Microbial Biofilms in Bioremediation and Wastewater Treatment

Editors

Y.V. Nancharaiah

Bhabha Atomic Research Centre
Kalpakkam - 603102, INDIA

Homi Bhabha National Institute
Mumbai - 400 094, INDIA

and

Vayalam P. Venugopalan

Bhabha Atomic Research Centre
Mumbai - 400 094, INDIA

Homi Bhabha National Institute
Mumbai - 400 094, INDIA

CRC Press
Taylor & Francis Group
Boca Raton London New York

CRC Press is an imprint of the
Taylor & Francis Group, an **informa** business

A SCIENCE PUBLISHERS BOOK

Cover illustration provided by the first editor, Dr. Y.V. Nancharaiah

CRC Press
Taylor & Francis Group
6000 Broken Sound Parkway NW, Suite 300
Boca Raton, FL 33487-2742

First issued in paperback 2021

Version Date: 20190916

ISBN 13: 978-1-03-208724-5 (pbk)
ISBN 13: 978-1-138-62639-3 (hbk)

Library of Congress Cataloging-in-Publication Data

Names: Nancharaiah, Y. V., 1970- editor.
Title: Microbial biofilms in bioremediation and wastewater treatment /
    editors, Y.V. Nancharaiah, Biofouling and Biofilm Processes Section,
    Water and Steam Chemistry Division, Bhabha Atomic Research Centre,
    Kalpakkam, Tamil Nadu, India and Vayalam P. Venugopalan, Homi Bhabha
    National Institute, Mumbai, India.
Description: Boca Raton : CRC Press, [2019] | Includes bibliographical
    references and index.
Identifiers: LCCN 2019027244 | ISBN 9781138626393 (hardcover)
Subjects: LCSH: Biofilms. | Bioremediation. |
    Water--Purification--Biological treatment.
Classification: LCC QR100.8.B55 M534 2019 | DDC 579/.17--dc23
LC record available at https://lccn.loc.gov/2019027244

Visit the Taylor & Francis Web site at
http://www.taylorandfrancis.com

and the CRC Press Web site at
http://www.crcpress.com

Preface

Microorganisms present in the environment provide a number of ecological services, including removal of pollutants that reach our aquatic systems. It is now fairly well-established that majority of microbes in nature reside in communities referred to as biofilms. Biofilms, therefore, can be assumed to play very important role in biogeochemical cycles and natural attenuation of toxic contaminants that reach our environment. However, natural remediation processes are slow and cannot be expected to remove all the pollutants emanating from a multitude of anthropogenic activities. But microbial communities can be engineered for various biotechnological applications such as biodegradation and biotransformation of organic and inorganic contaminants. Aerobic granular biomass is a relatively novel type of microbial community that allows setting up of efficient and compact biological wastewater treatment systems. Growth of microorganisms in the form of biofilms and biogranules in bioreactors simplifies separation of biomass from the treated water and achieves high biomass concentrations and high rate biological conversions. Using this strategy, we can efficiently decrease the load of most of the pollutants in wastewater to within acceptable levels.

The present book titled *Microbial Biofilms in Bioremediation and Wastewater Treatment* attempts to collate information dealing with bioremediation and wastewater treatment approaches based on microbial communities, particularly, biofilms and aerobic granular sludge. It provides useful information on basic and applied aspects of microbial communities, giving emphasis to their applications in removal/degradation of recalcitrant compounds and complex wastewaters. We hope that this book will serve as a useful source of information on biofilm based bioremediation and wastewater treatment, in particular biological removal of persistent organic pollutants, volatile organic compounds, antibiotics, oil, explosive compounds and selenium remediation. Extracellular polymeric substances (EPS) are key players in formation of both biofilms and biogranules and in remediation. Thus, role of EPS in attachment of cells and formation of aerobic granular sludge was presented in separate chapters. The editors hope that the book will be used by students of biotechnology, microbiology, environmental engineering, environmental biotechnology and wastewater treatment as well as teachers, researchers and operating personnel interested in biological wastewater treatment.

Contents

CHAPTER 1

Biofilms in Oil Bioremediation
Advances and Challenges

Debdeep Dasgupta[1,2,3] and *Tapas K. Sengupta*[1,*]

1. Introduction

1.1 Oil spillage and bioremediation: An overview

1.1.1 Petroleum and spillage of oil

Petroleum is a complex mixture of hydrocarbons with a wide range of molecular weights and various liquid organic compounds, which are found in geologic formation beneath earth crust. The liquids are flammable mixture of crude oil components, which are refined to produce a range of petroleum products. The crude oil is derived from the fossil fuels, which are formed by the intense heat and pressure of sedimentary rocks on dead organism like zooplankton and algae buried underneath the earth surface for ages. Petroleum is recovered from mines mainly by means of oil drilling (Hall and Cleveland 1981). The identification of mines takes place by detailed studies of structural geology analysis of sedimentary basin and characterization of reservoir (Guerriero et al. 2011, 2013). The most adopted process of refining the crude oil is by separation of components based on their boiling points (Figure 1). The products are collected into large number of consumer products, starting from petrol (or gasoline), fuel oil, liquid petroleum gas, kerosene, rocket fuel, asphalt and chemical reagents used in making plastics and a range of pharmaceutical products. The estimated consumption of petroleum across the world is approximately 71.7 million barrels each day (Asif and Muneer 2007). The use of fossil fuels such as petroleum has a detrimental impact on earth's biosphere, releasing a large amount of pollutants and greenhouse gases into the air and damaging ecosystems by catastrophic events such as oil spills. On the other hand, the depletion of earth's finite reserves of oil has a tremendous impact on world economy (Simmons 2005). Thus, the management of petroleum end products is an important aspect across the globe.

[1] Department of Biological Sciences, Indian Institute of Science Education and Research (IISER) Kolkata, Mohanpur-741246, India.

[2] Centre for Environmental Science & Engineering (CESE), Indian Institute of Technology Bombay, Powai, Mumbai-400076.

[3] Amity Institute of Biotechnology, Amity University Mumbai, Bhatan, Post - Somathne, Panvel, Mumbai (MH)-410206.

* Corresponding author: senguptk@iiserkol.ac.in

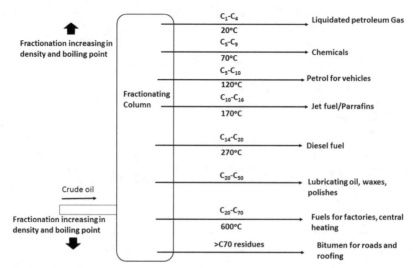

Figure 1. Distillation column for refining crude oil. Wide range of petroleum products are transported across the continents by shipping.

Oil refineries generally blend various solid waste feed-stocks with appropriate additives for providing short term storage and transfer by trucks, barges, product ships and railcars. However, the liquid fuels after refining, such as automotive and aviation grades of petrol, kerosene, various aviation turbine fuels and diesel fuels, lubricants such as motor oil, greases and machine oil are generally shipped by barge, rail, tanker ship, etc. Accidental spillages and leaks occur during production, transport and storage of petroleum goods. Recent data suggests natural oil seepage of 600,000 metric tons per year, with a sharp rise of 200,000 metric tons every year (Kvenvolden and Cooper 2003). One of the major causes of soil and water pollution is the oil and hydrocarbon spillage by human activities or by accidental means during transportation (Holliger et al. 1997). Water contamination with petroleum results in extensive damage, since the accumulation of toxic hydrocarbons may cause death or mutation in plant or animal tissues (Alvarez and Vogel 1991). The technology commonly used for remediation of oil spillage is by physical, chemical and biological means (Figure 2). However, the first two processes are expensive and time consuming.

Bioremediation is defined as the use of microorganisms for efficient removal and detoxification of pollutants by their diverse metabolic capabilities. Environmental pollutants including the products released from petroleum industries are effectively removed by this emerging technology (Medina-Bellver et al. 2005). The technology is not only cost effective but also non-invasive (April et al. 2000). Thus, biodegradation by natural population is one of the primary mechanisms (Ulrich 2000) being addressed currently by the environmentalists for effective removal of hydrocarbon pollution (Leahy and Colwell 1990). The bioremediation process is solely dependent on establishment and maintenance of conditions that favor oil biodegradation rates in the contaminated area. Numerous scientific reviews have addressed various factors that influence the rate of oil metabolism by microorganisms (Zobell 1946, Leahy and Colwell 1990, Zobell 1946, Atlas 1981, 1984, 1992, Foght and Westlake 1987). Adequate concentration of nutrient, oxygen and pH plays a crucial role in the rate of oil biodegradation. The success of bioremediation lies on two major modes: (a) bioaugmentation, where oil degrading bacteria are added to the existing microbial population; (b) biostimulation, where the growth of indigenous population is stimulated by the addition of growth limiting factors and nutrients. In a case study, 80% of the diesel was degraded in 40 hours when the substrate was readily available to microorganisms. Hence, bioaugmentation is more successful when pollutants are easily available and on the other hand, biostimulation would be more effective when the pollutants are limited (Moliterni et al. 2012). According to the study, the efficiency and bioavailability analyses showed that the best

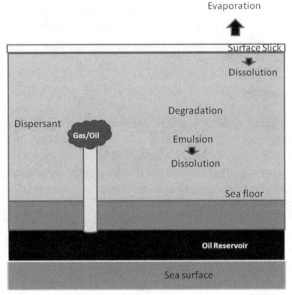

Figure 2. Fate of spilled oil in ocean and remediation efforts.

methodology to bioremediate the silty soil was biostimulation when native consortium was abundant, while bioaugmentation with a combination of native and exogenous consortia was effective to treat the clayey soil where the pollutants were easily available.

The potential of biodegradation and bioremediation technology got its first attention worldwide in the year of 1989 with the devastating oil spillage at Prince William and Gulf of Alaska from oil tanker Exxon Valdez (Atlas and Bartha 1998). Most of the studies so far focused on the factors altering the process and testing of favorable parameters in laboratory scale (Mearns 1997). Very few numbers of pilot scale and field studies have been carried out with convincing results (Prince 1993, Swanell et al. 1996, Venosa et al. 1996, 2002). From the international point of view, oil spill has taken place for more than one instance, including accident in small motor boat to mammoth tanker carrying mineral oil from one country to other. The major historical oil spillages include *Amoco Caldiz* on March 16, 1978 near coastline of Brittany, France; Arabian Gulf spill in January 1991; *Argo Merchant* in 1976 at Nantucket Island, Massachusetts; *Exxon Valdez* on Bligh Reef in Prince William Sound, Alaska, etc. Most recently, the *Deepwater Horizon* oil spill at the Gulf of Mexico has been considered as the largest accidental marine oil spill in the history of the petroleum industry (Cornwall 2015). Recent data suggest total mass of 0.18 ± 0.05 Tg hydrocarbons in the plume layers of Deepwater Horizon blowout was fully respired to CO_2 and 0.10 ± 0.08 Tg hydrocarbons incorporated into biomass (Du et al. 2012). The total discharge was estimated as 4.9 million barrels from the sea floor, gushing unabated for 87 days, before it was capped (Coordinator Report 2011).

Oil spillage has taken place in India in more than one instance. Notably, in early 2006, a Japanese tanker collided with a small Indian vessel 470 km west of the Nicobar and Andaman archipelago, spilling over 4,500 tons of oil into the Indian Ocean (Atroley 2006). In August 2009, an oil spill affected a vast area of 100-km off south Gujarat coast, threatening the marine biodiversity in the National Marine Park and Sanctuary. Close to the study area, another spillage took place on September 28, 2009 by a sunken Mongolian ship carrying iron ore. Most recently, a Panamanian ship carrying heavy oil and diesel tilted by an accident near Mumbai coast caused a disaster on the marine life and mangrove by devastating over thousands of natural habitats, affecting both flora and fauna (Sukhdhane et al. 2013). Kolkata port and nearby areas like Haldia port, part of Bay of Bengal close to the ports and Haldia refinery are major concerns for possible oil spillage due to everyday transport of fuel oil and other means, since it is a major shipping corridor for eastern region of India. In spite of possible threat of contamination of water sources by spilled oil in these areas, little work has been done on presence and characterization of oil degrading microorganisms naturally existing in water sources near Kolkata port and nearby areas. Current data suggests 28,220 tons of oily sludge is generated from the refineries in India (Dhote et al. 2013).

1.1.2 Microbial bioremediation

Research on microbial alkane degradation started close to a century ago, with a publication by Söhngen (1913) on microbes responsible for the disappearance of oil slicks on surface waters. Research carried out in the first half of the 20th century has provided a solid basis for our present extensive knowledge on the microbiology of alkane biodegradation, and many microorganisms capable of thriving on these highly reduced organic compounds were identified (Figure 3). In the case of Eubacteria, almost all of these belong to the *α-*, *β-*, and *γ-Proteobacteria* and the *Actinomycetales* (high G + C Gram-positives) (van Beilen et al. 2003, Coleman et al. 2006). More recently, isolates of *Bacillus, Geobacillus* (phylum Firmicutes) and *Thermus* (phylum Deinococcus-Thermus) were found to degrade alkanes (Marchant et al. 2006, Meintanis et al. 2006). Among the eukaryotes, many yeasts and fungi, and some algae, are known to thrive on alkanes (van Beilen et al. 2003).

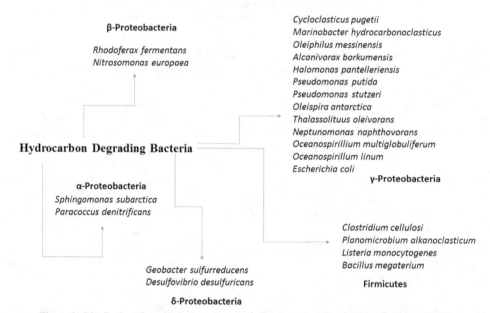

Figure 3. Distribution of aerobic hydrocarbon-degrading bacteria (Adopted from Head et al. 2006).

1.1.3 Hydrocarbon degradation by microorganisms

Biodegradation of petroleum hydrocarbons depends on chemical nature of the hydrocarbon components present in the oil sample. Petroleum based oils consists of four classes of hydrocarbons such as saturates, aromatics, asphaltenes and resins (Colwell et al. 1977). Asphaltenes include phenols, fatty acids, ketones, esters and porphyrins; resins consist of pyridines, quinolines, carbazoles, sulfoxides and amides. Rate of degradation of such hydrocarbons also depends on availability to microorganisms (Cooney et al. 1985). Susceptibility of microbial degradation is ranked as linear alkanes > branched alkanes > small aromatics > cyclic alkanes (Ulrichi 2000, Perry 1984), while some of the high molecular weight aromatics are not degraded at all (Atlas and Bragg 2009).

Microbial degradation of hydrocarbons is the sole natural mechanism by which cleanup of petroleum hydrocarbon was achieved (Atlas 1992, Amund and Nwokoye 1993, Lal and Khanna 1996). Four bacterial strains belonging to the family Bacillaceae, including *Bacillus licheniformis* STK08, *Geobacillus stearothermophilus* STM04, *Lysinibacillus sphaericus* STZ75 and *Bacillus firmus* STS84 were isolated from a refinery effluent (Hamed et al. 2013). The successful biodegradation of petroleum hydrocarbons in marine sediment was reported in a previous study by Jones et al. (1983). The study emphasized on the alkyl aromatics degradation prior to n-alkane by various organisms like *Sphingomonas, Burkholderia, Arthrobacter, Mycobacterium, Pseudomonas* and *Rhodococcus*. The study on petroleum hydrocarbon degradation in polluted tropical streams was also reported (Adebusoye et al. 2007). Thirteen cultivated and four wild species belonging to *Klebsiella, Microbacterium, Agrococcus, Kocuria, Arthrobacter, Bacillus, Planomicrobium* and *Rhodococcus* grow in presence of aromatic hydrocarbons, better than in presence of aliphatic hydrocarbons (Al-Awadhi et al. 2012). Nine different organisms including *Corynebacterium* sp., *Alcaligenes* sp., *Pseudomonas fluorescens, Micrococcus roseus, P. aeruginosa, Bacillus subtilis, Bacillus* sp., *Acinetobacter lwoffi* and *Flavobacterium* sp. were isolated from water samples with the ability of degrading crude oil (Adebusoye et al. 2007). Recently, the potential isolation and genome sequencing of two Oceanospirillales single cells has been documented in *Deepwater Horizon* oil spill, which revealed the presence of genes coding for n-alkane and cycloalkane degradation (Mason et al. 2012). *Micromonospora, Pseudomonas, Achromobacter* and *Bacillus* were isolated from a petroleum sludge sample with 82–88% degradation efficiency analyzed by gas chromatography (Gojgic-Cvijovic et al. 2012).

Hydrocarbon degradation in environment is achieved by yeasts, bacteria and fungi. Wild yeast, *Hansenula angusta* and *Rhodotorula minuta* were isolated from industrial effluents with the ability to uptake fluoranthene and carry out biotransformation (Romero et al. 2011). The reported degradation efficiency ranged from 0.003% to 100% by marine bacteria, 6% to 82% by soil fungi and 0.13% to 50% by soil bacteria (Jones et al. 1970, Pinholt et al. 1979, Hollaway et al. 1980). Microbial conversion of petroleum resins, aliphatic and aromatic compounds was observed among the three novel isolates of *Serratia* sp., *Raoultella* sp. and *Ochrobactrum* sp. (Ghollami et al. 2013). Many groups have also reported the application of broad enzymatic capacities of mixed bacterial strains for better degradation of hydrocarbons such as crude oil in soil, fresh water and marine environments (Bartha and Bossert 1984, Cooney 1984, Atlas 1984, Floodgate 1984).

Among the other organisms, bacteria are the most active agent for bioremediation in environment. Some of the bacteria prefer hydrocarbon meal exclusively (Rahman et al. 2003, Brooijmans et al. 2009, Yakimov et al. 2007). Twenty five genera of hydrocarbon degrading bacteria and fungi were listed by Floodgate (Floodgate 1984) and 22 genera of bacteria and 31 genera of fungi by Bartha and Bossert (Bartha and Bossert 1984). *Dietzia, Brevibacterium, Gordonia, Aeromicrobium, Burkholderia,* and *Mycobacterium* were isolated from petroleum contaminated soil (Chaillan et al. 2004). Twenty one isolates including *Selenomonas, Blastococcus, Pectinatus, Humicoccus, Methylophilus, Erysipelothrix, Microlunatus, Nevskia, Rubrimonas* and *Stenotrophomonas* were isolated from various environmental sources in Thailand, capable of producing surface active compounds (SACs) (Saimmai et al. 2012). Degradation of polyaromatic hydrocarbons by the genus *Sphingomonas* was also reported (Daugulis and McCracken 2003). *Acinetobacter* sp. was found to be capable of utilizing *n*-alkanes of chain length of C10-C40 (Throne-Holst et al. 2007).

Fungal genera including *Neosartorya, Amorphoteca, Talaromyces* and *Graphium* and yeast genera, namely, *Yarrowia, Candida,* and *Pichia* were isolated from petroleum contaminated soil. These organisms have proved their potential for hydrocarbon degradation (Chaillan et al. 2004). Terrestrial fungi, namely, *Cephalosporium, Aspergillus* and *Pencillium* were reported to be efficient degraders of crude oil hydrocarbons. The yeast species, namely, *Geotrichum* sp., *Candida lipolytica, Rhodotorula mucilaginosa,* and *Trichosporon mucoides* isolated from contaminated water are noteworthy as potential hydrocarbon degraders (Boguslawska and Dabrowski 2001).

Algae and protozoa are the important members of the microbial community in both aquatic and terrestrial ecosystems, although the reports are not sufficient for their potential in hydrocarbon biodegradation as they are scanty regarding their involvement in hydrocarbon biodegradation. The alga *Prototheca zopfi* was capable of utilizing crude oil and a mixed hydrocarbon substrates (Walker et al. 1975). This genus also exhibited degradation of n-alkanes and isoalkanes as well as some aromatic hydrocarbons. Some of the cyanobacteria, five green algae, one red alga, one brown alga and two diatoms could oxidize naphthalene (Cerniglia et al. 1980). But no report of biodegradation of crude oil by such organisms has yet been reported.

1.1.4 Mechanism of petroleum hydrocarbon degradation

Aerobic condition is the most favorable for rapid and complete biodegradation of hydrocarbons (Fritsche and Hofrichter 2000). Oxygenases and peroxidases are the key enzymes participating in intracellular attack on organic pollutants. A step by step peripheral degradation converts organic pollutants into intermediates of central metabolism, i.e., tri-carboxylic acid cycle. Co-metabolism approach has been reported in biodegradation, where the microorganisms such as *Pseudomonas* sp. assimilate different substrates from organic pollutants and survive using it. The microorganisms often fail to utilize the contaminant directly and thus it is termed as non-growth substrate. However, the potential of co-metabolism in bioremediation is not yet well understood (Nzila 2013). Mutation in *rpoB* gene which encodes for RNA polymerase beta subunit resulted in enhanced hydrocarbon degradation (Rachid 2012). *P. aeruginosa PA01* strain was able to degrade n-hexadecane, 1-undecene, 1-nonene, 1-decene, 1-dodecene and kerosene (Nisenbaum et al. 2013). The uptake of hydrocarbons takes place by attachment of microbial cells to oil/hydrocarbon droplets or by production of surface active agents called biosurfactants (Hommel 1990). The exact mechanism of uptake is still unknown, although the production of biosurfactants has been extensively studied. Batch experiments conducted with three microbial consortia to study the kinetics of diesel biodegradation showed that the consortia can function at high concentrations of hydrocarbons without significant growth inhibition, which is important for the design of bioreactors for oil-containing wastewater treatment (Moliterni et al. 2012).

Cytochrome P450 alkane hydroxylases plays a critical role in microbial degradation of hydrocarbons, oil, additives and chlorinated compounds (van Beilen and Funhoff 2007). Increased catabolic gene abundance with different changes in the microbial structure was observed as the adaptive response to the simulated pollutant rebound, which is due to functional redundancy in biodegrading microbial communities (Korotkevych et al. 2011). The ubiquitous heme-thiolate monooxygenases introduce oxygen into the substrate to initiate biodegradation. Cytochrome P450, integral membrane di-iron alkane hydroxylase, di-iron methane monooxygenases and copper containing methane monooxygenases are the diverse enzymes found in prokaryotes and eukaryotes participating in the biodegradation process (van Beilen and Funhoff 2005). Six aerobic alkanotrophs, including *Alcanivorax* sp. strains EPR7 and MAR14, *Marinobacter* sp. strain EPR21, *Nocardioides* sp. strains EPR26w, EPR28w, and *Parvibaculum hydrocarbonoclasticum* strain EPR92 isolated from deep-sea hydrothermal vents were characterized by using the substrate norcarane, where metalloenzyme plays a crucial role to oxidize alkanes (Bertrand et al. 2013). Potential of the metalloenzymes that catalyze alkane oxidation in the environment was recently reviewed (Austin and Groves 2011). For example, the capacity to degrade n-alkanes by yeast species lies on cytochome P450, which was isolated from *Candida tropicalis, Candida maltosa,* and *Candida apicola* (Scheuer et al. 1998).

1.1.5 Biodegradation by immobilized cells

Immobilized cells are essential tool of bioremediation. Cells entrapped inside the matrix not only protects them from external environments, but also mitigates the drawback of dispersion of cells at the spillage site. It also simplifies the separation and recovery process and thereby reduces the overall cost. It increases the contact between the cells and the hydrocarbon droplets and at the same time the rhamnolipid production also (Wilson and Bradley 1996). Immobilized cells of *Pseudoxanthomonas* sp. RN402 could maintain high efficacy and viability throughout 70 cycles of bioremedial treatment of diesel-contaminated water (Nopcharoenkul et al. 2013). Rhamnolipids trigger greater dispersion of water insoluble n-alkanes and reduce the interfacial surface tension of the oil water system. This in turn increases the microbial access to the hydrocarbon. The enhanced degradation of crude oil with immobilized cells was obtained at a wide range of salinity (Diaz et al. 2002). Immobilized cells were found most successful in degradation of diesel oil within 32 days. The cells were entrapped by polyvinyl alcohol (PVA) cryogelation (Cunningham et al. 2004). Microorganisms indigenous to diesel contaminated sites were utilized and entrapped by polyvinyl alcohol (PVA) cryogelation technique with successful removal of diesel after 32 days. A role for immobilized-cell bioaugmentation was studied for bioremediation in case of persistent fuel spills and/or in extreme environments. The repeated use of alginate beads with immobilized microorganisms was tested for biodegradation study and found promising too. The designed systems were found to be most successful with maximum removal in co-immobilization system where the PVA trapped microorganisms along with synthetic oil absorbent were examined.

1.1.6 Use of fertilizer and extremophiles in bioremediation

Bioremediation agents are classified as bioaugmentation agents and bio-stimulants. United States Environmental Protection Agency (USEPA) has enlisted 15 bioremediation agents that are required for Clean Water Act, Oil Pollution Act 1990, and National Contingency Plan (NCP) (Nichols 2001, ES EPA 2002). But these agents have shown very less activity in field (Mearns 1997, Venosa 1996, Venosa 2002, Lee et al. 1997). Multi-extremophilic microorganisms capable of surviving in environments with high temperature, pressure and salinity have been identified in petroleum reservoirs (Lenchi et al. 2013). *Pusillimonas* sp. strain T7-7, a novel cold tolerant strain capable of utilizing diesel oils (C5 to C30 alkanes) as the sole source of carbon and energy, was isolated recently (Li et al. 2013). Fertilizers are often used as an external source of N and P to improve the availability of these nutrients in the marine environment. Biodegradation is partially alleviated by incorporation of fertilizers and also being acceptable by environmental agencies as safe and cost effective method of stimulation. However, one limitation is that they are rapidly washed out due to the tide and wave action. Some of the oleophilic nutrient products like Inopol EAP22, Oil spill eater, Bioren 1 and Bioren 2 have been optimized as bioremediation agents during the catastrophe at Prince William, Alaska (US EPA 2002, Ladousse and Tramier 1991, Zwick et al. 1997).

1.1.7 Use of recombinant strains in bioremediation

Genetically engineered microorganisms (GEMs) have been reported with better bioremediation efficiency. Biomolecular engineering approaches including rational designing, suicidal GEMs and directed evolution have been employed to develop strains capable of degradation of recalcitrant organic pollutants like polyaromatic hydrocarbons, polychlorinated hydrocarbons and pesticides (Kumar et al. 2013). Genetically engineered bacteria had been a noted candidate for higher degrading capacity. *A. eutrophus* H850Lr, *P. putida* TVA8, *P. fluorescens* HK44, *B. cepacia* BRI6001L, *P. fluorescens* 10586s/pUCD607, *Pseudomonas* strain Shk1, *A. eutrophus* 2050 are some of the genetically engineered stains tested for their efficiency of degradation of polychlorinated biphenyl, trichloroethylene, benzene, toluene, ethylbenzene, and o-, m-, and p-xylenes (BTEX), 2,4-Dichlorophenoxyacetic acid, non-polar narcotic degradation (Dyke et al. 1996, Applegate et al. 1998, Sayler and Ripp 2000, Masson et al. 2002, Sousa et al. 1997, Kelly et al. 1999, Layton et al. 1999). Even though the use of GEMs is an effective cleanup process at lower cost, various factors including the biochemical mechanism, field engineering design, ecological, survival of the strain under field conditions and microbiological knowledge are sole requirements for its successful implementation.

1.1.8 Factors limiting degradation of hydrocarbons

A number of limiting factors are reported which affect the rate of biodegradation of petroleum hydrocarbons (Brusseau 1998). The composition of hydrocarbon mixture in petroleum products is the first and foremost factor influencing the suitability of remediation approach. Temperature is another crucial parameter which not only determines the chemical nature of hydrocarbons at spilled site but also the physiology and diversity of the microbial flora. At very low temperature, the viscosity of toxic low molecular weight hydrocarbons is reduced, thus delaying the onset of biodegradation process (Atlas 1975). Temperature also determines the solubility of hydrocarbons (Foght et al. 1996). The rate of biodegradation generally declines with low temperature. The best degradation takes place between 30–40°C in soil, 20–30°C in fresh water and 15–20°C in marine environment. Temperature affects both properties of spilled oil as well as the activity of microorganisms (Venosa and Zhu 2003). Psychrophilic microorganisms isolated from temperate region are reported to have significant potential of biodegradation of hydrocarbons (Pelletier et al. 2004).

Nutrients are also important requirements for successful biodegradation of hydrocarbons. Nitrogen, phosphorus and iron are sole inorganic nutrients which directly affect the biodegradation process (Cooney 1984). Oil spillage brings about significant rise in carbon source at the spilled site and corresponding reduction in nitrogen and phosphorus (Atlas 1985). In marine environment, the effect is more pronounced due to lower level of dissolved nitrogen and phosphorus (Floodgate 1984).

Additional supplement of inorganic nutrients is the key for successful biodegradation of oil pollutants (Choi et al. 2002, Kim et al. 2005). The fresh water wetlands are generally nutrient deficient by heavy pull of nutrients by plants (Mitsch and Gosselink 1993). On the other hand, excessive addition of nutrient is also detrimental to biodegradation (Chalillan et al. 2006). The higher level of NPK is detrimental for biodegradation rate (Chaineau et al. 2005, Oudot et al. 1988), especially for aromatics (Cramichael and Pfaender 1997). The use of fertilizer is another advent in the biodegradation research (Pelletier et al. 2004). Poultry manure and waste had been utilized for enhanced biodegradation (Okolo et al. 2005). Photo oxidation increases the biodegradation of petroleum products by increasing the bioavailability and thereby enhancing the microbial activity.

In situ bioremediation faces major obstacle for adequate mass transport of the electron acceptors. Oxygen is one of the most preferred electron acceptor and has to be ensured that it is not rate limiting (Cauwenberghe and Roote 1998). Major nutrient sources including nitrogen and phosphorus are rate limiting factors for successful bioremediation. A ratio of 100:10:1 for carbon, nitrogen and phosphorus is essential for efficient biodegradation of hydrocarbons. In most of the cases, a pH of 5.5–8.5 is optimum for effective bioremediation. High concentrations of heavy metal, chlorinated compounds, long chain n-alkanes and inorganic salts are reported as toxic for microorganisms. Also, the unavailability of hydrocarbon to the microorganism is a potential limitation in achieving bioremediation success. At a very low (ppm) concentration of hydrocarbons, the microbial activity is not stimulated even with the administration of stimulants.

Thus, the limitation of bioremediation is broadly classified in three major titles: (a) bioactivity, (b) bioconversion, and (c) bioavailability. Bioactivity is an important parameter for completion of bioremediation. The state of organism, capability of the organism to degrade recalcitrant compounds and the tolerance of organism to toxic compounds are crucial factors altering the process. The ability of organism to break down the contaminants into simpler form is a diverse process. The conversion can result in complete removal of contaminants, or by-products with increasing toxicity with release of toxic metabolite or production of new recalcitrant compound.

Finally, another major parameter for best bioremediation result lies in the ability and availability of reduced organic materials which serve as sole energy source. Thus, the oxidation state of the carbon source implies the effectiveness of biodegradation. Organism prefers the carbon source to be more in reduced state, which offers higher energy yield and thus effective energetic incentive.

The bioconversion of environmental contaminants also depends on the rate of substrate uptake and metabolism. The sole limiting factor is the rate of transfer of contaminants to the cell or mass transfer. As stated earlier, when mass transfer is a limiting factor, higher biotransformation is inhibited even if there is increased microbial conversion capacity (Bhopathy and Manning 1998). For example, when explosive containing soil undergoes rigorous mixing, the degradation was drastically stimulated (Manning et al. 1995). Thus, bioavailability of organic pollutants is a major drawback for bioremediation process. The problem was addressed by the addition of food grade surfactant which increased the availability of contaminants to some extent for microbial degradation (Bhoopathy and Maning 1999).

2. Advances in bioremediation technology

2.1 Role of biosurfactants

Biodegradation of hydrophobic compounds by *Pseudomonas aeruginosa* has long been reported. Recently, the role of surface active compounds in the biodegradation of hydrophobic contaminants was reported in Gram positive *Bacillus subtilis* DSVP23 (Pemmaraju et al. 2012). Bioavailability of hydrophobic compounds is one of the major drawbacks for the bioremediation of oil spills (Volkering et al. 1998, Angelova et al. 1999). The role of surfactants in enhanced degradation had been shown by rhamnolipid, a biosurfactant produced by the *Pseudomonas aeruginosa* (Hisatsuka et al. 1971, Zhang and Miller 1994). The microbial conversion of organic hydrocarbons is usually low and thus, the rhamnolipid enhanced bioavailability is one of the promising applications of microbial products in bioremediation (Noordman 2002). Recent report suggests enhanced biodegradation of hexadecane by stimulation with rhamnolipid (Noordman 2002) compared to other biosurfactants (Itoh and Suzuki 1972) or synthetic surfactants (Nakahara et al. 1981). However, the stimulation failed in case of rhamnolipid treated condition under mixed culture and other strains (Hisatsuka et al. 1971, Providenti et al. 1995). Energy dependent stimulation was observed in uptake of water insoluble compounds in the presence of rhamnolipid (Noordman 2002). Surface active compounds are essential for oil and hydrocarbon degradation (Muthusamy et al. 2008, Mahmound et al. 2008, Ilori et al. 2005, van Beilen et al. 2005, Ilori et al. 2005, Kiran et al. 2009, Obayori et al. 2009) which enhances the solubilization and biodegradation of hydrocarbons (Brusseau et al. 1995, Bai et al. 1997). Novel biosurfactant producing *Staphylococcus, Chrysomonas* and *Photobacterium* were described for the first time by isolating from seawater along with other isolates such as *Bacillus, Pseudomonas, Micrococcus, Neisseria,* and *Aeromonas* (Hamed et al. 2013). About 90% of hydrocarbons were degraded within a span of six weeks in liquid culture of *Pseudomonas aeruginosa* and *Rhodococcus erythropolis* isolated from oil contaminated soil (Cameotra and Singh 2008). Recently, gas-permeable microchannel filled with crude oil was developed in micro-scale environment to study the microbial processes where oil droplets degradation was carried out by the cells. In this study the production of biosurfactants attributed towards the reduction of interface tension and increase the dissolution of the crude oil (Wang et al. 2013). The addition of nutrient mixture or crude biosurfactant was also tested. Eleven different congeners of rhamnolipids were isolated from consortium members. The role of surfactant is equally applicable for oil contaminated soil. About 91–95% of depletion was reported in a time span of four weeks, being enhanced by crude surfactant treatment. About 98% of degradation was reported with additional supplement of nutrient and surfactant (Cameotra and Singh 2008). *Pseudomonas* sp. had been widely studied for production of biosurfactant (Cameotra and Singh 2008, Beal and Betts 2000, Pornsunthorntawee et al. 2008, Rahman et al. 2003). Biosurfactants form micelle and decrease surface tension of oil-water interface, thereby accelerating the biodegradation process forward. The micro droplets of oil/hydrocarbon are taken up by microbial cells. Surface active agents also increase the oil surface area, thereby increasing the amount of oil accessible to bacteria (Nikolopoulou and Kalogerakis 2009).

2.2 Biofilm and chemotaxis: Role in bioremediation

The major limitation of bioavailability of a compound to the bacterial cells can be improved by exploiting chemotactic bacteria (Paul et al. 2005, Pandey and Jain 2002, Stelmack et al. 1999). Cells displaying chemotaxis are able to sense the chemicals absorbed on the soil particles and thereby the mass transfer limitation can be overcome. Once the cells come in close contact to the surface, the biofilm formation starts as the cells attach and grow on their substrates. The biofilm formation and surfactant production together increase the bioavailability and biodegradation of oil components.

Chemotaxis is instrumental for swimming of organism towards the nutrient, thereby establishing the biofilm using the bacterial flagella (Pratt and Kolter 1999, O'Toole and Kolter 1998, Stelmack et al. 1999). The role of biofilm in bioremediation is documented earlier (Kargi and Eker 2005, Puhakka 1995). For removal of 2,4 dichlorophenol (DCP) from synthetic waste water, rotary perforated biofilm reactor was implemented and 100% degradation was achieved. Nitro aromatic compounds are xenobiotic in nature and mixed strain of culture were treated with those compounds in fluidized bed biofilm reactor with the success of 98% degradation (Lendenmann and Spain 1998). The biofilm supported remediation was also successful in the remediation of heavy metals and radio nucleotides (Barkay and Schaefer 2001, Lloyd 2003). In a report, 20–200 µM copper was found accumulating in the form of copper sulphide by sulphate reducing biofilm grown in continuous culture (White and Gadd 2000). Formation of sphalerite by members of aerotolerant *Desulfobacteriaceae* in natural biofilm was also reported (Labrenz et al. 2000). Sphalerites are abundant, micrometer-scale, spherical aggregates of 2- to 5-nanometer diameter ZnS particles formed within natural biofilms and are an interesting example of how microbes control metal concentrations in groundwater.

2.2.1 Biofilm: Role in bioremediation

Biofilm is multi cellular aggregation of cells attached to a surface. Biofilm development takes place on every moist environment where the nutrient flow is abundant and surface to cell attachment is feasible, as such film formation was observed in marine sediments at alkane–water interface (Klein et al. 2010). Biofilms can be formed by bacteria, fungi, algae and protozoa as single species or multi species together. As reported, 97% of biofilm matrix is made up of water, which in turn is bound to capsules of microbial cells. Biofilm surrounded by secreted polymer is essential for biofilm development and maturation processes (Gacesa 1998, Govan and Deretic 1996, Govan and Fyfe 1978). An interesting study described improvement of bioremediation by uptake of DNA harboring catabolic genes that facilitate biodegradation of selected pollutants by natural transformation within biofilms (Perumbakkam et al. 2006). One of the fascinating fields of development in the area includes engineering of secreted proteins of the biofilm matrix (Absalon et al. 2012). One can also use various combined processes, for example, biofilm trickling filters, with a set up for RBC or also with activated sludge, to enhance the efficiency of the process (Safa et al. 2014, Kaindl 2010).

2.2.2 Biofilm development

There are several steps which must be optimized for the formation of bacterial biofilm. First, the organism approaches the surface and motility gets slowed down. Next, the organism makes transient association with the surface and settles down. A nascent microcolony is developed and thereby the building of the three dimensional biofilm structure is developed and the cells occasionally detach from the biofilm (Figure 4).

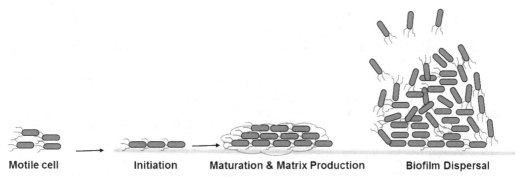

Figure 4. Development of biofilm microcolonies and dispersal of cells from biofilm-matrix.

2.2.3 Role of phosphate

Recently, the detachment process of biofilm associated cells has been correlated with nutrient deprivation including phosphate (Newell et al. 2011). High phosphate supports the production of high cyclic dimeric GMP which in turn binds to a diverse receptor. In a study, Newel et al. (2011) showed that at high cyclic GMP it binds to LapD and alters the conformation which in turn sequesters LapG. The LapG bound to LapA in outer membrane facilitates the binding or adherence of biofilm growth. In contrast, when the concentration of c-GMP in the surrounding is low, the conformation of LapD is altered, resulting in the release of LapG by cleaving out of LapA. This way, the bacteria detach from the surface and move in search of phosphate.

2.2.4 Role of flagella/pili

Genetic screening of biofilm defective mutants has shown that the surface interaction is mediated by type IV pili and flagella. Previous reports suggest that once the temporary contact is established, bacteria utilize type IV pili or flagella to move in two dimensional axes until other bacteria arrive and microcolonies are formed and enlarged to make biofilm (O'Toole and Kolter 1998, Pratt and Kolter 1998, Watnick and Kolkter 1999). The phenomenon is similar to fruiting body formation of *Myxobacteria xanthus* (Kuspa et al. 1992).

2.2.5 Role of signaling molecules

Intracellular communication between bacteria is generally carried out by extra cellular products which can diffuse away from one cell to another. The organism in planktonic stage gets limited access to these products as in natural and aquatic environment where they are generally carried off. But in case of diffusion limited environment like biofilm, the signaling by these compounds is ideally suited. The role of acyl homo serine lactones (HSL) has been documented in maturation of microcolonies into mushroom and pillar like structure with water channels in between and its separation and differentiation in three dimensional structures (Allison et al. 1998, Davies et al. 1993, McLean et al. 1997, Stickler et al. 1998). *Pseudomonas aeruginosa* mutant failing to produce acyl HSL produce closely packed biofilm, liable to easy disruption by surface active agents such as sodium dodecyl sulphate (SDS) treatment compared to control (Parsek et al. 2000).

2.2.6 Biofilm based reactors

Biofilms represent matrix enclosed population of organisms adhered to each other or with the substratum or interface (Costerton et al. 1999). Majority of microorganisms get associated on solid surface or microscopic particle or on the surface of animal/plant tissue to form the biofilms (Brusseau 1998). Biofilm has long been reported to be useful in environmental biotechnology processes in the treatment of potable and waste water. Growing urbanization has made extension of wastewater treatment plant difficult because of lack of space for pumping station and wastewater transfer. To avoid this limitation, biofilm based reactor with capability of high biomass concentration has been employed, which requires less land area and work efficiently. When organisms like bacteria floating in water attach to a surface, they slowly start aggregating, eventually forming biofilm. This is termed as fixed biofilm and it is the most common form of microbial life. Fixed-bed reactor has been previously employed successfully in bioremediation of mercury (Wagner Doobler 2003). Mercury resistant microorganisms grow on porous carrier material using contaminants as carbon source and obtain essential trace element from water.

In the enclosed matrix, organisms tend to create a small ecosystem which makes them resistant to environmental changes. In another report, similar biofilm mediated enhanced degradation of crude oil components was described using static biofilm based reactor (Dasgupta et al. 2013). Another type of biofilm reactor is known as trickle-bed biofilm bioreactor (TBR) where wastewater is allowed to trickle downward over the biofilm surface held on a fixed media. In case of fluidized-bed biofilm reactor, polluted water is slowly pumped upward though a column of biofilm coated beads (Shieh and Keenam 1986). This set up enables development of biofilms on a large surface area with high biomass and has been used for the treatment of various organic and inorganic compounds floating in streams (Shieh and Keenan 1986, Denac and Dunn 1988, Kumar and Saravanan 2009). Another important and successful biofilm reactor till date with potential of reducing both chemical oxygen demand and biochemical oxygen demand is Rotating Biological Contactors (RBC) or modified forms of RBC. It has so far been globally used for wastewater treatment for achieving nitrification and denitrification (Costley and Wallis 2001, Eker and Kargi 2008, 2010). The principle of RBC lies with disc being submerged in effluent and slowly rotating with exposure to air and concomitant formation of biofilm. RBCs have been used for treatment of water contaminated poly aromatic hydrocarbons, dyes, heavy metals and volatile compounds (Costley and Wallis 2001, Abraham 2003, Jeswani and Mukherji 2012, Sarayu and Sandhya 2012). However, no report on oil remediation is yet documented using biofilm based reactor except the use of membrane biofilm reactor (MBfR) (Li et al. 2014). In case of MBfR, biofilm is formed on membrane exterior where pressurized air or oxygen is provided though gas permeable membranes on the attached biofilms. It is an innovative technology where biofilm aided microbial platform for oil bioremediation was investigated in one finite system with proper oxygen supply. It is composed of biofilm and hollow fiber membrane component, where the microorganisms inside the biofilm counter oxygen and pollutant from aerobic to anaerobic layer in inside-out fashion (Casey et al. 1999, Semmens et al. 2003). Recently, membrane aerated biofilm reactor (MABR) system was also used in treatment of pharmaceutical waste water (Wei et al. 2012). Interestingly, the study by Li et al. (2014) actually showed the potential of MABR in oil removal. Oil contaminated seawater was treated with engineering bacteria ADB350 M with potent degradation ability to petroleum hydrocarbon, which was incubated in the MABR system supplemented with nutrient and oil spilled seawater pretreated with rhamnolipid. The MABR system showed significant removal of oil content and other pollutants during the study, and it would therefore be of great value in terms of scaling up for new startup venture for the oil-contained seawater bioremediation.

2.2.7 *Factors regulating biofilm*

Secondary metabolites are broad class of molecules produced at late exponential and stationary stages of microbial growth in laboratory cultures. Microbial secondary metabolites include antibiotics, receptor antagonists, effectors, pheromones, antitumor agents, pigments, toxins, enzyme inhibitors, immunomodulating agents, agonists, pesticides, antitumor agents, growth promoters, etc. They often have complex structures with various derivatives. Secondary metabolites also function as antibiotics and are thought to be produced in copious amounts for protecting the producer from competitors during the stationary phase of its growth (Firn and Jones 2003). The production and release of secondary metabolites are coded by both chromosomal DNA and plasmid DNA. Their formation is influenced by nutrients, growth rate, enzyme inactivation or enzyme induction and feedback mechanism. In recent years, however, the idea of 'secondary' metabolites has dramatically changed with the discovery of their role in controlling gene expression (Goh 2002). The compounds also play a critical role in iron acquisition and survival in microbial communities (Hernandez et al. 2004, Banin et al. 2005). Moreover, the secondary metabolism is resumed with the exhaustion of a nutrient as well as biosynthesis or external addition of inducer(s). Interestingly, a larger part of the microbes typically exists in nature as biofilm communities, where the organisms are in metabolically silent stage due to various reasons. The role of the secondary metabolite in gene expression and performing different function on sessile and biofilm adhered cells is one of the most interesting aspects in the research of modern microbiology. Phenazines are one of the most common compounds produced by various organisms including bacteria, actinomycetes and fungi as secondary metabolites. The phenazines are thus an illustrative example for deciphering their role for known and unknown functions.

Pseudomonas sp. is widespread in marine, freshwater and terrestrial environments and has the ability to infect hosts ranging from human, animal and plant. The compound phenazine was first documented as early as in 1860 as blue coloration in pus. Two strains of *P. aeruginosa* PAO1 and PA14 are well documented strains which produce a range of phenazine derivatives. Other strains include *Pseudomonas aureofaciens*, *P. fluorescens* and *P. chlororaphis*. Phenazine metabolites are also produced by *Brevibacterium, Burkholderia* and *Xanthomonas*, as well as the Gram-positive genus *Streptomyces* and archaeal genus *Methanosarcina* (Villavicencio 1998, Turner and Messenger 1986, Pirnay et al. 2005, Mavrodi et al. 2001, Delaney et al. 2001, Chin-A-Woeng et al. 2001, Beifuss and Tietze 2005, Rao and Sureshkumar 2000, Lau et al. 2005, Wilson et al. 1988).

Phenazines are large group of heterocyclic compounds secreted naturally by different bacterial species. More than 100 different phenazine structural derivatives have been identified as natural products and over 6,000 derivatives containing phenazine as a central moiety have been synthesized (Mavrodi et al. 2006). In recent years, the naturally occurring and synthetic phenazines are gaining significant interest because of their potential impact on bacterial interactions and biotechnological processes. The alteration of functional groups added to the phenazine ring determines the color, redox potential and solubility of compounds, thereby affecting their activities (Laursen and Nielsen 2004, Kerr 2000, Chin-A-Woeng et al. 1998). Phenazines are widely used as electron acceptors/donors, biosensors and reagents in fuel cells as well as antitumor agents. Phenazines are model compounds for studying biofilm and quorum sensing in *Pseudomonas aeruginosa* and other phenazine-excreting bacteria (Juhas et al. 2005, Hall-Stoodley 2004, Lazdunski et al. 2004). They modify cellular redox state by acting as electron shuttles and contribute to biofilm formation in oxygen depleted environment. Recently, the role of phenazine was studied on biofilm mediated biodegradation of crude oil components, where the addition of metabolite phenazine 1,6 di-carboxylic acid could increase the degradation of oil components by 10–20%, depending on the hydrocarbon (Dasgupta et al. 2015). The recent research work suggests that phenazines and other excreted compounds can react with common primary metabolites and also be potentially transformed by enzymes active in central metabolic pathways. Although many aspects of the primary function of the phenazines for producing organisms such as *Pseudomonas* are known, roles of these compounds in behavioral and ecological fitness of the producers are still unknown. Phenazine compounds can act as virulence factors during infection as well as directly as antibiotics (Price-Whelan et al. 2006).

3. Conclusion

The technology of bioremediation have few major challenges to be mitigated by taking interdisciplinary approach from different fields such as environmental microbiology, civil engineering and soil science. First of all, the process bioremediation is comparatively slow compared to chemical treatment for pollutant degradation. Second is the chance of formation of toxic metabolic products that could be generated during the biodegradation process. The process is generally favored in condition where the level of pollution is relatively low and chemical treatment process is not feasible. Moreover, the process is strictly driven by availability of essential nutrients which often limits the microbial growth. Finally, the bioavailability of pollutant to microbes is another crucial factor determining the rate and efficiency of bioremediation. Some of these issues can be addressed by using biofilms of natural isolates of microorganisms for bioremediation. Novel approaches are constantly being tested for better biodegradation, stable reactor design and designing new improved strains that energetically favor pollutants as the sole carbon source. Therefore, despite several challenges and limitations, the use of biofilm mediated bioremediation remains an attractive choice in mitigating environmental pollution and opens plethora of opportunities for environmental microbiologists.

Acknowledgments

Authors would like to thank Indian Institute of Science Education and Research (IISER), Kolkata for financial support.

References

Abraham, T.E., R.C. Senan, T.S. Shaffiqu, J.J. Roy, T.P. Poulose and P.P. Thomas. 2003. Bioremediation of textile azo dyes by an aerobic bacterial consortium using a rotating biological contactor. Biotechnology Progress 19(4): 1372–6.

Absalon, C., P. Ymele-Leki and P.I. Watnick. 2012. The bacterial biofilm matrix as a platform for protein delivery. MBio. 3: e00127–00112.

Adebusoye, S.A., M.O. Ilori, O.O. Amund, O.D. Teniola and S.O. Olatope. 2007. Microbial degradation of petroleum hydrocarbons in a polluted tropical stream. World Journal of Microbiology and Biotechnology 23: 1149–1159.

Al-Awadhi, H., D. Al-Mailem, N. Dashti, L. Hakam, M. Eliyas and S. Radwan. 2012. The abundant occurrence of hydrocarbon-utilizing bacteria in the phyllospheres of cultivated and wild plants in Kuwait. International Biodeterioration & Biodegradation 73: 73–79.

Allison, D.G., B. Ruiz, C. SanJose, A. Jaspe and P. Gilbert. 1998. Extracellular products as mediators of the formation and detachment of *Pseudomonas fluorescens* biofilms. FEMS Microbiol. Lett. 167: 179–184.

Alvarez, P.J.J. and T.M. Vogel. 1991. Substrate interactions of benzene, toluene, and para-xylene during microbial degradation by pure cultures and mixed culture aquifer slurries. Applied and Environmental Microbiology 57: 2981–2985.

Amund, O.O. and N. Nwokoye. 1993. Hydrocarbon potentials of yeast isolates from a polluted Lagoon. Journal of Scientific Research and Development 1: 65–68.

Angelova, B. and H.P. Schmauder. 1999. Lipophilic compounds in biotechnology interactions with cells and technological problems. Journal of Biotechnology 67: 13–32.

Applegate, B.M., S.R. Kehrmeyer and G.S. Sayler. 1998. A chromosomally based tod-lux CDABE whole-cell reporter for benzene, toluene, ethybenzene, and xylene (BTEX) sensing. Applied and Environmental Microbiology 64: 2730–2735.

April, T.M., J.M. Foght and R.S. Currah. 2000. Hydrocarbon degrading filamentous fungi isolated from flare pit soils in northern and western Canada. Canadian Journal of Microbiology 46: 38–49.

Asif, M. and T. Muneer. 2007. Energy supply, its demand and security issues for developed and emerging economies. Renewable and Sustainable Energy Reviews 11(7): 1388–1413.

Atlas, R. and J. Bragg. 2009. Bioremediation of marine oil spills: when and when not—the Exxon Valdez experience. Microbial Biotechnology 2: 213–221.

Atlas, R.M. 1981. Microbial degradation of petroleum hydrocarbons: an environmental perspective. Microbiological Reviews 45: 180–209.

Atlas, R.M. 1984. Petroleum Microbiology, Macmillion, New York, NY, USA.

Atlas, R.M. 1992. Petroleum Microbiology. Encyclopedia of Microbiology Academic Press, Baltimore, Md, USA, 363–369.

Atlas, R.M. and R. Bartha. 1998. Fundamentals and Applications. Microbial Ecology Benjamin Cummings, San Francisco, Calif, USA, 4th Edition, 523–530.

Atroley, A. 2006. Available from Oil Spill Off India's Andaman and Nicobar Islands, http://www.wwfindia.org/news_facts/infocus/oil.cfm.

Austin, R.N. and J.T. Groves. 2011. Alkane-oxidizing metalloenzymes in the carbon cycle. Metallomics 3: 775–787.

Bai, G., M.L. Brusseau and R.M. Miller. 1997. Biosurfactant enhanced removal of residual hydrocarbon from soil. Journal of Contaminant Hydrology 25: 157–170.

Banin, E., M.L. Vasil and E.P. Greenberg. 2005. Iron and *Pseudomonas aeruginosa* biofilm formation. Proceedings of the National Academy of Sciences. USA 102: 11076–11081.

Barkay, T. and J. Schaefer. 2001. Metal and radionuclide bioremediation: issues, considerations and potentials. Current Opinion in Microbiology 4: 318–323.

Bartha, R. and I. Bossert. 1984. The treatment and disposal of petroleum wastes. Petroleum Microbiology. R.M. Atlas, Ed. Macmillan, New York, NY, USA, 553–578.

Beal, R. and W.B. Betts. 2000. Role of rhamnolipid biosurfactants in the uptake and mineralization of hexadecane in *Pseudomonas aeruginosa*. Journal of Applied Microbiology 89: 158–168.

Beifuss, U. and M. Tietze. 2005. Methanophenazine and other natural biologically active phenazines. Topics in Current Chemistry 244: 77–113.

Bertrand, E.M., R. Keddis, J.T. Groves, C. Vetriani and R.N. Austin. 2013. Identity and mechanisms of alkane-oxidizing metalloenzymes from deep-sea hydrothermal vents. Frontiers in Microbiology 4: 109.

Bogusławska-Was, E. and W. Dabrowski. 2001. The seasonal variability of yeasts and yeast-like organisms in water and bottom sediment of the Szczecin Lagoon. International Journal of Hygiene and Environmental Health 203: 451–458.

Boopathy, R. and J. Manning. 1998. A laboratory study of the bioremediation of 2,4,6-trinitrotoluene-contaminated soil using aerobic anaerobic soil slurry reactor. Water Environment Research 70: 80–86.

Brooijmans, R.J.W., M.I. Pastink and R.J. Siezen. 2009. Hydrocarbon-degrading bacteria: the oil-spill clean-up crew. Microbial Biotechnology 2: 587–594.

Brusseau, M.L., R.M. Miller, Y. Zhang, X. Wang and G.Y. Bai. 1995. Biosurfactant and cosolvent enhanced remediation of contaminated media. ACS Symposium Series 594: 82–94.

Brusseau, M.L. 1998. The impact of physical, chemical and biological factors on biodegradation. pp. 81–98. *In*: R. Serra (ed.). Proceedings of the International Conference on Biotechnology for Soil Remediation: Scientific Bases and Practical Applications. C.I.P.A. S.R.L., Milan, Italy.

Cameotra, S.S. and P. Singh. 2008. Bioremediation of oil sludge using crude biosurfactants. International Biodeterioration and Biodegradation 62: 274–280.

Carmichael, L.M. and F.K. Pfaender. 1997. Polynuclear aromatic hydrocarbon metabolism in soils: Relationship to soil characteristics and preexposure. Environmental Toxicology and Chemistry 16: 666–675.

Casey, E., B. Glennon and G. Hamer. 1999. Review of membrane aerated biofilm reactors. Resources, Conservation and Recycling 27: 203–215.

Cauwenberghe, L.V. and D.S. Roote. 1998. *In situ* bioremediation technology overview report. Ground Water Technology Analysis Centre O Series. TO-98-01.

Cerniglia, C.E., D.T. Gibson and C. Van Baalen. 1980. Oxidation of naphthalene by cyanobacteria and microalgae. Journal of General Microbiology 116: 495–500.

Chaillan, F., A. Le Flèche and E. Bury. 2004. Identification and biodegradation potential of tropical aerobic hydrocarbon degrading microorganisms. Research in Microbiology 155: 587–595.

Chaillan, F., C.H. Chaîneau, V. Point, A. Saliot and J. Oudot. 2006. Factors inhibiting bioremediation of soil contaminated with weathered oils and drill cuttings. Environmental Pollution 144: 255–265.

Chaîneau, C.H., G. Rougeux, C. Yéprémian and J. Oudot. 2005. Effects of nutrient concentration on the biodegradation of crude oil and associated microbial populations in the soil. Soil Biology and Biochemistry 37: 490–1497.

Chin-A-Woeng, T.F.C, G.V. Bloemberg, A.J. van der Bij, K.M.G.M. van der Drift, J. Schripsema, B. Kroon, R.J. Scheffer, C. Keel, P.A.H.M. Bakker, H. Tichy, F.J. de Bruijn, J.E. Thomas-Oates and B.J.J. Lugtenberg. 1998. Biocontrol by phenazine-1-carboxamide-producing *Pseudomonas chlororaphis* PCL1391 of tomato root rot caused by *Fusarium oxysporum* f. sp. *radicis-lycopersici*. Molecular Plant-Microbe Interactions 11: 1069–1077.

Chin-A-Woeng, T.F.C., J.E. Thomas-Oates, B.J.J. Lugtenberg and G.V. Bloemberg. 2001. Introduction of the phzH gene of *Pseudomonas chlororaphis* PCL1391 extends the range of biocontrol ability of phenazine-1-carboxylic acid-producing *Pseudomonas* spp. strains. Molecular Plant-Microbe Interactions 14: 1006–1015.

Choi, S.C., K.K. Kwon, J.H. Sohn and S.J. Kim. 2002. Evaluation of fertilizer additions to stimulate oil biodegradation in sand seashore mesocosms. Journal of Microbiology and Biotechnology 12: 431–436.

Coleman, N.V., N.B. Bui and A.J. Holmes. 2006. Soluble di-iron monooxygenase gene diversity in soils, sediments and ethene enrichments. Environmental Microbiology 7: 1228–39.

Colwell, R.R., J.D. Walker and J.J. Cooney. 1977. Ecological aspects of microbial degradation of petroleum in the marine environment. Critical Reviews in Microbiology 5: 423–445.

Cooney, J.J. 1984. The fate of petroleum pollutants in fresh water ecosystems. Petroleum Microbiology, R.M. Atlas, Ed. Macmillan, New York, NY, USA, 399–434.

Cooney, J.J., S.A. Silver and E.A. Beck. 1985. Factors influencing hydrocarbon degradation in three freshwater lakes. Microbial Ecology 11: 127–137.

Cornwall, W. 2015. Deepwater Horizon: After the oil. Science 348(6230): 22–29.

Costerton, J.W., P.S. Stewart and E.P. Greenberg. 1999. Bacterial biofilms: a common cause of persistent infections. Science 284: 1318–1322.

Costley, S.C. and F.M. Wallis. 2001. Bioremediation of heavy metals in a synthetic wastewater using a rotating biological contactor. Water Research 35: 3715–3723.

Cunningham, C.J., I.B. Ivshina, V.I. Lozinsky, M.S. Kuyukina and J.C. Philp. 2004. Bioremediation of diesel contaminated soil by microorganisms immobilised in polyvinyl alcohol. International Biodeterioration and Biodegradation 54: 167–174.

Dasgupta, D., R. Ghosh and T.K. Sengupta. 2013. Biofilm-mediated enhanced crude oil degradation by newly isolated *Pseudomonas* species. ISRN Biotechnology 5: 250749. http://dx.doi.org/10.5402/2013/250749.

Dasgupta, D., A. Kumar, B. Mukhopadhyay and T.K. Sengupta. 2015. Isolation of phenazine 1,6-di-carboxylic acid from *Pseudomonas aeruginosa* strain HRW.1-S3 and its role in biofilm-mediated crude oil degradation and cytotoxicity against bacterial and cancer cells. Applied Microbiology and Biotechnology 99(20): 8653–65.

Daugulis, A.J. and C.M. McCracken. 2003. Microbial degradation of high and low molecular weight polyaromatic hydrocarbons in a two-phase partitioning bioreactor by two strains of *Sphingomonas* sp. Biotechnology Letters 25: 1441–1444.

Davies, D.G., A.M. Chakrabarty and G.G. Geesey. 1993. Exopolysaccharide production in biofilms: substratum activation of alginate gene expression by *Pseudomonas aeruginosa*. Applied Environmental Microbiology 59: 1181–1186.

Delaney, S.M., D.V. Mavrodi, R.F. Bonsall and L.S. Thomashow. 2001. phzO, a gene for biosynthesis of 2-hydroxylated phenazine compounds in *Pseudomonas aureofaciens* 30–84. Journal of Bacteriology 183: 318–327.

Denac, M. and I.J. Dunn. 1988. Packed- and fluidized-bed biofilm reactor performance for anaerobic wastewater treatment. Biotechnology and Bioengineering 32: 159–173.

Dhote, M., A. Juwarkar and A. Kumar. 2013. Microbial assisted bioremediation of oil sludge: Present and future. ABS Journal of Sustainable Biotechnology, 1–18.

Díaz, M.P., K.G. Boyd, S.J.W.W. Grigson and J.G. Burgess. 2002. Biodegradation of crude oil across a wide range of salinities by an extremely halotolerant bacterial consortium MPD-M, immobilized onto polypropylene fibers. Biotechnology and Bioengineering 79: 145–153.

Du, M. and J.D. Kessler. 2012. Assessment of the spatial and temporal variability of bulk hydrocarbon respiration following the deepwater horizon oil spill. Environmental Science and Technology 46: 10499–10507.

Eker, S. and F. Kargi. 2008. Biological treatment of 2,4-dichlorophenol containing synthetic wastewater using a rotating brush biofilm reactor. Bioresource Technology 99: 2319–2325.

Eker, S. and F. Kargi. 2010. COD, para-chlorophenol and toxicity removal from synthetic wastewater using rotating tubes biofilm reactor (RTBR). Bioresource Technology 101: 9020–9024.

Firn, R.D. and C.D. Jones. 2003. Natural products—a simple model to explain chemical diversity. Natural Products Reports 20: 382–391.

Floodgate, G. 1984. The fate of petroleum in marine ecosystems. Petroleum Microbiology, R.M. Atlas, Ed. Macmillion, New York, NY, USA, 355–398.

Foght, J.M. and D.S.W. Westlake. 1987. Biodegradation of hydrocarbons in freshwater oil. pp. 217–230. *In*: J.H. Vandermeulen and S.R. Hrudey (eds.). Freshwater: Chemistry, Biology, Countermeasure Technology. Pergamon Press, New York, NY, USA.

Foght, J.M., D.S.W. Westlake, W.M. Johnson and H.F. Ridgway. 1996. Environmental gasoline-utilizing isolates and clinical isolates of *Pseudomonas aeruginosa* are taxonomically indistinguishable by chemotaxonomic and molecular techniques. Microbiology 142: 2333–2340.

Fritsche, W. and M. Hofrichter. 2000. Aerobic degradation by microorganisms. pp. 146–155. *In*: J. Klein (ed.). Environmental Processes-Soil Decontamination. Wiley-VCH, Weinheim, Germany.

Gacesa, P. 1998. Bacterial alginate biosynthesis: recent progress and future prospects. Microbiology 144: 1133–43.

Ghollami, M., M. Roayaei, F. Ghavipanjeh and B. Rasekh. 2013. Bioconversion of heavy hydrocarbon cuts containing high amounts of resins by microbial consortia. Journal of Petroleum and Environmental Biotechnology 4: 139.

Goh, E.B, G. Yim, W. Tsui, J. McClure, M.G. Surette and J. Davies. 2002. Transcriptional modulation of bacterial gene expression by subinhibitory concentrations of antibiotics. Proceedings of the National Academy of Sciences. USA 99: 17025–17030.

Gojgic-Cvijovic, G.D., J.S. Milic, T.M. Solevic, V.P. Beskoski, M.V. Ilic, L.S. Djokic, T.M. Narancic and M.M. Vrvic. 2012. Biodegradation of petroleum sludge and petroleum polluted soil by a bacterial consortium: a laboratory study. Biodegradation 23(1): 1–14.

Govan, J.R.W. and J.A.M. Fyfe. 1978. Mucoid *Pseudomonas aeruginosa* and cystic fibrosis: resistance of the mucoid form to carbenicillin, flucloxacillin, and tobramycin and the isolation of mucoid variants *in vitro*. Journal of Antimicrobial Chemotherapy 4: 233–40.

Govan, J.R.W. and V. Deretic. 1996. Microbial pathogenesis in cystic fibrosis: mucoid *Pseudomonas aeruginosa* and *Burkholderia cepacia*. Microbiology Reviews 60: 539–74.

Guerriero, V., S. Vitale, S. Ciarcia and S. Mazzoli. 2012. Improved statistical multi-scale analysis of fractured reservoir analogues. Tectonophysics 504: 1–4.

Guerriero, V., S. Mazzoli, A. Iannace, S. Vitale, A. Carravetta and C. Strauss. 2013. A permeability model for naturally fractured carbonate reservoirs. Marine and Petroleum Geology 40: 115–134.

Hall-Stoodley, L. 2004. Bacterial biofilms: From the natural environment to infectious diseases. Nature Reviews in Microbiology 2: 95–108.

Hall, C.A. and C.J. Cleveland. 1981. Petroleum drilling and production in the United States: yield per effort and net energy analysis. Science 6; 211(4482): 576–9.

Hamed, S.B., L. Smii, A. Ghram and A. Maaroufi. 2012. Screening of potential biosurfactant-producing bacteria isolated from seawater biofilm. African Journal of Biotechnology 11: 14153–14158.

Hamed, S.B., A. Maaroufi, A. Ghram, B.A.G. Zouhaier and M. Labat. 2013. Isolation of four hydrocarbon effluent-degrading Bacillaceae species and evaluation of their ability to grow under high-temperature or high-salinity conditions. African Journal of Biotechnology 12: 1636–1643.

Head, I.M., D.M. Jones and W.F. Röling. 2006. Marine microorganisms make a meal of oil. Nature Reviews in Microbiology 4(3): 173–82.

Hernandez, M.E., A. Kappler and D.K. Newman. 2004. Phenazines and other redox-active antibiotics promote microbial mineral reduction. Applied and Environmental Microbiology 70: 921–928.

Hisatsuka, K., T. Nakahara, N. Sano and K. Yamada. 1971. Formation of rhamnolipid by *Pseudomonas aeruginosa* and its function in hydrocarbon fermentation. Agricultural and Biological Chemistry 35: 686–692.

Hollaway, S.L., G.M. Faw and R.K. Sizemore. 1980. The bacterial community composition of an active oil field in the Northwestern Gulf of Mexico. Marine Pollution Bulletin 11: 153–156.

Holliger, C., S. Gaspard and G. Glod. 1997. Contaminated environments in the subsurface and bioremediation: organic contaminants. FEMS Microbiology Reviews 20: 517–523.

Hommel, R.K. 1990. Formation and phylogenetic role of biosurfactants. Journal of Applied Microbiology 89: 158–119.

Ilori, M.O., S.A. Adebusoye and A.C. Ojo. 2008. Isolation and characterization of hydrocarbon-degrading and biosurfactant-producing yeast strains obtained from a polluted lagoon water. World Journal of Microbiology and Biotechnology 24: 2539–2545.

Itoh, S. and T. Suzuki. 1972. Effect of rhamnolipids on growth of *Pseudomonas aeruginosa* mutant deficient in n-paraffin-utilizing ability. Agricultural and Biological Chemistry 36: 2233–2235.

Jeswani, H. and S. Mukherji. 2012. Degradation of phenolics, nitrogen-heterocyclics and polynuclear aromatic hydrocarbons in a rotating biological contactor. Bioresource Technology 111: 12–20.

Jones, D.M., A.G. Douglas, R.J. Parkes, J. Taylor, W. Giger and C. Schaffner. 1983. The recognition of biodegraded petroleum-derived aromatic hydrocarbons in recent marine sediments. Marine Pollution Bulletin 14: 103–108.

Jones, S.C., W.O. Roszelle and M.A. Svaldi. 1970. High water content oil-external micellar dispersions. Date: 2/24/1970 US 3497006. https://www.osti.gov/biblio/7301198.

Juhas, M., L. Eberl and B. Tummler. 2005. Quorum sensing: the power of cooperation in the world of *Pseudomonas*. Environmental Microbiology 7: 459–471.

Kaindl, N. 2010. Upgrading of an activated sludge wastewater treatment plant by adding a moving bed biofilm reactor as pre-treatment and ozonation followed by biofiltration for enhanced COD reduction: design and operation experience. Water Science and Technology 62: 2710–2719.

Kargi, F. and S. Eker. 2005. Removal of 2,4-dichlorophenol and toxicity from synthetic wastewater in a rotating perforated tube biofilm reactor. Process Biochemistry 40: 2105–2111.

Kelly, C.J., C.A. Lajoie, A.C. Layton and G.S. Sayler. 1999. Bioluminescent reporter bacterium for toxicity monitoring in biological wastewater treatment systems. Water Environment Research 71: 31–35.

Kerr, J.R. 2000. Phenazine pigments: antibiotics and virulence factors. Infectious Disease Review 2: 184–194.

Kim, S.J., D.H. Choi, D.S. Sim and Y.S. Oh. 2005. Evaluation of bioremediation effectiveness on crude oil-contaminated sand. Chemosphere 59: 845–852.

Kiran, G.S., T.A. Hema and R. Gandhimathi. 2009. Optimization and production of a biosurfactant from the sponge-associated marine fungus *Aspergillus ustus* MSF3. Colloids and Surfaces B Biointerfaces 73: 250–256.

Klein, B., P. Bouriat, P. Goulas and R. Grimaud. 2010. Behavior of Marinobacter hydrocarbonoclasticus SP17 cells during initiation of biofilm formation at the alkane-water interface. Biotechnology and Bioengineering 105: 461–468.

Korotkevych, O., J. Josefiova, M. Praveckova, T. Cajthaml, M. Stavelova and M.V. Brennerova. 2011. Functional adaptation of microbial communities from jet fuel-contaminated soil under bioremediation treatment: simulation of pollutant rebound. FEMS Microbiology Ecology 78: 137–149.

Kumar, S., V.K. Dagar, Y.P. Khasa and R.C. Kuhad. 2013. Genetically Modified Microorganisms (GMOs) for bioremediation. Biotechnology for Environmental Management and Resource Recovery, 191–218.

Kumar, T.A. and S. Saravanan. 2009. Treatability studies of textile wastewater on an aerobic fluidized bed biofilm reactor (FABR): a case study. Water Science Tehnology 59: 1817–1821.

Kuspa, A., L. Plamann and D. Kaiser. 1992. A signaling and the cell density requirement for *Myxococcus xanthus* development. Journal of Bacteriology 174: 7360–69.

Kvenvolden, K.A. and C.K. Cooper. 2003. Natural seepage of crude oil into the marine environment. Geo-Marine Letters 23: 140–146.

Labrenz, M., G.K. Druschel, T. Thomsen-Ebert, B. Gilbert, S.A. Welch, K.M. Kemner, G.A. Logan, R.E. Summons, G. De Stasio, P.L. Bond, B. Lai, S.D. Kelly and J.F. Banfield. 2000. Formation of sphalerite (ZnS) deposits in natural biofilms of sulfate-reducing bacteria. Science 290: 1744–1747.

Ladousse, A. and B. Tramier. 1991. Results of 12 years of research in spilled oil bioremediation: inipol EAP 22. In Proceedings of the International Oil Spill Conference American Petroleum Institute, Washington, DC, USA, 577–581.

Lal, B. and S. Khanna. 1996. Degradation of crude oil by *Acinetobacter calcoaceticus* and *Alcaligenes odorans*. The Journal of Applied Bacteriology 81(4): 355–62.

Lau, G.W., D.J. Hassett and B.E. Britigan. 2005. Modulation of lung epithelial functions by *Pseudomonas aeruginosa*. Trends in Microbiology 13: 389–397.

Laursen, J.B. and J. Nielsen. 2004. Phenazine natural products: biosynthesis, synthetic analogues, and biological activity. Chemical Reviews 104: 1663–1686.

Layton, A.C., B. Gregory, T.W. Schultz and G.S. Sayler. 1999. Validation of genetically engineered bioluminescent surfactant resistant bacteria as toxicity assessment tools. Ecotoxicology and Environmental Safety 43: 222–228.

Lazdunski, A.M., I. Ventre and J.N. Sturgis. 2004. Regulatory circuits and communication in Gram-negative bacteria. Nature Reviews Microbiology 2: 581–592.

Leahy, J.G. and R.R. Colwell. 1990. Microbial degradation of hydrocarbons in the environment. Microbiological Reviews 54: 305–315.

Lee, K., G.H. Tremblay, J. Gauthier, S.E. Cobanli and M. Griffin. 1997. Bioaugmentation and biostimulation: a paradox between laboratory and field results. Proceedings of the International Oil Spill Conference, American Petroleum Institute, Washington, DC, USA, 697–705.

Lenchi, N., Ö. İnceoğlu, S. Kebbouche-Gana, M.L. Gana and M. Llirós. 2013. Diversity of microbial communities in production and injection waters of algerian oilfields revealed by 16S rRNA gene amplicon 454 pyrosequencing. PLoS ONE 8(6): e66588.

Lendenmann, U. and J.C. Spain. 1998. Simultaneous biodegradation of 2,4-dinitrotoluene and 2,6-dinitrotoluene in an aerobic fluidized-bed biofilm reactor. Environmental Science and Technology 32: 82–87.

Li, P., L. Wang and L. Feng. 2013. Characterization of a novel rieske-type alkane monooxygenase system in *Pusillimonas* sp. strain T7-7. Journal of Bacteriology 195: 1892–1901.

Li, P., Y. Zhang, M. Li and B. Li. 2015. Bioremediation of oil containing seawater by membrane-aerated biofilm reactor. Industrial and Engineering Chemical Research 54: 13009−13016.

Lloyd, J.R. 2003. Microbial reduction of metals and radionuclides. FEMS Microbiol. Rev. 27: 411–425.

Mahmound, A., Y. Aziza, A. Abdeltif and M. Rachida. 2008. Biosurfactant production by Bacillus strain injected in the petroleum reservoirs. Journal of Industrial Microbiology & Biotechnology 35: 1303–1306.

Manning, J., R. Boopathy and C.F. Kulpa. 1995. A laboratory study in support of the pilot demonstration of a biological soil slurry reactor. Report no. SFIM-AEC-TS-CR-94038. US Army Environmental Center, Aberdeen Proving Ground, MD.

Marchant, R., F.H. Sharkey, I.M. Banat, T.J. Rahman and A. Perfumo. 2006. The degradation of n-hexadecane in soil by thermophilic geobacilli. FEMS Microbiology Ecology 56(1): 44–54.

Mason, O.U., T.C. Hazen, S. Borglin, P.S.G. Chain, E.A. Dubinsky, J.L. Fortney, J. Han, H.Y.N. Holman, J. Hultman, R. Lamendella, R. Mackelprang, S. Malfatti, L.M. Tom, S.G. Tringe, T. Woyke, J. Zhou, E.M. Rubin and J.K. Jansson. 2012. Metagenome, metatranscriptome and single-cell sequencing reveal microbial response to Deepwater Horizon oil spill. The ISME Journal 6: 1715–1727.

Masson, L., B.E. Tabashnik and A. Mazza. 2002. Mutagenic analysis of a conserved region of domain III in the Cry1ac toxin of *Bacillus thuringiensis*. Applied and Environmental Microbiology 68: 194–200.

Mavrodi, D.V., R.F. Bonsall, S.M. Delaney, M.J. Soule, G. Phillips and L.S. Thomashow. 2001. Functional analysis of genes for biosynthesis of pyocyanin and phenazine-1-carboxamide from *Pseudomonas aeruginosa* PAO1. Journal of Baceteriology 183: 6454–6465.

Mavrodi, D.V., W. Blankenfeldt and L.S. Thomashow. 2006. Phenazine compounds in fluorescent *Pseudomonas* spp. biosynthesis and regulation. Annual Reviews of Phytopathology 44: 417–45.

McLean, R.J., M. Whiteley, D.J. Strickler and W.C. Fuqua. 1997. Evidence of autoinducer activity in naturally occurring biofilms. FEMS Microbiol. Lett. 154: 259–263.

Mearns, A.J. 1997. Cleaning oiled shores: putting bioremediation to the test. Spill Science and Technology Bulletin 4: 209–217.

Medina-Bellver, J.I., P. Marín and A. Delgado. 2005. Evidence for *in situ* crude oil biodegradation after the Prestige oil spill. Environmental Microbiology 7: 773–779.

Meintanis, C., K.I. Chalkou, K.A. Kormas and A.D. Karagouni. 2006. Biodegradation of crude oil by thermophilic bacteria isolated from a volcano island. Biodegradation 17: 105–11.

Mitsch, W.J. and J.G. Gosselink. 1993. Wetlands. John Wiley & Sons, New York, NY, USA, 2nd Edition.

Moliterni, E., R.G. Jiménez-Tusset, M.V. Rayo, L. Rodriguez, F.J. Fernández and J. Villaseñor. 2012. Kinetics of biodegradation of diesel fuel by enriched microbial consortia from polluted soils. International Journal of Environmental Science and Technology 9: 749–758.

Muthusamy, K., S. Gopalakrishnan, T.K. Ravi and P. Sivachidambaram. 2008. Biosurfactants: properties, commercial production and application. Current Science 94: 736–747.

Nakahara, T., K. Hisatsuka and Y. Minoda. 1981. Effect of hydrocarbon emulsification on growth and respiration of microorganisms in hydrocarbon media. Journal of Fermentation Technology 59: 415–418.

Newell, P.D., C.D. Boyd, H. Sondermann and G.A. O'Toole. 2011. A c-di-GMP effector system controls cell adhesion by inside-out signaling and surface protein cleavage. PLoS Biol. 9: e1000587.

Nichols, W.J. 2001. The U.S. Environmental Protect Agency: National Oil and Hazardous Substances Pollution Contingency Plan, Subpart J Product Schedule (40 CFR 300.900). In Proceedings of the International Oil Spill Conference, American Petroleum Institute, Washington, DC, USA, 1479–1483.

Nikolopoulou, M. and N. Kalogerakis. 2009. Biostimulation strategies for fresh and chronically polluted marine environments with petroleum hydrocarbons. Journal of Chemical Technology and Biotechnology 84: 802–807.

Nisenbaum, M., G.H. Sendra, G.A.C. Gilbert, M. Scagliola, J.F. González and S.E. Murialdo. 2013. Hydrocarbon biodegradation and dynamic laser speckle for detecting chemotactic responses at low bacterial concentration. Journal of Environmental Sciences 25: 613–625.

Noordman, W.H. and D.B. Janssen. 2002. Rhamnolipid stimulates uptake of hydrophobic compounds by *Pseudomonas aeruginosa*. Applied and Environmental Microbiology 68(9): 4502–8.

Nopcharoenkul, W., P. Netsakulnee and O. Pinyakong. 2013. Diesel oil removal by immobilized *Pseudoxanthomonas* sp. RN402. Biodegradation 24: 387–397.

Nzila, A. 2013. Update on the cometabolism of organic pollutants by bacteria. Environmental Pollution 178: 474–482.

Obayori, O.S., M.O. Ilori, S.A. Adebusoye, G.O. Oyetibo, A.E. Omotayo and O.O. Amund. 2009. Degradation of hydrocarbons and biosurfactant production by *Pseudomonas* sp. strain LP1. World Journal of Microbiology and Biotechnology 25: 1615–1623.

Okolo, J.C., E.N. Amadi and C.T.I. Odu. 2005. Effects of soil treatments containing poultry manure on crude oil degradation in a sandy loam soil. Applied Ecology and Environmental Research 3: 47–53.

On Scene Coordinator Report on Deepwater Horizon Oil Spill. 2011. http://www.uscg.mil/foia/docs/dwh/fosc_dwh_report.pdf.

O'Toole, G.A. and R. Kolter. 1998. Flagella and twitching motility are necessary for *Pseudomonas aeruginosa* biofilm development. Molecular Microbiology 30: 295–304.

Oudot, J., F.X. Merlin and P. Pinvidic. 1998. Weathering rates of oil components in a bioremediation experiment in estuarine sediments. Marine Environmental Research 45: 113–125.

Pandey, G. and R.K. Jain. 2002. Bacterial chemotaxis toward environmental pollutants: role in bioremediation. Applied and Environmental Microbiology 68: 5789–5795.

Parsek, M.R. and G.P. Greenberg. 2000. Acyl-homoserine lactone quorum sensing in gram-negative bacteria: a signaling mechanism involved in associations with higher organisms. Proceeding of National Academy of Science USA 97: 8789–93.

Paul, D., G. Pandey, J. Pandey and R.K. Jain. 2005. Accessing microbial diversity for bioremediation and environmental restoration. Trends in Biotechnol. 23: 135–142.

Pelletier, E., D. Delille and B. Delille. 2004. Crude oil bioremediation in sub-Antarctic intertidal sediments: chemistry and toxicity of oiled residues. Marine Environmental Research 57: 311–327.

Pemmaraju, S.C., D. Sharma, N. Singh, R. Panwar, S.S. Cameotra and V. Pruthi. 2012. Production of microbial surfactants from oily sludge-contaminated soil by *Bacillus subtilis* DSVP23. Applied Biochemistry and Biotechnology 167(5): 1119–31.

Perry, J.J. 1984. Microbial metabolism of cyclic alkanes. Petroleum Microbiology, R.M. Atlas, Ed., Macmillan, New York, NY, USA, 61–98.

Perumbakkam, S., T.F. Hess and R.L. Crawford. 2006. A bioremediation approach using natural transformation in pure-culture and mixed-population biofilms. Biodegradation 17: 545–557.

Pinholt, Y., S. Struwe and A. Kjøller. 1979. Microbial changes during oil decomposition in soil. Holarct. EcoL. 2: 195.

Pirnay, J.P., S. Matthijs, H. Colak, P. Chablain, F. Bilocq, J. Van Eldere, D. De Vos, M. Ziz, L. Triest and P. Cornelis. Global *Pseudomonas aeruginosa* biodiversity as reflected in a Belgian river. Environmental Microbiology 7: 969–980.

Pornsunthorntawee, O., P. Wongpanit, S. Chavadej, M. Abe and R. Rujiravanit. 2008. Structural and physicochemical characterization of crude biosurfactant produced by *Pseudomonas aeruginosa* SP4 isolated from petroleum contaminated soil. Bioresource Technology 99: 1589–1595.

Pratt, L.A. and R. Kolter. 1999. Genetic analysis of bacterial biofilm formation. Curr. Opin. Microbiol. 2: 598–603.

Price-Whelan, A., L.E. Dietrich and D.K. Newman. 2006. Rethinking 'secondary' metabolism: physiological roles for phenazine antibiotics. Nat. Chem. Biol. 2(2): 71–8.

Prince, R.C. 1993. Petroleum spill bioremediation in marine environments. Critical Reviews in Microbiology 19: 217–242.

Providenti, M.A., C.W. Greer, H. Lee and J.T. Trevors. 1995. Phenanthrene mineralization by *Pseudomonas* sp. UG14. World Journal of Microbiology 11: 271–279.

Puhakka, J.A. 1995. Fluidized bed biofilms for chlorophenol mineralization. Water Science and Technology 31: 227–235.

Rachid, S. 2012. Mutation of RNA polymerase beta subunit confers in both enhancement of environmental bioremediation and antibiotic production in a *Streptomyces* species, petroleum and mineral resources. WIT Transaction on Engineering Sciences 81: 279.

Rahman, K.S.M., T.J. Rahman, Y. Kourkoutas, I. Petsas, R. Marchant and I.M. Banat. 2003. Enhanced bioremediation of n-alkane in petroleum sludge using bacterial consortium amended with rhamnolipid and micronutrients. Bioresource Technology 90: 159–168.

Rao, Y.M. and G.K. Sureshkumar. 2000. Oxidative-stress-induced production of pyocyanin by *Xanthomonas campestris* and its effect on the indicator target organism, *Escherichia coli*. J. Journal of Industrial Microbiology and Biotechnology 25: 266–272.

Romero, M.C., M.I. Urrutia, E.H. Reinoso and A.M. Kiernan. 2011. Effects of the sorption/desorption process on the fluoranthene degradation by wild strains of *Hansenula angusta* and *Rhodotorula minuta*. International Research Journal of Microbiology, 2230–236.

Safa, M., I. Alemzadeh and M. Vossoughi. 2014. Biodegradability of oily wastewater using rotating biological contactor combined with an external membrane. Journal of Environmental Health Science and Engineering 12: 117.

Saimmai, A., O. Rukadee, T. Onlamool, V. Sobhon and S. Maneerat. 2012. Characterization and phylogenetic analysis of microbial surface active compound-producing bacteria. Applied Biochemistry and Biotechnology 168: 1003–1018.

Sarayu, K. and S. Sandhya. 2012. Rotating biological contactor reactor with biofilm promoting mats for treatment of benzene and xylene containing wastewater. Applied Biochemistry and Biotechnology 168: 1928–1937.

Sayler, G.S. and S. Ripp. 2000. Field applications of genetically engineered microorganisms for bioremediation processes. Current Opinion in Biotechnology 11: 286–289.

Scheuer, U., T. Zimmer, D. Becher, F. Schauer and W. Schunck. 1998. Oxygenation cascade in conversion of n-alkanes to α,ω-dioic acids catalyzed by cytochrome P450 52A3. Journal of Biological Chemistry 273: 32528–32534.

Semmens, M.J., K. Dahm, J. Shanahan and A. Christianson. 2003. COD and nitrogen removal by biofilms growing on gas permeable membranes. Water Research 37: 4343−4350.

Shieh, W. and J. Keenan. 1986. Fluidized Bed Biofilm Reactor for Wastewater Treatment. Bioproducts, Springer Berlin Heidelberg, pp. 131–169.

Simmons, M.R. 2005. Twilight in the Desert: The Coming Saudi Oil Shock and the World Economy. Unites States. Wiley.

Sohngen, N.L. 1913. Benzin, Petroleum, Paraffino und Paraffin als Kohlenstoff—und Energiequelle, fur Mikroben. Zentl. Bakt. Parasitek Aby. II Vol. 37: 595–609.

Sousa, C., V. De Lorenzo and A. Cebolla. 1997. Modulation of gene expression through chromosomal positioning in *Escherichia coli*. Microbiology 143: 2071–2078.

Stelmack, P.L., M.R. Gray and M.A. Pickard. 1999. Bacterial adhesion to soil contaminants in the presence of surfactants. Applied Environmental Microbiology 65: 163–168.

Stickler, D.J., N.S. Morris, R.J. McLean and C. Fuqua. 1998. Biofilms on indwelling urethral catheters produce quorum-sensing signal molecules *in situ* and *in vitro*. Applied Environmental Microbiology 64: 3486–3490.

Sukhdhane, K.S., E.R. Priya, M.R. Shailendra and J. Teena. 2013. Status of oil pollution in Indian coastal waters. Fishing Chimes 33(5): 53–54.

Swannell, R.P.J., K. Lee and M. Mcdonagh. 1996. Field evaluations of marine oil spill bioremediation. Microbiological Reviews 60: 342–365.

Throne-Holst, M., A. Wentzel, T.E. Ellingsen, H.K. Kotlar and S.B. Zotchev. 2007. Identification of novel genes involved in long-chain n-alkane degradation by *Acinetobacter* sp. strain DSM 17874. Applied and Environmental Microbiology 73(10): 3327–32.

Turner, J.M. and A.J. Messenger. 1986. Occurrence, biochemistry and physiology of phenazine pigment production. Advances in Microbial Physiology 27: 211–275.

Ulrici, W. 2000. Contaminant soil areas, different countries and contaminant monitoring of contaminants in Environmental Process II. Soil Decontamination Biotechnology, H.J. Rehm and G. Reed, Eds. 11: 5–42.

U.S. EPA, Spill NCP Product Schedule. 2002. http://www.epa.gov/oilspill.

Van Beilen, J.B., Z. Li, W.A. Duetz, T.H.M. Smits and B. Witholt. 2003. Diversity of alkane hydroxylase systems in the environment. Oil & Gas Science and Technology 58: 427–440.

Van Beilen, J.B. and E.G. Funhoff. 2005. Expanding the alkaneoxygenase toolbox: new enzymes and applications. Current Opinion in Biotechnology 16: 308–314.

Van Beilen, J.B. and E.G. Funhoff. 2007. Alkane hydroxylases involved in microbial alkane degradation. Applied Microbiology and Biotechnology 74: 13–21.

Van Dyke, M.I., H. Lee and J.T. Trevors. 1996. Survival of luxAB-marked *Alcaligenes eutrophus* H850 in PCB-contaminated soil and sediment. Journal of Chemical Technology and Biotechnology 65: 15–122.

Venosa, A.D., M.T. Suidan and B.A. Wrenn. 1996. Bioremediation of an experimental oil spill on the shoreline of Delaware Bay. Environmental Science and Technology 30: 1764–1775.

Venosa, A.D., D.W. King and G.A. Sorial. 2002. The baffled flask test for dispersant effectiveness: a round Robin evaluation of reproducibility and repeatability. Spill Science and Technology Bulletin 7: 299–308.

Venosa, A.D. and X. Zhu. 2003. Biodegradation of crude oil contaminating marine shorelines and freshwater wetlands. Spill Science and Technology Bulletin 8: 163–178.

Villavicencio, R.T. 1998. The history of blue pus. Journal of the American College of Surgeons 187: 212–216.

Vlamakis, H., Y. Chai, P. Beauregard, R. Losick and R. Kolter. 2013. Sticking together: building a biofilm the *Bacillus subtilis* way. Nature Reviews in Microbiology 11(3): 157–68.

Volkering, F., A.M. Breure and W.H. Rulkens. 1998. Microbiological aspects of surfactant use for biological soil remediation. Biodegradation 8: 401–417.

Wagner-Dobler, I. 2003. Pilot plant for bioremediation of mercury-containing industrial wastewater. Applied Microbiology and Biotechnology 62: 124–133.

Walker, J.D., R.R. Colwell, Z. Vaituzis and S.A. Meyer. 1975. Petroleum degrading achlorophyllous alga *Prototheca zopfii*. Nature 254: 423–424.

Wang, L., Y.Q. Tang, P. Guo, Y Luo, X.L. Wu and H. Wang. 2013. Microbial functioning on crude oil in a gas-permeable single microfluidic channel. Journal of Petroleum Science and Engineering 104: 38–48.

Watnick, P.I. and R. Kolter. 1999. Steps in the development of a *Vibrio cholerae* biofilm. Molecular Microbiology 34: 586–595.

Wei, X., B. Li, S. Zhao, L. Wang, H. Zhang, C. Li and S. Wang. 2012. Mixed pharmaceutical wastewater treatment by integrated membrane-aerated biofilm reactor (MABR) system—a pilot-scale study. Bioresource Technology 122: 189–95.

White, C. and G.M. Gadd. 2000. Copper accumulation by sulfate-reducing bacterial biofilms. FEMS Microbiology Letters 183: 313–318.

Wilson, N.G. and G. Bradley. 1996. The effect of immobilization on rhamnolipid production by *Pseudomonas fluorescens.* Journal of Applied Bacteriology 81: 525–530.

Wilson, R., D.A. Sykes, D. Watson, A. Rutman, G.W. Taylor and P.J. Cole. 1988. Measurement of *Pseudomonas aeruginosa* phenazine pigments in sputum and assessment of their contribution to sputum sol toxicity for respiratory epithelium. Infection and Immunity 56: 2515–2517.

Yakimov, M.M., K.N. Timmis and P.N. Golyshin. 2007. Obligate oil-degrading marine bacteria. Current Opinion in Biotechnology 18: 257–266.

Zhang, Y. and R.M. Miller. 1994. Effect of a *Pseudomonas* rhamnolipid biosurfactant on cell hydrophobicity and biodegradation of octadecane. Applied and Environmental Microbiology 60: 2101–2106.

Zobell, C.E. 1946. Action of microorganisms on hydrocarbons. Bacteriological Review 10: 1–49.

Zwick, T.C., E.A. Foote and A.J. Pollack. 1997. Effects of nutrient addition during bioventing of fuel contaminated soils in an arid environment. *In-Situ* and On-Site Bioremediation. Battelle Press, Columbus, Ohio, USA, 403–409.

CHAPTER **2**

Aerobic Granulation Technology for Wastewater Treatment

Joo-Hwa Tay[1],* and *Saurabh Jyoti Sarma*[1,2],*

1. Introduction

Aerobic granulation technology (AGT) is an emerging technology of wastewater treatment, which has been mostly explored by the researchers from around 1998–2000 (Dangcong et al. 1999, Tay et al. 2001a). Owing to its potential advantages over the conventional wastewater treatment processes, it has received immense attention after its discovery and has the potential to replace the existing wastewater treatment processes (Tay et al. 2001b). Aerobic granules are formed during wastewater treatment in sequencing batch reactor (SBR) by self-immobilization of different types of microorganisms present in the wastewater (Gonzalez-Gil and Holliger 2014). These granules are denser and heavier than the loose microbial aggregates (e.g., activated sludge) commonly found in the modern day wastewater treatment processes. Therefore, they can settle down very quickly and separation of the sludge or biomass from the liquid phase becomes easy (Ni 2012). The amount of sludge produced by the AGT is significantly lower than that of a conventional activated sludge process (Nor Anuar et al. 2006). Reduction in land and infrastructure related investment requirement is considered as another benefit of this technology. Apart from the aforementioned advantages, the AGT is capable of simultaneously removing carbon, nitrogen, phosphorus, various organic pollutants and heavy metals from both municipal and industrial wastewaters (Wang et al. 2010, Ahn and Hong 2015, Jiang et al. 2016).

The microbial and molecular biological mechanisms behind granule formation, such as quorum sensing, are the topics for further investigations on this technology. This chapter will provide an outline of the roles of SBR in granule formation and various hypotheses proposed to explain the granulation process. Based on authors' own experience and literature data, a new mechanism of aerobic granule formation has been proposed. Importance and major challenges of this technology have been discussed and new research directions have been outlined. Over the years, the technology has been going through modifications to improve its efficiency. A short discussion on the recent advances and modifications of this technology has also been included.

---

[1] Department of Civil Engineering, Schulich School of Engineering, University of Calgary, 2500 University Drive NW, Calgary (Alberta), Canada, T2N 1N4.

[2] Department of Biotechnology, School of Engineering and Applied Sciences, Bennett University, Plot No 8-11, TechZone II, Greater Noida, Uttar Pradesh 201310, India.

* Corresponding authors: jhtay@ucalgary.ca; saurabh.sarma@bennett.edu.in

2. Sequencing batch reactors for granule formation

Sequencing batch reactors (SBR) are the most suitable bioreactors for aerobic granulation. High biomass retention capacity of SBR and its cyclic feast and famine operation mode are two possible reasons behind this observation. Both stirred tank based SBR and airlift type SBR could be used for this purpose (Rezaei et al. 2012, Mosquera-Corral et al. 2011, Di Bellaa and Torregrossa 2014). Typically, SBR operation is divided into different phases such as (i) feeding phase, (ii) reaction or mixing phase, (iii) biomass settling phase and (iv) effluent draining phase (Jungles et al. 2014, Hwan Oh 2018). In the feeding phase, the reactor is allowed to fill with influent wastewater. Reaction or mixing phase is for homogeneous distribution of the biomass and substrate, and for efficient oxygen transfer. It is achieved either by airflow or by combined aeration and stirring. The biomass settling phase is crucial for SBR as it allows separation of solid and liquid phase and prevents the biomass from washing out. As a part of the final phase of a SBR cycle, a definite amount of the liquid is drained from the top of the reactor. The time allocated to each phase is predetermined and is operated in cycles. Now-a-days, different phases of SBR are controlled by programmable controllers and they are very useful in precisely controlling the duration of the phases and cycles (Jungles et al. 2014, Boon 2003). Intermittent flow and continuous flow SBR are two broad variations of SBR technology. In the case of intermittent flow SBR, influent wastewater flow (feeding phase) is periodically stopped and remainder of the phases are allowed to complete before the next cycle. However, in the case of continuous flow SBR, the reactor is continuously fed and drained, and reaction/mixing and settling phases keep on repeating in cycle (Morgenroth and Wilderer 1998). According to Singh and Srivastava (2011), SBR allows a single reactor to have both aerobic and anaerobic phases. This feature reduces the aeration cost (by nearly 25%) as well as overall sludge production (Singh and Srivastava 2011). This is one of the reasons behind relatively low sludge production by AGT.

SBR has an excellent biomass retention capacity (Strous et al. 1998). Due to high biomass retention, high cell density can be quickly achieved and the process quickly enters into the stationary phase of growth. As the accumulated biomass/sludge is not frequently wasted, the SBR process has the advantage of using the microorganisms mostly in their stationary phase of growth (Williams et al. 2017). In the stationary phase, microbial growth rate is the minimum; this is another reason for the relatively low sludge production during wastewater treatment by AGT.

It has been demonstrated that in some microorganisms, the production of extracellular polymeric substances (EPS) is highest in the stationary phase of the growth (Hu et al. 2012, Petry et al. 2000, Liu and Buskey 2000). High biomass retention in SBR has direct effect on microbial growth phase and in turn, growth phase is related to EPS production. Thus, EPS production and granulation are relatively more common when the process is operated under SBR mode.

Granule formation in the SBR is commonly promoted by introducing a very short biomass settling phase. It allows relatively heavier aggregate of biomass to quickly settle down, whereas the light-weight microbial flocs are washed out with the effluent liquid drained from the top of the reactor (Morgenroth et al. 1997). Thus, the microorganisms capable of forming granules or heavier aggregates of biomass gradually dominate the reactor. Beun et al. (1999) have demonstrated that short hydraulic retention time (HRT) and high shear force are the two SBR parameters which are crucial for granule formation (Beun et al. 1999).

Organic loading rate (OLR) is another SBR parameter which has influence on granule development. For instance, a relatively low OLR is applied at the beginning of the process. This is helpful to maintain alternating feast and famine condition. OLR is increased after appearance of the granules (Inizan et al. 2005).

Height to diameter ratio of the SBR tank is also a factor to influence aerobic granule formation. A high height to diameter ratio is usually preferred. According to Jungles et al. (2014), a height to diameter ratio of 19.5 was found to be suitable for granule formation. Tay et al. (2001a) have used a height to diameter ratio of around 13.3. SBR used for wastewater treatment by AGT is known to be operated under a wide range of temperature. It could be as high as 16 to 27°C (Jungles et al. 2014, Su and Yu 2005).

A SBR can be used to produce aerobic granules under alternating aerobic and anoxic conditions. Aerobic granules can be used for alternating nitrification and denitrification of the wastewater by this type of SBR operation (Jang et al. 2003). Alternating feast and famine periods applied by the SBR operation is considered to be crucial for the stability of aerobic granules (Schwarzenbeck et al. 2005). Organic loading rate ranging from 2.5 kg COD/m³ d to 15 kg COD/m³ d can be used for the cultivation of aerobic granules and they usually can tolerate a relatively high COD load (Thanh et al. 2009, Arrojo et al. 2004, Long et al. 2015). The COD of municipal wastewater is usually less than 2.5 kg/m³. Therefore, for efficient granule formation, the COD of such wastewater should be increased by addition of a carbon source. In this context, volatile fatty acids (VFAs) produced by fermentation of primary sludge can be considered. In traditional enhanced biological nutrient removal process, VFAs produced by this method are commonly used to increase the COD of the wastewater. Therefore, the same strategy can be used for AGT.

3. Aerobic granule formation mechanisms

Many hypotheses have been proposed to explain the mechanisms behind microbial granule formation. However, experimental evidences are not sufficient to support any such hypotheses. Barr et al. (2010) have concluded that two distinct types of granules could be formed. One type of granule was formed from a single colony of microorganisms and it was usually dominated by a single type of bacterium. The second type of granule was formed by many individual micro-colonies grown as a single granule (Barr et al. 2010).

Based on a previous report, Lv et al. (2014) have concluded that there are four distinct steps in granule formation. First, the cells attach to each other and then they form smaller aggregates of cells. These two steps are followed by EPS production by the cells in the aggregates. Finally, compact granules are formed where the hydrodynamic forces of the reactor play a crucial role (Lv et al. 2014).

According to another theory, a nucleating agent is necessary for granule formation. According to this hypothesis, the nucleating agent acts as a substratum for the microorganisms to attach and grow in the form of biofilms (Show 2006). Precipitated calcium phosphate found within the granules is often considered as one of the nucleating agents (Xin et al. 2016, Wan et al. 2015). Likewise, Vlyssides et al. (2008) have demonstrated that precipitated ferrous sulfide can play the role of a nucleating agent for granule formation. Linlin et al. (2005) have investigated development of aerobic granules by using anaerobic granular sludge as the seed sludge. The authors observed that previously disintegrated granules act as precursors for development of new aerobic granules. Verawaty et al. (2012) have used crushed granules as seed and concluded that addition of crushed granules can accelerate the granule formation, where the crushed granules act as a nucleating agent. Yellow earth is another nucleating agent found to be capable of rapid granule formation (He et al. 2016). He et al. (2016) have used yellow earth and found that aerobic granules could be developed as early as 4th day of operation. Li et al. (2015) have used dried and ground municipal sewage sludge as nucleating agent for the development of aerobic granules. The authors have concluded that the dried sludge powders of 20 to 250 μm were capable of significantly reducing the time required for granulation (Li et al. 2015). Particulate matter present in the wastewater might have a role as nucleating agent during granulation. Therefore, real wastewater containing particulate materials should be used to investigate the role of these materials on aerobic granule formation (Weissbrodt et al. 2013).

The strategy of allowing the microorganisms capable of granule formation to dominate the process is known as 'selection pressure strategy' for granule formation (Liu et al. 2003). According to this theory, by applying a short settling time, the microorganisms capable of attaching to each other and quickly settling down are allowed to remain in the reactor. The planktonic microorganisms or the aggregate of microorganisms which cannot settle down quickly are washed out with the liquid periodically drained from the upper section of the reactor (Xiong and Liu 2012). Since every day it is done in many cycles, gradually the microorganisms capable of forming granules dominate the process.

Dehydration related hydrophobic interaction among the cell surfaces initiated by proton translocation across cell membranes is another theory proposed for cell-cell attachment and granule formation (Liu et al. 2003, Tay et al. 2000). According to this hypothesis, due to fermentation of the substrate by the microorganisms, intracellular proton concentration increases. It activates the proton pumps in the cell membrane to translocate the protons to the outer surface of the cell. The cell surfaces are dehydrated by proton translocation and they attain slightly hydrophobic property. Hydrophobic interaction among microbial cell surfaces is one of the possible mechanisms behind cell-cell attachment and granule formation (Sarma et al. 2017).

Quorum sensing (QS) observed among biofilm forming microorganisms could be one of the molecular mechanisms responsible for granule formation (Sarma et al. 2016, Zhou et al. 2015). QS is a cell-cell communication mechanism used by the microorganisms. As a response to the environmental signals, microorganisms release autoinducer molecules to their surroundings. These molecules are detected by the members of the microbial community and start regulating their metabolic activities in response to the signal being propagated in the form of these molecules. A detailed molecular mechanism of quorum sensing is still a subject of research; however, its role in biofilm formation and pathogenicity is widely accepted. Aerobic granules are a special kind of biofilm and the role of QS on aerobic granule formation is relatively unexplored (Li and Zhu 2014).

Alternative feast and famine conditions created by different cyclic phases of sequencing batch reactor might have a role in granule formation. EPS production may be improved by applying a starvation period (Gao et al. 2011). Similarly, starvation period with no air flow and organic loading creates a temporary anaerobic environment which is considered to be beneficial for granule formation (Gao et al. 2011). A relatively short cycle time and a short starvation phase were found to be useful for rapid formation of aerobic granules (Liu and Tay 2008). However, a slightly longer starvation period was found to be suitable for forming stable aerobic granules (Li et al. 2014a).

4. Extracellular polymeric substances (EPS)

EPS produced by the microbial cells may include proteins, polysaccharides, and nucleic acids among others. It is widely believed that they play a crucial role in attachment of microbial cells to the substratum and each other facilitating formation of biofilm and bio-aggregates. According to a theory of granule formation, multivalent cations can form ionic bridges to crosslink the polymers of EPS (Kończak et al. 2014). These ionic bridges are considered to be one of the factors behind stable granule formation (Liu et al. 2003). Li et al. (2014b) have investigated the role of the ionic bridges formed by the EPS and multivalent cations in aerobic granule formation (Li et al. 2014b). In order to prove the role of this cation based gelation in aerobic granule formation, they have used latex particles and biopolymers such as alginate and peptone for bacterial cells and EPS, respectively. Multivalent cations such as, Ca^{2+}, Mg^{2+} and Fe^{3+} were added to the solution to achieve gelation by crosslinking the polymers by ionic bridges. The authors have demonstrated that at low alginate concentration, small flocs were formed. By increasing the alginate concentration, granules similar to that of the microbial granules could be formed (Li et al. 2014b).

Exopolysaccharides, proteins, lipids and nucleic acid are the major components of the EPS found in aerobic granules. Synthesis of exopolysaccharides could be extracellular or intracellular. Intracellular exopolysaccharides are secreted by the cells (Rabin et al. 2015). The EPS matrix developed outside the cell membrane protects the cell from environmental stresses and helps the cells in attaching to one another. In *Bacillus subtilis*, the genes of the epsA-O operon are responsible for the EPS synthesis (Mielich-Süss and Lopez 2015). BslA is a protein found in the EPS matrix and it is involved in developing a hydrophobic outer layer of a biofilm. These types of proteins are usually known as hydrophobins and they provide hydrophobic property to the biofilm (Mielich-Süss and Lopez 2015). EPS allows the microorganisms to attach to any substratum or to each other and thus protects them from washing out of the reactor. Moreover, by restricting the mobility of the cells and by increasing the cell density, biofilm environment facilitates horizontal gene transfer. Plasmid exchanged by horizontal gene transfer may spread the characteristic such as antibiotic resistance and also catabolic pathways among the microbial population of biofilm (Rabin et al. 2015).

5. Aerobic granules versus fungal pellets

Filamentous fungus can form large visible microspheres known as fungal pellets. The diameter of a mature fungal pellet could be around 2.1 mm to 3.3 mm (Zhang and Zhang 2016). Fungal pellet formation has been a subject of research for quite some time. There are many hypotheses behind fungal pellet formation and after many years of research, two distinct mechanisms have been established. As shown in Figure 1, a single fungal spore can germinate and form a pellet. The pellet formed from a single spore is known as non-coagulative type of pellet (Ryoo and Choi 1999). The pellets of *Penicillium* come under this category (Papagianni and Mattey 2006). Similarly, as shown in the same figure, there is another type of pellet which is known as coagulative type of pellet (Vechi-Lifshitz et al. 1990). In this case, as high as 500 spores first coagulate and then germinate to form a single pellet (Zhang and Zhang 2016). For example, the pellets of *Aspergillus niger* are of coagulative type (Papagianni and Mattey 2006, Vechi-Lifshitz et al. 1990, Braun and Vecht-Lifshitz 1991). According to a hypothesis of fungal spore coagulation, the spores have a 5 to 10 nm thick layer of a protein called hydrophobin. The hydrophobicity conferred by the hydrophobin layer is considered to be responsible for spore aggregation. However, according to another hypothesis, salt bridges between the polysaccharides present on the fungal spore surface are the main driving forces for spore coagulation (Zhang and Zhang 2016).

Based on this observation, we have proposed a new hypothesis for aerobic granule formation. According to this hypothesis, germinated fungal spores can serve as a backbone of aerobic granule formation. As shown in Figure 1, similar to fungal pellet formation, there could be two possible scenarios for aerobic granule formation from newly germinated fungal spores. In the first case, single spore can germinate to give rise to a small fungal hyphae matrix which is immediately colonized by bacteria. Agitation of the process increases the chance of the bacteria to come into contact with the newly developed fungal hyphae matrix. Factor such as cell surface hydrophobicity and Van der Waals forces may be responsible for initial attachment of the bacteria to newly developed fungal hyphae matrix. Relatively high temperature, fast growth rate and high bacterial cell density will help the bacteria to dominate over the fungus. Therefore, instead of becoming a proper fungal pellet, the structure will eventually developed into a mature aerobic granule dominated by bacteria. Bacteria will use their QS and EPS production ability to ensure the granule development process. Ionic bridges involving multivalent cations will be the other crucial factor in this type of granule formation.

As shown in Figure 1, the second type of aerobic granule may be formed from coagulated fungal spores. Similar to the first case, coagulated spores will germinate to develop a small granule like structure (newly developed fungal hyphae matrix). The matrix of the fungal hyphae will entrap bacterial cells, which will later dominate to develop into an aerobic granule. In both the cases, due to their filamentous nature the newly developed fungal hyphae, matrixes which are not colonized by bacteria, will find it difficult to settle down. Therefore, they are most likely to wash out from the SBR used for granule development. Likewise, apart from Van der Waals forces, hydrophobic interactions, EPS synthesis, ionic bridges and mechanical entrapment of the bacterial cells within the fungal hyphae matrix are also responsible for attachment of the bacterial cells to the embryonic fungal pellets. Hydrodynamic forces of the reactor are some of the factors responsible for mechanical entrapment of bacterial cells within the fungal hyphae matrix.

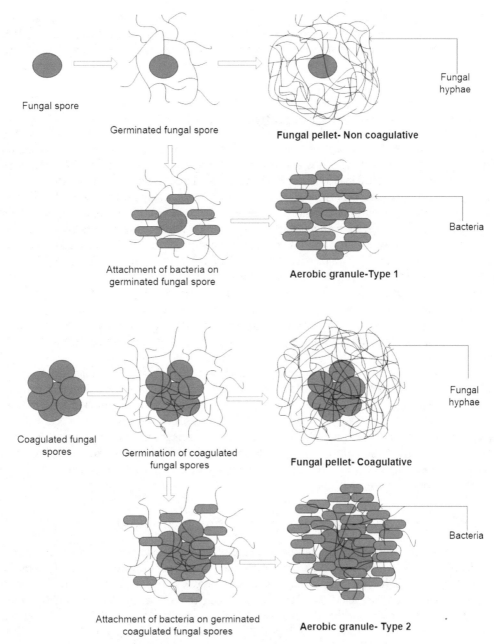

Figure 1. Proposed fungal pellet mediated aerobic granule formation mechanisms.

Color version at the end of the book

Li et al. (2010) have investigated the effect of COD on fungal growth on aerobic granules (Li et al. 2010). The authors have initially used bacterial granules containing black fungi like filamentous microorganism as seed granules.

It was observed that at an organic loading rate of 2 g COD/L/day, the seed granules could develop into mature aerobic granules dominated by rod shaped bacteria (Li et al. 2010). However, at an organic loading rate of 0.5 g COD/L/day, the seed granules grown in another reactor developed into black fungal granules dominated by filamentous fungi (Li et al. 2010). This study highlights that immature or newly developed fungal hyphae matrix has the potential to develop into both fungi dominated or bacteria dominated granules. The process conditions play an important role to decide whether it will become bacteria dominated stable aerobic granule or filamentous fungi dominated granule with lesser mechanical stability. Also, it supports our new hypothesis of fungal pellet mediated aerobic granule formation.

6. Importance of aerobic granulation technology

As already discussed, AGT is a new generation of efficient microbial technology for wastewater treatment. It has many obvious advantages over conventional wastewater treatment technologies. Aerobic granules are more dense and heavier than microbial flocs found in activated sludge process of conventional wastewater treatment plants. Therefore, they quickly settle down and it is easy to separate the sludge (granules) from the liquid phase. Thus, the entire treatment process becomes faster and dedicated sludge settling tanks are not required for this technology. It helps to construct a wastewater treatment plant within one-fifth of the space occupied by a conventional activated sludge based plant (Sarma et al. 2017). Likewise, AGT is a high cell density process. Since an SBR allows the biomass to settle down before releasing the effluent, microbial cells get a chance to remain within the reactor and the cell density of the process gradually increases. Thus, the AGT can use three to five times more microbial cells to treat the same amount of wastewater (Sarma et al. 2017). AGT can avoid frequent sludge wasting. It helps to reduce overall sludge production by the process. It is a special advantage because the energy requirement and cost of sludge dewatering could be reduced by reducing the amount of sludge produced by a process.

Apart from all the advantages of AGT outlined above, low process cost and relatively less investment required for the infrastructure are the other benefits of this technology. Similarly, compared to an activated sludge process, the energy consumption of an aerobic granulation process is around 30% less. Additionally, compared to activated sludge process, aerobic granules can deal with high organic loading rate and shows high degree of tolerance towards the toxic compounds present in the wastewater (Li et al. 2014a). Moreover, AGT allows simultaneous carbon, nitrogen and phosphorus removal within a single reactor. Thus, it is considered to be an effective tool for the treatment of industrial and municipal wastewater.

7. Challenging areas of aerobic granulation technology

Experience shows that 100% of the activated sludge of a process cannot be converted into aerobic granules. Usually, only around 30–50% of the activated sludge can be converted to aerobic granules. Remainders of the sludge are found as dense microbial flocs. Co-existence of as high as 47% of the biomass as aerobic granules and 53% of the biomass as microbial flocs has been reported (Carvalho et al. 2006). It is also known that the process could be operated even at 10% granular and 90% floccular sludge condition. However, if it is possible to achieve 100% sludge granulation, it would be easier to operate the process and a more compact system with better efficiency could be developed.

Development of aerobic granules in a continuous process would be the next big challenge for this technology. Presently, intermittent flow SBR is used for granule development. However, for its real application in large volume (few hundred million litres per day) of municipal and industrial wastewater treatment, a continuous process would be more appropriate. In this context, further investigation on granule formation in continuous flow SBR would be more appropriate.

Long incubation time required for granule development and the uncertainty involving the appearance of stable granules are the other concerns of this technology. Fundamental principles behind aerobic granule formation are still the subject of research. Further information on the granule formation mechanism would be useful to deal with these challenges.

8. Modifications of aerobic granulation technology

8.1 Bioaugmentation

Bioaugmentation is a strategy to enhance the performance of a bioprocess by externally adding one or more microbial strains to the process. Bioaugmentation has been evaluated to enhance the performance of AGT. Ivanov et al. (2006) first enriched the aerobic granular sludge with microorganisms with high auto-aggregation index. The authors enriched the sludge by repeatedly selecting and cultivating well settling granular sludge. Then a strain of *Klebsiella pneumoniae* and a strain of *Pseudomonas veronii* with high auto-aggregation index were isolated from the enriched granular sludge. These two strains were used as the bio-augmenting agents for subsequent aerobic granule formation study using activated sludge as the inoculum. The authors have noted that addition of these two strains could accelerate aerobic granule formation and the granules were observed within 8 days (Ivanov et al. 2006). In another investigation, Nancharaiah et al. (2008) have used *Pseudomonas putida* KT2442 strain containing TOL (pWWO) plasmid for bio-augmentation of aerobic granules. This strain can degrade benzyl alcohol and transmit the plasmid to other bacteria present in the granule. The authors have noted that compared to non-augmented granules, degradation of benzyl alcohol was significantly increased in the case of the granules augmented with the strain (Nancharaiah et al. 2008). Duque et al. (2011) have investigated bio-augmentation of aerobic granules for 2-fluorophenol (2-FP) degradation. The aerobic granules initially used by the authors were not capable of degrading 2-FP. However, after bioaugmentation of the granules with a microbial strain capable of degrading 2-FP, complete degradation of the pollutant was achieved (Duque et al. 2011).

8.2 Acclimatization

Acclimatization is a strategy where microorganisms of a process are gradually exposed to increasing concentration of a pollutant, a substrate or to an environmental condition so that they can adapt themselves to survive and function in such situations. Aerobic granules can be acclimatized for the treatment of certain wastewater which requires special treatment. Liu et al. (2012) have used this strategy for the treatment of polyacrylamide wastewater. The authors have used aerobic granules grown in glucose fed SBR. The granules were gradually exposed to increasing influent concentrations of polyacrylamide for around 43 days. The acclimatized granules obtained by this method were used for polyacrylamide biodegradation (Liu et al. 2012). Likewise, Lee et al. (2011) have used phenol acclimatized aerobic granules for cresol degradation. For the first two months, enriched activated sludge, which was used as inoculum, was cultivated in a medium containing 500 mg/L of phenol. Phenol concentration was gradually increased from 500 to 3000 mg/L over the next four months. Phenol acclimatized granules obtained by this method were used for the degradation of cresol isomers (Lee et al. 2011).

9. Concluding remarks

Biological treatment of wastewater is a successful multimillion litre scale process presently being operated all over the world. Over the years, the technology has been going through modernization. Enhanced biological nutrient removal process involving activated sludge is a modern version of biological wastewater treatment process. Although this technology is hugely successful and widely used in the developed world, there are some drawbacks of this technology. AGT has been projected as a replacement of presently used wastewater treatment technology. In this chapter, the strengths and weaknesses of the AGT have been discussed. Better settle-ability, low footprint, lesser energy consumption, high toxicity tolerance and reduction in the initial investment are some of the benefits of this technology. Long incubation times required for granule development and its dependence on a

SBR for granule production are considered as its major challenges. Additionally, granule formation mechanism is a poorly understood aspect of this technology. Thus, a new hypothesis of fungal pellet mediated aerobic granule formation has been discussed in this chapter. Further investigations on this subject would be useful in understanding the mechanisms behind granule formation.

Acknowledgements

Authors are thankful to NSERC and City of Calgary, Canada for financial assistance in the form of industrial research chair granted to Prof. J.H. Tay.

References

Ahn, K.H. and S.W. Hong. 2015. Characteristics of the adsorbed heavy metals onto aerobic granules: isotherms and distributions. Desalination and Water Treatment 53(9): 2388–2402.

Arrojo, B., A. Mosquera-Corral, J.M. Garrido and R. Méndez. 2004. Aerobic granulation with industrial wastewater in sequencing batch reactors. Water Research 38(14-15): 3389–3399.

Barr, J.J., A.E. Cook and P.L. Bond. 2010. Granule formation mechanisms within an aerobic wastewater system for phosphorus removal. Applied and Environmental Microbiology 76(22): 7588–7597.

Beun, J., A. Hendriks, M.C.M. van Loosdrecht, E. Morgenroth, P.A. Wilderer and J.J. Heijnen. 1999. Aerobic granulation in a sequencing batch reactor. Water Research 33(10): 2283–2290.

Boon, A.G. 2003. Sequencing batch reactors: A review. Water and Environment Journal 17(2): 68–73.

Braun, S. and S.E. Vecht-Lifshitz. 1991. Mycelial morphology and metabolite production. Trends in Biotechnology 9(1): 63–68.

Carvalho, G., R.L. Meyer, Z. Yuan and J. Keller. 2006. Differential distribution of ammonia- and nitrite-oxidising bacteria in flocs and granules from a nitrifying/denitrifying sequencing batch reactor. Enzyme and Microbial Technology 39(7): 1392–1398.

Dangcong, P., N. Bernet, J.P. Delgenes and R. Moletta. 1999. Aerobic granular sludge—A case report. Water Research 33(3): 890–893.

Di Bellaa, G. and M. Torregrossa. 2014. Aerobic granular sludge for leachate treatment. Chemical Engineering Transactions 38: 493–498.

Duque, A.F., V.S. Bessa, M.F. Carvalho, M.K. de Kreuk, M.C.M. van Loosdrecht and P.M. Castro. 2011. 2-Fluorophenol degradation by aerobic granular sludge in a sequencing batch reactor. Water Research 45(20): 6745–6752.

Gao, D., L. Liu, H. Liang and W.-M. Wu. 2011. Aerobic granular sludge: characterization, mechanism of granulation and application to wastewater treatment. Critical Reviews in Biotechnology 31(2): 137–152.

Gonzalez-Gil, G. and C. Holliger. 2014. Aerobic granules: microbial landscape and architecture, stages, and practical implications. Applied and Environmental Microbiology 80(11): 3433–3441.

He, Q.L., S.L. Zhang, Z.C. Zou and H.Y. Wang. 2016. Enhanced formation of aerobic granular sludge with yellow earth as nucleating agent in a sequencing batch reactor. IOP Conference Series: Earth and Environmental Science 39(1): 012025.

Hu, W., J. Wang, I. McHardy, R. Lux, Z. Yang, Y. Li and W. Shi. 2012. Effects of exopolysaccharide production on liquid vegetative growth, stress survival, and stationary phase recovery in *Myxococcus xanthus*. The Journal of Microbiology 50(2): 241–248.

Hwan Oh, J. 2018. Fundamental and application of aerobic granulation technology for wastewater treatment. http://home.eng.iastate.edu/~tge/ce421-521/Jin%20Hwan%20Oh.pdf (accessed on 08/05/2017).

Inizan, M., A. Freval, J. Cigana and J. Meinhold. 2005. Aerobic granulation in a sequencing batch reactor (SBR) for industrial wastewater treatment. Water Science and Technology 52(10-11): 335–343.

Ivanov, V., X.-H. Wang, S.T.L. Tay and J.-H. Tay. 2006. Bioaugmentation and enhanced formation of microbial granules used in aerobic wastewater treatment. Applied Microbiology and Biotechnology 70(3): 374–381.

Jang, A., Y.-H. Yoon, I.S. Kim, K.-S. Kim and P.L. Bishop. 2003. Characterization and evaluation of aerobic granules in sequencing batch reactor. Journal of Biotechnology 105(1-2): 71–82.

Jiang, Y., Y. Shang, H. Wang and K. Yang. 2016. Rapid formation and pollutant removal ability of aerobic granules in a sequencing batch airlift reactor at low temperature. Environmental Technology 37(23): 3078–3085.

Jungles, M., J. Campos and R. Costa. 2014. Sequencing batch reactor operation for treating wastewater with aerobic granular sludge. Brazilian Journal of Chemical Engineering 31(1): 27–33.

Kończak, B., J. Karcz and K. Miksch. 2014. Influence of calcium, magnesium, and iron ions on aerobic granulation. Applied Biochemistry and Biotechnology 174(8): 2910–2918.

Lee, D.-J., K.-L. Ho and Y.-Y. Chen. 2011. Degradation of cresols by phenol-acclimated aerobic granules. Applied Microbiology and Biotechnology 89(1): 209–215.

Li, A.-J., T. Zhang and X.-Y. Li. 2010. Fate of aerobic bacterial granules with fungal contamination under different organic loading conditions. Chemosphere 78(5): 500–509.

Li, J., L.-B. Ding, A. Cai, G.-X. Huang and H. Horn. 2014a. Aerobic sludge granulation in a full-scale sequencing batch reactor. BioMed Research International 2014: 268789.

Li, J., J. Liu, D. Wang, T. Chen, T. Ma, Z. Wang and W. Zhuo. 2015. Accelerating aerobic sludge granulation by adding dry sewage sludge micropowder in sequencing batch reactors. International Journal of Environmental Research and Public Health 12(8): 10056–10065.

Li, Y.-C. and J.-R. Zhu. 2014. Role of N-acyl homoserine lactone (AHL)-based quorum sensing (QS) in aerobic sludge granulation. Applied Microbiology and Biotechnology 98(17): 7623–7632.

Li, Y., S.-F. Yang, J.-J. Zhang and X.-Y. Li. 2014b. Formation of artificial granules for proving gelation as the main mechanism of aerobic granulation in biological wastewater treatment. Water Science and Technology 70(3): 548–554.

Linlin, H., W. Jianlong, W. Xianghua and Q. Yi. 2005. The formation and characteristics of aerobic granules in sequencing batch reactor (SBR) by seeding anaerobic granules. Process Biochemistry 40(1): 5–11.

Liu, H. and E.J. Buskey. 2000. Hypersalinity enhances the production of extracellular polymeric substance (EPS) in the Texas brown tide alga, *Aureoumbra lagunensis* (Pelagophyceae). Journal of Phycology 36(1): 71–77.

Liu, L., Z. Wang, K. Lin and W. Cai. 2012. Microbial degradation of polyacrylamide by aerobic granules. Environmental Technology 33(9): 1049–1054.

Liu, Y., H.-L. Xu, S.-F. Yang and J.-H. Tay. 2003. Mechanisms and models for anaerobic granulation in upflow anaerobic sludge blanket reactor. Water Research 37(3): 661–673.

Liu, Y.-Q. and J.-H. Tay. 2008. Influence of starvation time on formation and stability of aerobic granules in sequencing batch reactors. Bioresource Technology 99(5): 980–985.

Long, B., C.-Z. Yang, W.-H. Pu, J.-K. Yang, F.-B. Liu, L. Zhang, J. Zhang and K. Cheng. 2015. Tolerance to organic loading rate by aerobic granular sludge in a cyclic aerobic granular reactor. Bioresource Technology 182: 314–322.

Lv, Y., C. Wan, D.-J. Lee, X. Liu and J.-H. Tay. 2014. Microbial communities of aerobic granules: Granulation mechanisms. Bioresource Technology 169: 344–351.

Mielich-Süss, B. and D. Lopez. 2015. Molecular mechanisms involved in *Bacillus subtilis* biofilm formation. Environmental Microbiology 17(3): 555–565.

Morgenroth, E., T. Sherden, M.C.M. van Loosdrecht, J.J. Heijnen and P.A. Wilderer. 1997. Aerobic granular sludge in a sequencing batch reactor. Water Research 31(12): 3191–3194.

Morgenroth, E. and P.A. Wilderer. 1998. Sequencing batch reactor technology: concepts, design and experiences (abridged). Water and Environment Journal 12: 314–320.

Mosquera-Corral, A., B. Arrojo, M. Figueroa, J. Campos and R. Méndez. 2011. Aerobic granulation in a mechanical stirred SBR: treatment of low organic loads. Water Science and Technology 64(1): 155–161.

Nancharaiah, Y.V., H.M. Joshi, M. Hausner and V.P. Venugopalan. 2008. Bioaugmentation of aerobic microbial granules with *Pseudomonas putida* carrying TOL plasmid. Chemosphere 71(1): 30–35.

Ni, B.J. 2012. Formation, Characterization and Mathematical Modeling of the Aerobic Granular Sludge. Springer, Berlin Heidelberg.

Nor Auuar, A., Z. Ujang, M.C.M. van Loosdrecht and M. de Kreuk. 2006. Aerobic granular sludge technology for wastewater treatment—an overview. http://www.academia.edu/3054336/Aerobic_Granular_Sludge_Technology_For_Wastewater_Treatment_An_Overview (accessed on 08/05/2017).

Papagianni, M. and M. Mattey. 2006. Morphological development of *Aspergillus niger* in submerged citric acid fermentation as a function of the spore inoculum level. Application of neural network and cluster analysis for characterization of mycelial morphology. Microbial Cell Factories 5(1): 3.

Petry, S., S. Furlan, M.-J. Crepeau, J. Cerning and M. Desmazeaud. 2000. Factors affecting exocellular polysaccharide production by *Lactobacillus delbrueckii* subsp. *bulgaricus* grown in a chemically defined medium. Applied and Environmental Microbiology 66(8): 3427–3431.

Rabin, N., Y. Zheng, C. Opoku-Temeng, Y. Du, E. Bonsu and H.O. Sintim. 2015. Biofilm formation mechanisms and targets for developing antibiofilm agents. Future Medicinal Chemistry 7(4): 493–512.

Rezaei, L.S., B. Ayati and H. Ganjidoust. 2012. Cultivation of aerobic granules in a novel configuration of sequencing batch airlift reactor. Environmental Technology 33(20): 2273–2280.

Ryoo, D. and C.-S. Choi. 1999. Surface thermodynamics of pellet formation in *Aspergillus niger*. Biotechnology Letters 21(2): 97–100.

Sarma, S.J., J.H. Tay and A. Chu. 2017. Finding knowledge gaps in aerobic granulation technology. Trends in Biotechnology 35(1): 66–78.

Schwarzenbeck, N., J.M. Borges and P.A. Wilderer. 2005. Treatment of dairy effluents in an aerobic granular sludge sequencing batch reactor. Applied Microbiology and Biotechnology 66(6): 711–718.

Show, K.-Y. 2006. Chapter 1 Mechanisms and models for anaerobic granulation. pp. 1–33. *In*: Joo-Hwa Tay ST-LTYLK-YS and I. Volodymyr (eds.). Waste Management Series, Volume 6. Elsevier.

Singh, M. and R.K. Srivastava. 2011. Sequencing batch reactor technology for biological wastewater treatment: a review. Asia-Pacific Journal of Chemical Engineering 6(1): 3–13.

Strous, M., J.J. Heijnen, J. Kuenen and M. Jetten. 1998. The sequencing batch reactor as a powerful tool for the study of slowly growing anaerobic ammonium-oxidizing microorganisms. Applied Microbiology and Biotechnology 50(5): 589–596.

Su, K.-Z. and H.-Q. Yu. 2005. Formation and characterization of aerobic granules in a sequencing batch reactor treating soybean-processing wastewater. Environmental Science & Technology 39(8): 2818–2827.

Tay, J.-H., H.-L. Xu and K.-C. Teo. 2000. Molecular mechanism of granulation. I: H^+ translocation-dehydration theory. Journal of Environmental Engineering 126(5): 403–410.

Tay, J.H., Q.S. Liu and Y. Liu. 2001a. The effects of shear force on the formation, structure and metabolism of aerobic granules. Applied Microbiology and Biotechnology 57(1-2): 227–233.

Tay, J.H., Q.S. Liu and Y. Liu. 2001b. The role of cellular polysaccharides in the formation and stability of aerobic granules. Letters in Applied Microbiology 33(3): 222–226.

Thanh, B.X., C. Visvanathan and R.B. Aim. 2009. Characterization of aerobic granular sludge at various organic loading rates. Process Biochemistry 44(2): 242–245.

Vecht-Lifshitz, S.E., S. Magdassi and S. Braun. 1990. Pellet formation and cellular aggregation in *Streptomyces tendae*. Biotechnology and Bioengineering 35(9): 890–896.

Verawaty, M., M. Pijuan, Z. Yuan and P.L. Bond. 2012. Determining the mechanisms for aerobic granulation from mixed seed of floccular and crushed granules in activated sludge wastewater treatment. Water Research 46(3): 761–771.

Vlyssides, A., E.M. Barampouti and S. Mai. 2008. Granulation mechanism of a UASB reactor supplemented with iron. Anaerobe 14(5): 275–279.

Wan, C., D.-J. Lee, X. Yang, Y. Wang, X. Wang and X. Liu. 2015. Calcium precipitate induced aerobic granulation. Bioresource Technology 176: 32–37.

Wang, S., S. Teng and M. Fan. 2010. Interaction between heavy metals and aerobic granular sludge. Environmental Management. Edited by Santosh Kumar Sarkar, SCIYO, Croatia, 173–188.

Weissbrodt, D., T. Neu, U. Kuhlicke, Y. Rappaz and C. Holliger. 2013. Assessment of bacterial and structural dynamics in aerobic granular biofilms. Frontiers in Microbiology 4(175).

Williams, C., C.D. Goldsmith, T. Smith and J.J. Classen. Alternative natural technologies sequencing batch reactor performance verification. https://www.bae.ncsu.edu/topic/waste-mgmt-center/smithfield-project/phase2/A.5.pdf (accesed on 08/05/2017).

Xin, X., H. Lu, L. Yao, L. Leng and L. Guan. 2016. Rapid formation of aerobic granular sludge and its mechanism in a continuous-flow bioreactor. Applied Biochemistry and Biotechnology, 1–10.

Xiong, Y. and Y. Liu. 2012. Essential roles of eDNA and AI-2 in aerobic granulation in sequencing batch reactors operated at different settling times. Applied Microbiology and Biotechnology 93(6): 2645–2651.

Zhang, J. and J. Zhang. 2016. The filamentous fungal pellet and forces driving its formation. Critical Reviews in Biotechnology 36(6): 1066–1077.

Zhou, Y., H. Xu and Y. Liu. 2015. Molecular mechanisms governing aerobic granular sludge processes. Water Practice and Technology 10(2): 277–281.

CHAPTER 3

The EPS Matrix of Aerobic Granular Sludge

Y.V. Nancharaiah,[1,2,]* M. Sarvajith[1,2] and V.P. Venugopalan[2,3]

1. Introduction

In nature, microbial cells predominantly live in the form of microbial communities. Common examples of microbial communities are biofilms, flocs, aggregates and granules (Figure 1). Biofilms are the microbial communities that develop at solid-liquid, liquid-liquid or air-liquid interfaces and are embedded in a slimy matrix. Biofilms are developed as a result of attachment and progressive growth on a solid substratum (carrier material). On the other hand, microbial aggregates such as bioflocs and biogranules are not attached to any substratum (intentionally added carrier material) but share the characteristics of biofilms. The key components of these microbial communities are microbial cells and the slimy matrix. The quantity and the composition of the matrix vary widely between the biofilms and suspended microbial aggregates (bioflocs or biogranules). The microbial cells account for about < 10% dry mass while the matrix accounts for > 90% in most biofilms (Flemming and Wingender 2010). Since the matrix contains a mixture of biopolymers, it is referred to as extracellular polymeric substances (EPS). The EPS of microbial communities consists of polysaccharides, proteins, nucleic acids, lipids and other biomolecules (Flemming and Wingender 2010). In the case of biofilms, the EPS matrix is responsible for the adhesion of microbial cells to the substratum, formation of microcolonies, biofilm development and cohesion in the biofilms. The EPS matrix of biofilms has been relatively well investigated (Flemming and Wingender 2010) to identify its structural and functional role. The EPS matrix of bioflocs and biogranules has attracted research interest as it plays a pivotal role in microbial aggregation, development of microbial aggregates (flocs and granules), stability of microbial aggregates and wastewater treatment.

Activated sludge process (ASP) is widely applied for treating sewage and industrial wastewaters. It is a century old process originally developed for removing organic carbon from wastewaters (van

[1] Biofouling and Biofilm Processes Section, Water and Steam Chemistry Division, Bhabha Atomic Research Centre, Kalpakkam-603102, Tamil Nadu, India.
[2] Homi Bhabha National Institute, Anushakti Nagar, Mumbai 400094.
[3] Nuclear Agriculture and Biotechnology Division, Bhabha Atomic Research Centre, Mumbai 400094.
* Corresponding author: yvn@igcar.gov.in, venkatany@gmail.com

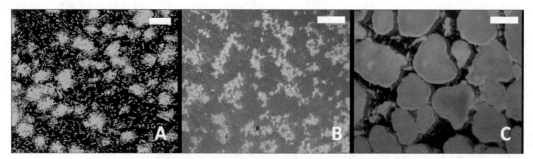

Figure 1. Examples of microbial communities. (A) *P. putida* biofilm on glass substratum. Bar = 20 μm. (B) Activated sludge flocs, bar = 2 mm. (C) Aerobic granular biomass, bar = 2 mm.

Color version at the end of the book

Loosdrecht and Brdjanovic 2014). ASP systems are applied worldwide for removing contaminants from wastewaters. In general, ASP systems are equipped with separate aeration tank and settling tanks. Biological treatment of wastewater takes place in the aeration tank while separation of activated sludge (microbial biomass) from treated water takes place in the settling tank (Figure 2). Aeration tank is further divided if nutrient (nitrogen and phosphorus) removal is considered important for wastewater treatment. Returning of settled activated sludge (from settling tank to aeration tank) and integration of biological nitrogen removal requires re-circulation flows. Therefore, ASP systems require large land footprint and energy for re-circulation flows. Apart from these drawbacks, ASP systems also suffer from sludge bulking leading to more biosolids in the treated water and poor biological purification of wastewater.

To overcome the above mentioned drawbacks of ASP, aerobic granular biomass was invented (Morgenroth et al. 1997). Aerobic granular biomass (also referred to as aerobic granular sludge, aerobic granules or granular activated sludge) is a new form of microbial community developed from activated sludge inoculum for new generation wastewater treatment systems. Unlike ASP systems, aerobic granular biomass needs only a single tank in place of both aeration and settling tanks (Figure 3). This single sludge system saves the energy incurred on re-circulation flows of ASP systems. Moreover, lower aeration (through on/off and lower aeration rate) is used in operating an aerobic granular biomass system for simultaneously removing organic carbon, reactive nitrogen and phosphorus from wastewater. Therefore, aerobic granular biomass systems are promising due to advantages like lower footprint, lower capital and operational costs. Aerobic granular biomass systems maintain microbial community in the form of compact granules and high biomass concentrations are achieved in the bioreactor. These features of aerobic granular biomass not only offer efficient removal of contaminants from wastewater but also exhibit higher tolerance to pollutants and perturbations in environmental conditions.

But, it is still a new technology which mainly relies on the growth of microorganisms of activated sludge in the form of compact and dense microbial granules. Aerobic granules are distinct, compact and dense particles which separate out from the water on their own by sedimentation under idle conditions. Thus, aerobic granules are distinct from that of activated sludge in terms of physical characteristics, microbial community structure and the EPS matrix (Nancharaiah and Kiran Kumar Reddy 2018).

Ever since the invention of aerobic granular biomass, it is conceived that the EPS matrix plays a key role in the aggregation of microbes, granulation of activated sludge and maintaining structural stability of developed biogranules (McSwain et al. 2005). Recent studies have identified additional roles for the EPS matrix of aerobic granules in removing nutrients (nitrogen and phosphorus) and contaminants (i.e., dyes, phenolic compounds) from wastewater. In spite of key roles in the structure and function of aerobic granules, the EPS matrix is not fully characterized owing to complex nature of the matrix. This chapter is aimed to provide an overview of the research performed on the EPS

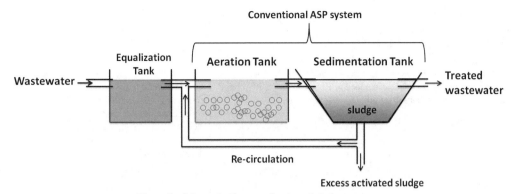

Figure 2. Schematic diagram of activated sludge process.

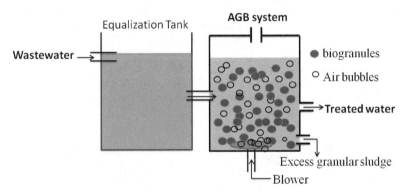

Figure 3. Schematic diagram of aerobic granular biomass system.

matrix of aerobic granular biomass. This chapter provides an insight into the EPS matrix of aerobic granular biomass in terms of extraction methods, composition of the EPS matrix, role of EPS in granule formation, stability of granules, contaminant and nutrient removal. Special emphasis was given to the extraction of bacterial alginate-like polymer as bio-based product from the excess aerobic granular biomass formed during wastewater treatment.

2. The EPS matrix

2.1 Ex situ EPS extraction methods

Different methods have been applied for extracting the EPS from aerobic granular biomass. Summary of different methods applied for extraction of EPS from aerobic granular biomass is shown in Table 1. The yields of EPS obtained from aerobic granular biomass varied among the extraction methods. The EPS yield was dependant on the reagents and the incubation conditions of extraction procedures. Due to compact and dense physical structure, the extraction methods find it difficult to extract the EPS form aerobic granules (Adav et al. 2008). Therefore, the reported values for the yield of EPS from the aerobic granules are highly dependent on the applied extraction protocol. Thus, the differences in the reported values of the EPS were not likely due to the differences in the real EPS quantities present in the aerobic granules.

Tay et al. (2001) extracted polysaccharides and proteins separately from the aerobic granules. Exopolysaccharides (PS) were extracted from the granular sludge using alkaline heating method (1 M NaOH at 80°C for 30 min). For obtaining extracellular proteins (PN), the granular sludge was suspended in treatment buffer (0.0625 M Tris HCl buffer, pH 6.8, 2% SDS, 10% glycerol and 5%

Table 1. Summary of extraction methods applied for harvesting EPS from aerobic granular biomass.

Extraction method	Extraction procedure	EPS harvesting	References
Alkaline treatment with heat	Homogenized or non-homogenized aerobic granules were added to 1 M NaOH with pH 11 and incubated at 80°C for 30 min	After NaOH extraction, the samples were centrifuged at 10000 rpm for 1 min. The supernatant was transferred to clean tubes and centrifuged again for 10 min and analyzed for different EPS components	McSwain et al. 2005, Tay et al. 2001
Cation exchange resin method	0.5 g of granules were added to 35 g of cation exchange resin (Dowex 50x8, Na+ form) and stirred at 750 rpm for 4 h in the dark at 4°C	After the Dowex extraction, the samples were centrifuged to harvest EPS in the supernatant	McSwain et al. 2005
Formamide–sodium hydroxide extraction	3 g of granules were added to 50 ml demineralized water in a 100 ml glass bottle. To this, add 0.3 ml of 99% formamide and store the glass bottle at 4°C for 1 h. Then add 20 ml of 1 M NaOH and again store the glass bottle at 4°C for 3 h	Centrifuge the samples at 4000 × g for 20 min to harvest EPS in the supernatant	Adav et al. 2008
Formaldehyde–sodium hydroxide extraction	Transfer known weight (3 g) of granules to 50 ml demineralized water in a 100 ml glass bottle and add 0.37 ml of 37% formaldehyde. Incubate the bottle in refrigerator at 4°C for 1 h. Add 20 ml of 1 M NaOH and again incubate the bottle in refrigerator at 4°C for 3 h	Centrifuge the sample at 4000 × g for 20 min to harvest EPS in the supernatant	Felz et al. 2016
Centrifugation extraction	Transfer 3 g (wet weight) granules in 50 ml demineralized water in a centrifuge tube and subject to centrifugation at 4000 x g for 20 min at 4°C	Harvest the supernatant for EPS analysis	Felz et al. 2016
Sonication extraction	Transfer 3 g granules to 50 ml demineralized water in a centrifuge tube and subject to pulse sonication on ice for 2.5 min at 40 W	Centrifuge the samples at 4000 × g for 20 min at 4°C to harvest EPS in the supernatant	Felz et al. 2016
EDTA extraction	Transfer 3 g granules to 100 ml glass bottle containing 50 ml 2% (w/v) EDTA solution. Mix and incubate the samples in refrigerator 4°C for 3 h	Subject the samples to centrifugation at 4000 × g for 20 min at 4°C to collect EPS in the supernatant	Felz et al. 2016
Sodium carbonate extraction	Transfer 3 g granules to 250 ml baffled flask containing 50 ml demineralized water. Add 0.25 g Na_2CO_3 anhydrous to the flask to obtain 0.5% (w/v) Na_2CO_3 solution. Place the flask in a water bath on a magnetic stirrer at 80°C. Stir the mixture at 400 rpm for 35 min at 400 rpm and 80°C	Subject the samples to centrifugation 4000 × g for 20 min at 4°C to harvest EPS in the supernatant	Lin et al. 2008, 2013, Felz et al. 2016

2-mercaptoethanol) and heated to 100°C for 10 min. This study showed that the polysaccharide (PS) content was always higher in the EPS than the PN content in the aerobic granules.

Subsequently, McSwain et al. (2005) have applied two different extraction methods, namely cation exchange resin (Dowex) and alkaline (NaOH) treatment with heat for extracting the EPS from aerobic granules. These authors have used both homogenized and non homogenized aerobic granules for extracting EPS. The NaOH extraction yielded more EPS as compared to the cation exchange resin method. Homogenization has increased the yield of EPS extraction by NaOH method. Moreover, the extracellular protein (PN) content was found to be greater than the polysaccharide content in the EPS extracts. The protein to polysaccharide (PN/PS) ratio was found to be 11 in the EPS extracts of aerobic granules developed in lab scale sequencing batch reactor by feeding acetate-containing synthetic wastewater. The EPS extracts obtained with alkaline heating method contained higher TOC, PN and PS than the extracts with the cation exchange resin method.

Caudan et al. (2012) have applied a multi-method protocol involving sequential mechanical and chemical extractions for efficient extraction of the EPS from aerobic granular biomass. Sonication and Tween 20 along with EDTA were used for mechanical and chemical extraction methods, respectively. This multi-method protocol extracted both polysaccharides (PS) and proteins (PN) from aerobic granular biomass. The PS to PN ratio in the extracted EPS was found to be 3.6. The polysaccharides were mostly recovered in the mechanical (sonication) extraction step. Lower recovery of polysaccharides was seen in the Tween 20 and EDTA extracts. This could be due to lower fraction of anionic polysaccharides in the EPS of aerobic granular sludge. However, extracellular proteins were recovered during both physical and chemical extraction steps. The recovery of proteins was higher in chemical extraction steps using Tween 20 and EDTA. PS to PN ratios were found to be 8.3 and 38.8, respectively, for extraction using Tween 20 and EDTA. Anion exchange chromatography of the extracted EPS also confirmed the presence of highly anionic proteins in the EPS of aerobic granules.

Felz et al. (2016) have evaluated six different methods for extracting gel-forming EPS from the aerobic granular sludge harvested from a pilot-scale reactor treating real sewage. Methods such as centrifugation, sonication, EDTA, formamide with NaOH, sodium carbonate with heat and constant mixing were used for extracting structural components of the EPS matrix. Among these methods, sodium carbonate method was found to be effective for disassembling the granular sludge structure by solubilizing the hydrogel EPS matrix. Based on the results, combination of physical and chemical treatments was recommended for solubilizing the EPS matrix and to extract the structural components of the EPS matrix.

2.2 *In situ characterization of the EPS matrix*

In situ methods refer to characterization of EPS in microbial communities without extraction or with no or minimal sample preparation. Spectroscopic techniques like infrared spectroscopy and nuclear magnetic resonance spectroscopy allow identification of chemical functional groups, i.e., carbonyl, aromatic and peptide bonds in the EPS. Multicolour fluorescence and confocal microscopy has received attention for *in situ* localization of different components of EPS.

Fluorophores specific for carbohydrates and proteins are used to selectively stain the EPS matrix components in intact and hydrated granules (Table 2). Fluorescence microscope is then used for visualization of the EPS components. McSwain et al. (2005) have applied three fluorophores to visualize the distribution of cells, proteins and polysaccharides. Fluorescein iosthiocyanate (FITC), Concanavalin A with texas red and Syto 63 were used, respectively, to stain proteins (amino acids of proteins and amino sugars of EPS), carbohydrates (α-mannopyranosyl and α-glucopyranosyl sugar residues) and cells. After staining, confocal laser scanning microscope (CLSM) was used for visualizing differential distribution of cells, polysaccharides and proteins in the intact granules.

In order to determine distribution of cells and EPS components, Chen et al. (2007a,b) applied a quadruple staining protocol to intact granules. The intact and hydrated aerobic granules were sequentially stained with Syto 60 (cells), FITC (an amine reactive dye that stains proteins and other amine-containing compounds such as amino-sugars), Con A tagged with tetramethylrhodamine (carbohydrates with α-mannopyranosyl and α-glucopyranosyl sugar residues) and calcofluor white (β-D-glucopyranose polysaccharides). CLSM was used to visualize stained cells and EPS components in the intact granules. The fluorescence of calcofluor white was excited using 400 nm laser line and the emission was detected at 410–480 nm (blue). The FITC probe was excited using 488 nm laser line and the emission was collected between 500–540 nm (green). Con A was excited with 543 nm laser line and emission was collected at 550–600 nm (red). SYTO 63 was excited using 633 nm laser line and the emission was collected at 650–700 nm.

A six-fold staining procedure was employed for *in situ* localization of total cells, dead cells, proteins, lipids, α-polysaccharides and β-polysaccharides in intact aerobic granules (Chen et al.

Table 2. Summary of the studies on the use of fluorophores for *in situ* visualization of the EPS matrix of aerobic granules.

Flurophores	Target	References
Fluorescein isothiocyanate (FITC)	All proteins and amino-sugars	McSwain
Concanavalin A (conA) conjugated with texas red	α-mannopyranosyl and α-glucopyranosyl sugar residues	et al. 2005
Syto 63	Cell permeative nucleic acid stain visualizes cells	
Fluorescein isothiocyanate (FITC)	All proteins and amino-sugars	Adav et al.
Concanavalin A conjugated with TRITC	α-mannopyranosyl and α-glucopyranosyl sugar residues	2008
Syto 63	Cell permeative nucleic acid	
Calcofluor white	β-D-glucopyranose polysaccharides	
Nile Red	Lipids	
Aleuria aurantia lectin	Glyco conjugates	Wagner
Syto 60	Nucleic acids	et al. 2009
Sypro orange	matrix proteins	
Acridine orange	Nucleic acids	Basuvaraj
Concanavalin A conjugated with Alexa flour 633	Polysaccharides (α-Mannose and glucose)	et al. 2015
Sypro orange	Proteins	
Coomassie brilliant blue	Proteins	Lin et al.
Periodic acid-Schiff (PAS)	Vicinal hydroxyl groups in glycoconjugates	2018
Alcian blue at pH 2.5	Carboxyl-rich and sulphated glycoconjugates	
Alcian blue at pH 1.0	Sulphated glycoconjugates	

2007b). Two additional fluorophores such as Sytox Blue and Nile Red for dead cells and lipids, respectively, were included to the quadruple staining procedure. Later, Adav et al. (2008) also applied similar staining procedure involving Syto 63 (cells), Syto Blue (dead cells), FITC (proteins), Con A Rhodamine (α-D-glucopyranose polysaccharides), calcofluor white (β-D-glucopyranose polysaccharides) and Nile Red (lipids) for determining distribution of cells and EPS components of an individual aerobic granule.

2.3 EPS matrix composition

The EPS matrix has generated tremendous interest among researchers because of its prominent role in formation and stability of aerobic granules. Various studies on the characterization of EPS matrix of aerobic granular sludge are summarized in Table 2.

Zhu et al. (2012) used Fourier-transform infrared spectroscopy and three dimensional excitation emission matrix (3D-EEM) fluorescence spectroscopy for characterizing major components of EPS extracted from aerobic granules. Signatures for aromatic protein-like substances, particularly tyrosine, were found in the stable aerobic granular sludge. Caudan et al. (2012) have found high recovery of proteins in Tween and EDTA extraction methods and detected anionic proteins in the EPS extracted from aerobic granules.

Proteins, carbohydrates and lipids accounted for 240 ± 13, 61 ± 9.4 and 51.1 ± 7.8 mg g^{-1} VSS, respectively, in the EPS extracted from phenol-degrading aerobic granules using formamide plus NaOH method (Adav et al. 2008). The PN/PS ratio was found to be around 3.9. A non-cellular region was proposed at the centre of aerobic granules. The non-cellular centre of aerobic granules was proposed to be containing proteins and β-polysaccharides. Selective hydrolysis of aerobic granules using enzymes revealed that β-polysaccharides are important in forming a network-like outer layer enmeshed with proteins, lipids, α-polysaccharides and microbial cells for providing mechanical stability of aerobic granules.

Lin et al. (2010) have extracted alginate-like polymeric substances from aerobic granular sludge formed in a pilot-scale sequencing batch reactor treating real sewage. For this, aerobic granules having size > 2 mm were dried, homogenized for 5 min and extracted in 0.2 M Na_2CO_3 (80 mL) at 80°C for 60 min. The supernatant was harvested by centrifugation at 15000 rpm for 20 min. The pH of the supernatant was adjusted by adding 0.1 M HCl. The precipitate formed while adjusting the pH was collected by another round of centrifugation at 15000 rpm for 30 min. The polymer pellet was washed with de-ionized water until the pH of the discarded supernatant reached 7.0. Finally, the precipitate was dissolved in 0.1 M NaOH. The polymer was precipitated further by adding cold absolute alcohol to a final concentration of 80% (v/v). The precipitated polymer could be recovered in the form of precipitate by centrifugation at 15000 rpm for 30 min. The yield of the polymer obtained from the aerobic granules was 160 ± 4 mg/g VSS suggesting that they are one of the dominant polysaccharides in the EPS matrix of aerobic granules. The polymer was identified by following the FAO/WHO alginate identification tests. The polymer was further characterized using biochemical tests, gelation with Ca^{2+}, fractionation, UV-visible, FT-IR, MALDI-TOF MS and electrophoresis techniques. The UV-visible and MALDI-TOF MS spectra of alginate-like exopolysaccharide resembled that of alginate commercially extracted from seaweed. High percentage of poly guluronic acid blocks (68%) was found in the alginate-like exopolysaccharide. The isolated alginate-like polymer formed rigid non-deformable gels in $CaCl_2$ solution. Polymers are responsible for highly hydrophobic, compact, strong and elastic structure of aerobic granular sludge.

Pronk et al. (2017) reported extraction of an acid soluble EPS from an acetate fed aerobic granular sludge. For cultivating aerobic granular sludge, a sequencing batch reactor was operated at 35°C by feeding with acetate as the sole carbon source. The reactor was operated with 3 h cycles containing 1 h anaerobic feeding from the bottom. Granules collected from the lab scale sequencing batch reactor were added to sodium carbonate solution and heated for 30 min at 80°C. After centrifugation, the supernatant was used for ALE extraction. The pellet was washed in demineralized water and re-suspended in 0.4 M acetic acid, stirred and heated to 95 C in a water bath for 2 min. The suspension was then centrifuged for 20 min. The supernatant was filtered through 0.2 μm filter. Ethanol (98%) was then added to the filtrate in 1:1 ratio. The precipitate was lyophilized and referred to as acid soluble EPS.

3. Role of EPS in aerobic granular sludge formation

Both *in situ* microscopy and *ex situ* chemical analysis revealed that concentrations and distribution of EPS components such as proteins and polysaccharides were different between the flocs and aerobic granules (McSwain et al. 2005, Seviour et al. 2009). McSwain et al. (2005) were the earliest to employ both *in situ* and *ex situ* methods for determining the spatial distribution and components of extracted EPS, respectively. These authors have reported that protein content was higher than the carbohydrates in the EPS matrix of aerobic granules. Moreover, the protein content of aerobic granules was about 50% higher than the EPS of activated sludge when extracted using cation-exchange resin (Dowex) method.

Application of specific fluorophores and confocal laser scanning microscope (CLSM) allowed obtaining information on the spatial distribution of the EPS matrix and microbial cells. These experiments have shown that microbial cells and polysaccharides were mainly distributed in the outer layers of the aerobic granules, while the centre of the aerobic granules was devoid of cells and mostly comprised of proteins. Based on this differential distribution pattern, authors have predicted that the protein core at the centre is likely to play a role both in granulation and stability of aerobic granules (McSwain et al. 2005). Almost at the same time, Wang et al. (2005) reported that β-polysaccharides are mainly found in the outer layer of aerobic granules and contribute to cohesive force. Zhang et al. (2007) also suggested that non-cellular protein core contributes to the stability of aerobic granules. Thus, better understanding of the EPS matrix composition, spatial distribution and role in structural stability is needed.

Chen et al. (2007b, c) and Adav et al. (2008) have attempted to determine the contribution of individual components of the EPS to the structural stability of aerobic granules. Selective hydrolysis of the EPS components with specific enzymes or chemicals (i.e., α-amylase, pronase, EDTA treatment) was applied to determine the components responsible for structural stability of phenol-degrading aerobic granules. The structural changes in the EPS distribution brought in by selective hydrolysis were probed using specific fluorophores and CLSM. The 3-dimensional structure of the aerobic granule was intact even after the selective hydrolysis of proteins, lipids and α-polysaccharides. However, fragmentation of aerobic granule was not noticed when selective hydrolysis of β-polysaccharides of aerobic granules implied their prominent role in structural stability (Adav et al. 2008). The EPS extracted from the aerobic granules has been reported to be more adhesive (sticky) than the EPS extracted from activated sludge (Seviour et al. 2009). Therefore, the reactor operating conditions promote EPS production and induce compositional changes in the EPS which in turn facilitate microbial aggregation and formation of aerobic granules. Negatively charged polysaccharides and proteins were identified in the EPS extracted from aerobic granules (Caudan et al. 2012, Seviour et al. 2012).

The gel forming exopolysaccharides are involved in providing the structural stability in case of aerobic granules (Seviour et al. 2010, 2011, Lin et al. 2010). Until now, tow gel forming exopolysaccharides have been identified and their role in granulation was proposed. Alginate-like exopolysaccharide having relatively high-levels of glucuronic acid (69%) was identified in the extracted EPS of aerobic granules (Lin et al. 2010). On the other hand, Seviour et al. (2010, 2011) have identified a novel heteropolysaccharide named granulan in the extracted EPS. Both alginate-like exopolysaccharide and granulan exhibit gel-forming properties and form structural gels *in vitro*. Although gel-forming properties of exopolysaccharides may contribute to structural stability of aerobic granules, the biochemical composition of the EPS matrix is under-studied; hence, the role of other components of EPS in putative structural role is envisaged. The role of proteins and other EPS components (e.g., DNA) in the structural stability of aerobic granules is yet to be determined. Moreover, the link between the EPS components and the microorganisms as well as reactor operating conditions needs to be ascertained.

Feeding of SBRs with metals (Ca^{2+}, Mg^{2+}, and zerovalent iron (ZVI)) in the synthetic wastewater was reported to decrease the time required for formation of aerobic granules (Jiang et al. 2003, Kong et al. 2014, Yilmaz et al. 2017, Kończak et al. 2014). Multivalent cations can assist in bacterial aggregation by neutralizing negative charges on the surface of microbial cells. Therefore, addition of Ca^{2+}, mg^{2+} or ZVI not only speeds up aerobic granulation but the formed granules have a compact and dense microbial structure. In addition to releasing Fe^{2+}, ZVI itself can act as nuclei for the growth for biofilms (Kong et al. 2014). Recently, Li et al. (2017) showed that formation of aerobic granules from activated sludge was faster when fed with synthetic saline sewage prepared by mixing with different proportions of real seawater. Production of alginate-like exopolysaccharides and formation of Ca^{2+} and Mg^{2+} phosphate in the core of granules was suggested to be responsible for accelerated aerobic granulation. The metal cations aid in the aggregation and granule formation and also help in granule stability through increased EPS production and EPS-metal interactions.

Deng et al. (2016) found a linear relationship between the EPS content and SVI of aerobic granules. A 3 L volume sequential air-lift bioreactor was inoculated with anaerobic sludge and fed with acetate-containing synthetic wastewater. The reactor was operated with 6 h cycle and 50% volume exchange ratio for cultivating aerobic granules. The developed aerobic granules had diameters between 0.8 and 1.1 mm within 35 days. Authors reported that aerobic granules had good settling abilities when the EPS was higher than 200 mg/g MLVSS. The increased EPS content, particularly the protein content of the EPS matrix, was correlated with settling characteristics of aerobic granules.

Xiong et al. (2013) investigated the role of extracellular proteins in the structural integrity of aerobic granules. Aerobic granules cultivated in a lab scale sequencing batch reactor with a synthetic wastewater containing acetate and ethanol as carbon sources were used for determining the action

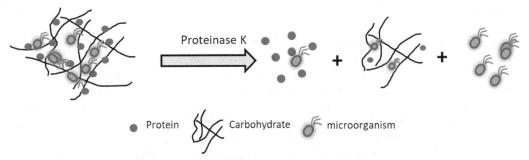

Figure 4. Diagram on mechanism of action of proteinase K on a single aerobic granule.

Color version at the end of the book

of proteinase K (Figure 4). The granules (0.5 g l⁻¹) were supplemented with 1 mg l⁻¹ Proteinase K in synthetic wastewater on rotary shaker set at 150 rpm and 37°C. After 2 days of incubation, the size of granules decreased from 562 to 320 μm in the presence of Proteinase K. The decrease in size of granules was associated with an increase in turbidity from 0.5 to 22.7 NTU of the medium. The PN content of the granules decreased by 40% due to exposure to Proteinase K. These results showed that Proteinase K decreased the protein content of aerobic granules and induced break-up of aerobic granules. Authors proposed that extracellular proteins are essential for maintaining the structural stability of aerobic granules.

4. Functional role of the EPS matrix in wastewater treatment

The roles of EPS matrix in wastewater treatment is shown in Figure 5. The EPS matrix can contribute to functional role in remediation by acting as a source of extracellular enzymes, redox centres, and nucleation centres in transformation and biomineralization of metal(loid)s and radionuclides. Moreover, the functional groups of the EPS matrix can also aid in the sequestration of ions through adsorption and ion exchange processes. The large number of functional groups present in the EPS matrix contributes to biosorption of organic compounds (i.e., dyes), ammonium and metal ions. The EPS matrix is also capable of biotransformation of redox-active metal(loid)s.

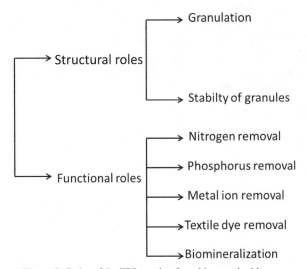

Figure 5. Roles of the EPS matrix of aerobic granular biomass.

4.1 Nitrogen removal

Ammonium nitrogen is released to the environment through various activities such as production of chemical fertilizers, release of unutilized reactive nitrogen in the agricultural runoff, sewage and animal manure. Landfill leachate, human urine, swine liquor and liquor from anaerobic digester are some of the examples of ammonium rich waste streams. Ammonium and its transformation products such as nitrite and nitrate are collectively called as reactive nitrogen because such higher levels of ammonium and nitrite nitrogen are toxic to living organisms. Nitrate contamination of ground water is widespread and consumption of nitrate contaminant water is linked to blue baby syndrome in children (Fowler et al. 2013, Nancharaiah and Venugopalan 2011). Therefore, pollution of waters with reactive nitrogen is a public health issue. Additionally, release of reactive nitrogen to surface waters (i.e., ponds and lakes) is associated with excessive growth of algae and aquatic plants and eutrophication. Removal of reactive nitrogen from waste streams is necessary to protect public health and to avoid eutrophication of water bodies.

Nitrogen removal is generally achieved by combination of nitrification and denitrification processes (Nancharaiah et al. 2016). In nitrification, NH_4^+ is oxidized to NO_2^- and then to NO_3^- by ammonia oxidizing bacteria and nitrite oxidizing bacteria, respectively. NO_3^- is reduced to N_2 gas in a multi-step reduction process by denitrifying bacteria. Due to the large size of aerobic granules, nitrogen removal via simultaneous nitrification and denitrification is possible even in the presence of oxygen in bulk liquid.

The measured products (nitrite and nitrate) formed by nitrification were found to be higher than the amount of ammonium removed from the medium. This was attributed to the occurrence of other processes like simultaneous nitrification of NH_4^+ produced by ammonification, biomass decay or to analytical issues (Nielsen 1996). Therefore, other processes including adsorption of NH_4^+ by biomass should be taken into account while tracking the flow of nitrogen compounds.

Ammonium adsorption is likely as both the EPS and microbial cell surfaces carry a net negative charge offering interactions with cations. The EPS matrix provides sites for interactions with cations (e.g., Ca^{2+}, Mg^{2+} and NH_4^+) as well as heavy metals. Bassin et al. (2011) were the first to quantify adsorptive removal of ammonium by aerobic granular sludge. Batch adsorption tests have shown that the NH_4^+ adsorption capacity of aerobic granular sludge is much higher as compared to activated sludge and anammox granules. The ammonium adsorption capacity was estimated to be approximately 18–24% towards the end of anaerobic feeding phase when the incoming wastewater had 50–100 mg l^{-1} NH_4^+-N. These high adsorption values were noticed when the pilot scale reactor contained granular sludge at 8 g l^{-1} VSS. Adsorption of ammonium to activated sludge was very fast and complete within 5 min. In contrast, the adsorption rates were slower for aerobic granules suggesting mass transfer limitations. Rapid adsorption was observed at the beginning of the experiments. The adsorption rate gradually decreased until the equilibrium concentration was reached. About 90% of this adsorbed ammonium could be desorbed by placing the aerobic granules in ammonium-free medium. These results suggested that the EPS matrix acts as an ion exchanger for cations. Effect of parameters such as temperature and saline condition were also determined on NH_4^+ adsorption by aerobic granular sludge. Addition of salt showed a considerable decrease in ammonium adsorption by aerobic granules. However, no difference in NH_4^+ adsorption was caused by changes in temperature (20°C and 30°C).

Yan et al. (2015) have determined nitrate and nitrite contents in the EPS of aerobic granular sludge along with ammonium. The content of these three nitrogen forms were measured in the EPS of aerobic granular sludge during cycle to understand the dynamic change in different forms. At the beginning of the cycle, the EPS mainly contained ammonium nitrogen and little nitrite and nitrate. The ammonium nitrogen quickly reached to 48.65 mg l^{-1} within 2 min of aeration and then decreased to 10 mg/l within 15 min of aeration. Nitrite nitrogen concentration increased up to 9.48 mg l^{-1} during prolonged aeration during cycle. However, nitrate nitrogen concentrations remained low at < 3 mg l^{-1}, throughout the cycle. The sum of the three nitrogen forms in the EPS was significantly lower at the end of aeration phase as compared to beginning. Based on these findings, authors suggested that the EPS acts as a temporary repository by adsorbing nitrogen forms. Thereby, the EPS matrix plays an important role in biological nitrification and denitrification and contributes to total nitrogen removal pathways of aerobic granular sludge. The large amount of functional groups (i.e., –OH, –COOH) available in the EPS of aerobic granules sequester reactive nitrogen through adsorption and ion exchange processes.

4.2 Phosphorus removal

Phosphorus is an essential element in all organisms. It is mined from phosphate rocks for manufacturing chemical fertilizers for application in modern agriculture. Therefore, phosphate availability is linked to agriculture and food security. The used phosphorus is lost to the environment via animal manure, sewage and agriculture drainage. Phosphorus is a liming nutrient in natural water bodies and any input of phosphorus (i.e., sewage and agricultural drainage) into these water bodies affects natural phosphorus cycle. Lost phosphorus can promote eutrophication of water bodies, i.e., lakes, reservoirs, estuaries and oceans.

As the phosphate rocks are non-renewable and finite (Cordell et al. 2009), removal and recovery of phosphorus from secondary sources allows sustainable phosphorus management. In this context, waste streams (i.e., sewage and industrial effluents) are attractive as secondary sources for recovering phosphorus. Moreover, many countries are following strict discharge limits for phosphorus release through effluent discharge (i.e., sewage and industrial effluents) to surface waters. The stringent limits are in force to essentially avoid nutrient pollution and eutrophication in receiving waters.

The strategies for achieving phosphorus sustainability includes utilization of technology for removing and reusing phosphorus from sewage and minimizing losses of organic phosphorus in the food chain. Physical, chemical and biological technologies are considered for improved management of nutrients. Addition of iron salts is practiced to achieve stringent discharge limits of 0.3 mg l^{-1} phosphorus-P in sewage (Wilfert et al. 2015). Biological methods are less energy intensive and concentrate phosphorus in biosolids. Generally, concentration of phosphorus is relatively low (< 10 mg l^{-1}) in sewage and industrial wastewaters (Yuan et al. 2012). Microbial cells can remove phosphorus from the water medium to meeting their cellular requirement. Specialized bacteria can store phosphorus in the form of polyphosphate inside the cells and use a different orchestrated mechanism for removing phosphorus from the extracellular environment. These specialized bacteria are called polyphosphate accumulating organisms. P removal by these PAOs is called enhanced biological P removal to distinguish it from normal bio-P removal for meeting cellular requirement. The alternating anaerobic and aerobic conditions of sequencing batch reactors are favourable for enrichment of PAOs and to establish EBPR for removing P from wastewater. EBPR has been successfully established in activated sludge and aerobic granular sludge by operating the system in fill-draw mode.

Studies on activated sludge systems have shown that EPS play an important role in phosphorus removal process. SEM-EDS analysis identified presence of significant quantities of phosphorus in the EPS extracted from EBPR performing activated sludge (Cloete and Oosthuizen 2001). Phosphorous content of EPS accounted for about 13–30% of total phosphorus removed by the activated sludge (Cloete and Oosthuizen 2001, Li et al. 2010, Zhang et al. 2013). Zhang et al. (2013) have applied ^{31}P NMR and found that orthophosphate, pyrophosphate and polyphosphate were the main P species in the EPS extracted from activated sludge. All these studies show that EPS of activated sludge indeed contributes to P removal process.

Likewise, the EPS content of aerobic granular sludge has been suggested to play a prominent role in phosphorus removal from wastewater. This can be due to the fact that the EPS content of aerobic granular sludge is much higher than in activated sludge (McSwain et al. 2005). Moreover, the sludge ages are longer in aerobic granular sludge systems than in conventional activated sludge process (Winkler et al. 2012). Higher EPS content and longer sludge ages of aerobic granular sludge systems would offer more scope for phosphorus accumulation in the EPS matrix of aerobic granules.

Researchers have anticipated that the role of EPS would be different and higher in phosphorus removal by aerobic granular sludge than in activated sludge (Wang et al. 2014).

The success of EBPR process depends on enrichment of PAOs that take up inorganic phosphate from the wastewater and convert to polyphosphate inside the cells. But, recent studies showed that extracellular polymeric substances (EPS), an important component of activated sludge and granular sludge, plays a crucial role in the EBPR process. Li et al. (2010, 2015) have reported that about 30% of the total phosphorus removed by activated sludge was accumulated in the EPS. Orthophosphate, pyrophosphate and polyphosphate were found to be main P species detected by ^{31}P nuclear magnetic resonance spectroscopy in the EPS extracted from activated sludge (Zhang et al. 2013). Inorganic phosphorus along with cations such as K^+, Mg^{2+} and Ca^{2+} were retained in the EPS matrix of aerobic granules (Wang et al. 2014). The rate of phosphorus release from the EPS matrix into the bulk liquid in the anaerobic phase was much faster than the phosphorus uptake into the EPS in the aerobic phase. Based on these, the authors have proposed that EPS matrix plays a crucial role in facilitating phosphorus accumulation by PAOs in the granular sludge (Wang et al. 2014).

Recently, Sarma and Tay (2018) reviewed the mechanisms of phosphorus removal by aerobic granular biomass. Apart from EBPR by microbial cells, precipitation and accumulation in the EPS matrix contributes to phosphorus removal mechanisms. For example, alginate, a prominent component of the EPS matrix, is proposed to be involved in accumulation and precipitation of phosphorus in the aerobic granules. A proposed mechanism for uptake and accumulation of phosphorus in the EPS matrix of aerobic granules is shown in Figure 6. Divalent cations like Ca^{2+} and Mg^{2+} can form ionic bridges by interacting with the carboxyl groups of alginate molecules and help in the accumulation of phosphorus.

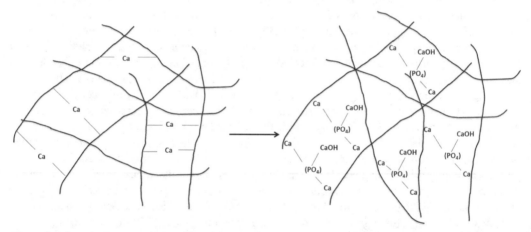

Figure 6. Schematic diagram on phosphorus accumulation in the EPS matrix of aerobic granular biomass. (A) Ionic bridges in the polysaccharides of EPS matrix. (B) Modified ionic bridges in the presence of phosphorus. Modified and re-drawn from Sarma and Tay 2018.

4.3 Biosorption of metal ions and other contaminants

Contamination of natural water resources with metals is a serious concern because metal pollutants are not biodegraded unlike organic pollutants and persist in the environment. Many of the metal pollutants can transfer across trophic levels in the food chain and accumulate in the living organisms. Metal pollutants commonly found in aqueous environments come under either cationic metals or oxyanionic metals. The cationic metal pollutants include cadmium(II), lead(II), nickel(II), copper(II) and zinc(II). Common oxyanionic metal pollutants encountered in aqueous environments are arsenic(III), arsenic(V), antimony(V) and chromium(VI). Many of these metal ions are essential in trace amounts in the living organisms. However, at higher concentrations they cause acute and chronic toxicity. As many as 13 metals, i.e., Ag, As, Be, Cd, Cr, Cu, Hg, Ni, Pb, Sb, Se, Tl, and Zn are listed as priority pollutants (US EPA) due to their toxicity. Release of these metal pollutants to environment (soil and water) needs to be controlled to minimize pollution and protect environmental and human health. Stringent discharge limits have been put forth for various metal pollutants to minimize toxicity and environmental pollution. Physical, chemical and biological methods are available for removing metal ions from water. Comprehensive reviews are available on the potential applications of physical and chemical methods for removing heavy metals from wastewaters (Fu and Wang 2011).

Microbial cells are known to interact with a broad range of metal ions, metal(loid)s and radionuclides. These microbe-metal interactions result in effecting the mobility of metal(loid) ions in both natural and engineered environments (Francis and Nancharaiah 2015). The basic mechanisms of metal ion removal by microbes include biosorption, bioaccumulation, bioreduction and biomineralization. Biotransformation of certain metal(loid)s and radionuclides by microbes is a useful strategy in *in situ* bioremediation of contaminated waters and wastewaters (Francis and Nancharaiah 2015).

Recently, Wang et al. (2018) reviewed the advances made in the biosorption of various sorbates, i.e., heavy metals, nuclides, dyes and other inorganic ions by aerobic granular biomass. The biosorption capacities of aerobic granules in removing various cationic metal contaminants (Cd(II), Pd(II), Ni(II), Zn(II), Cu(II), Co(II), Be(II), Mn(II), Fe(II), Ce(II) and Cr(III)) is summarized. General consensus is that the EPS of aerobic granules play a prominent role in the biosorption mechanisms. EPS extracted from the aerobic granules was found to show impressive adsorption capacities for removing the metal ions from the aqueous medium due to the presence of various functional groups.

The role of EPS in dye (methylene blue) removal was reported by Wei et al. (2015). Methylene blue was selected as the model dye pollutant for determining the functional role of EPS of aerobic granules in dye removal process. Pre-cultivated aerobic granules were evaluated for biosorption of methylene blue under batch conditions. The dye removal efficiencies of the EPS and granules were determined to be 9 and 80%, respectively, suggesting that the EPS matrix contributes to dye removal.

5. Recovery, re-use and commercial potential of EPS

The EPS content of granular sludge is much higher than in activated sludge. Role of bacterial alginate was proposed in activated sludge floc formation via gelation of alginate by metal cations like Ca^{2+}. Since the EPS content is higher in aerobic granular sludge, a much higher proportion of bacterial alginate is expected in the EPS of aerobic granules. Indeed, recent studies reported that alginate-like polymers account for almost 15–25% of aerobic granules by dry weight basis. Our results showed that alginate-like polymer account for about 27% of aerobic granular sludge cultivated in a pilot-scale sequencing batch reactor treating real sewage. Na_2CO_3 extraction was used successfully applied for extracting alginate-like polymer from aerobic granular sludge (Felz et al. 2016). This alginate-like polymer extracted from aerobic granules can thicken *in vitro* in the presence of cations (Ca^{2+}). It means that the extracted EPS from the aerobic granules forms gels or beads *in vitro*. At this juncture, it is considered a valuable bio-based product that can be extracted from waste aerobic granular sludge (Lin et al. 2015). But, it can be a raw material for applications in sectors like chemical, paper and textile industries. The polymer extracted from aerobic granular sludge can also be used as a soil-enhancer to improve water retention in semi-arid environments. The other advantage associated with polymer extraction is a decrease in biosolids volume by almost 27%. This would help in decreasing the biosolids volume to be treated by anaerobic digestion for further disposal.

The biological treatment step of wastewater treatment plants generates a constant stream of excess sludge which demands further treatment, i.e., anaerobic digestion, dewatering and disposal. Dewatering and anaerobic digestion are applied to recover energy, decrease pathogens and reduce and the bio-solids volume. Recovery of alginate-like polymer from aerobic granules offers alternative approaches for excess sludge management. Sludge remaining after extracting alginate can be processed by anaerobic digestion and dewatering. Studies in this direction have shown that extraction of alginate-like polymer not only decreased sludge volume significantly (by about 30%) but also improved digestibility and dewatering of remaining sludge. Recovery of alginate-like polymer for commercial applications of excess sludge would be an important step towards converting wastewater treatment systems into energy and resource facilities. This can be seen as a step towards development of bio-based products from waste microbial biomass and bioeconomy.

6. Conclusions

- Like biofilms, EPS matrix is an important component of aerobic granular biomass. It has a distinct structural role in formation of granules and maintenance of granular structure. The EPS content of granules is dependent on wastewater composition, microbial community and cultivation conditions.

- Different methods such as centrifugation, sonication, EDTA, alkaline heating, formamide with NaOH, sodium carbonate with heat and constant mixing have been applied for extracting the EPS matrix of aerobic granules. The EPS yield from aerobic granules varied with the extraction procedure (reagents and incubation conditions). Sodium carbonate method was found to be effective in disassembling the structure of granules and solubilizing the EPS matrix.

- Staining of hydrated granules with fluorophores specific for EPS components (i.e., carbohydrates and proteins) followed by visualization with confocal laser scanning microscope allowed to obtain spatial distribution of the EPS matrix and microbial cells *in situ* in intact and hydrate granules.

- Two gel forming exopolysaccharides, i.e., alginate like exopolysaccharide and granulan have been identified to contribute to the structural stability of aerobic granules.

- Apart from structural role, it is now becoming more evident that the EPS matrix can contribute to functional role in wastewater treatment by acting as a source of extracellular enzymes, redox centres and nucleation centres in transformation and biomeralization of metal(loid) contaminants. Additionally, the functional groups of the EPS matrix help in the sequestration of reactive nitrogen (ammonium, nitrate and nitrite), phosphorous, metal ions, dyes and other pollutants through biosorption.

Acknowledgements

Department of Atomic Energy is acknowledged for the financial support.

References

Adav, S.S., D.J. Lee and J.H. Tay. 2008. Extracellular polymeric substances and structural stability of aerobic granule. Water Research 42: 1644–1650.

Bassin, J.P., M. Pronk, R. Kraan, R. Kleerebezem and M.C.M. van Loosdrecht. 2011. Ammonium adsorption in aerobic granular sludge, activated sludge and anammox granules. Water Research 45: 5257–5265.

Basuvaraj, M., J. Fein and S.N. Liss. 2015. Protein and polysaccharide content of tightly and loosely bound extracellular polymeric substances and the development of a granular activated sludge floc. Water Research 82: 104–117.

Caudan, C., A. Filali, D. Lefebvre, M. Spérandio and E. Girbal-Neuhauser. 2012. Extracellular polymeric substances (EPS) from aerobic granular sludges: extraction, fractionation, and anionic properties. Applied Biochemistry and Biotechnology 166: 1685–1702.

Chen, M., D.J. Lee and J.H. Tay. 2007a. Distribution of extracellular polymeric substances in aerobic granules. Applied Microbiology and Biotechnology 73(6): 1463–1469.

Chen, M.Y., D.J. Lee, J.H. Tay and K.Y. Show. 2007b. Staining of extracellular substances and cells in bioaggregates. Applied Microbiology and Biotechnology 72(2): 467–474.

Cloete, T.E. and D.J. Oosthuizen. 2001. The role of extracellular exopolymers in the removal of phosphorus from activated sludge. Water Research 35(15): 3595–3598.

Cordell, D., J.O. Drangert and S. White. 2009. The story of phosphorus: global food security and food for thought. Global Environmental Change 19(2): 292–305.

Deng, S., L. Wang and H. Su. 2016. Role and influence of extracellular polymeric substances on the preparation of aerobic granular sludge. Journal of Environmental Management 173: 49–54.

Felz, S., S. Al-Zuhairy, O.A. Aarstad, M.C.M. van Loosdrecht and Y.M. Lin. 2016. Extraction of structural extracellular polymeric substances from aerobic granular sludge. Journal of Visualized Experiments 115: 54534, doi:10.3791/54534.

Flemming, H.C. and J. Wingender. 2010. The biofilm matrix. Nature Reviews in Microbiology 8: 623–633.

Fowler, D., M. Coyle, U. Skiba, M.A. Sutton, J.N. Cape, S. Reis, L.J. Sheppard, A. Jenkins, B. Grizzetti, J.N. Galloway, P. Vitousek, A. Leach, A.F. Bouwman, K. Butterbach-Bahl, F. Dentener, D. Stevenson, M. Amann and M. Voss. 2013. The global nitrogen cycle in the twenty-first century. Philosophical Transactions of the Royal Society B Biological Sciences 368(1621): 20130164.

Francis, A.J. and Y.V. Nancharaiah. 2015. *In situ* and *ex situ* bioremediation of radionuclide-contaminated soils at nuclear and norm sites. pp. 185–236. *In*: L. van Velzen (ed.). Environmental Remediation and Restoration of Contaminated Nuclear and Norm Sites, 978-1-78242-231-0, Woodhead Publishing Series in Energy.

Fu, F. and Q. Wang. 2011. Removal of heavy metal ions from wastewaters: a review. Journal of Environmental Management 92(3): 407–418.

Jiang, H.L., J.H. Tay, Y. Liu and S.T.L. Tay. 2003. Ca^{2+} augmentation for enhancement of aerobically grown microbial granules in sludge blanket reactors. Biotechnology Letters 25(2): 95–99.

Kong, Q., H.H. Ngo, L. Shu, R.S. Fu, C.H. Jiang and M.S. Miao. 2014. Enhancement of aerobic granulation by zero-valent iron in sequencing batch airlift reactor. Journal of Hazardous Materials 279: 511–517.

Kończak, B., J. Karcz and K. Miksch. 2014. Influence of calcium, magnesium, and iron salts on aerobic granulation. Applied Microbiology and Biotechnology 174(8): 2910–2918.

Li, X., J. Luo, G. Guo, H.R. Mackey, T. Hao and G. Chen. 2017. Seawater-based wastewater accelerates development of aerobic granular sludge: A laboratory proof-of-concept. Water Research 115: 210–219.

Li, N., N.Q. Ren, X.H. Wang and H. Kang. 2010. Effect of temperature on intracellular phosphorus absorption and extra-cellular phosphorus removal in EBPR process. Bioresource Technology 101(15): 6265–6268.

Li, W.W., H.L. Zhang, G.P. Sheng and H.Q. Yu. 2015. Roles of extracellular polymeric substances in enhanced biological phosphorus removal process. Water Research 86: 85–95.

Lin, Y.M., L. Wang, Z.M. Chi and X.Y. Liu. 2008. Bacterial alginate role in aerobic granular bio-particles formation and settleability improvement. Separation Science and Technology 43(7): 1642–1652.

Lin, Y.M., M. de Kreuk, M.C.M. van Loosdrecht and A. Adin. 2010. Characterization of alginate-like exopolysaccharides isolated from aerobic granular sludge in pilot-plant. Water Research 44: 3355–3364.

Lin, Y.M., P.K. Sharma and M.C.M. van Loosdrecht. 2013. The chemical and mechanical differences between alginate-like exopolysaccharides isolated from aerobic flocculent sludge and aerobic granular sludge. Water Research 47: 57–65.

Lin, Y.M., K.G.J. Nierop, E. Girbal-Neuhauser, M. Adriaanse and M.C.M. van Loosdrecht. 2015. Sustainable polysaccharide-based biomaterial recovered from waste aerobic granular sludge as a surface coating material. Sustainable Materials and Technologies 4: 24–29.

Lin, Y.M., C. Reino, J. Carrera, J. Pérez and M.C.M. van Loosdrecht. 2018. Glycosylated amyloid-like proteins in the structural extracellular polymers of aerobic granular sludge enriched with ammonium-oxidizing bacteria. Microbiology Open 7(6): e00616. doi:10.1002/mbo3.616.

McSwain, B.S., R.L. Irvine, M. Hausner and P.A. Wilderer. 2005. Composition and distribution of extracellular polymeric substances in aerobic flocs and granular sludge. Applied and Environmental Microbiology 71(2): 1051–1057.

Morgenroth, E., T. Sherden, M.C.M. van Loosdrecht, J.J. Heijnen and P.A. Wilderer. 1997. Aerobic granular sludge in a sequencing batch reactor. Water Research 31(12): 3191–3194.

Nancharaiah, Y.V. and V.P. Venugopalan. 2011. Denitrification of synthetic concentrated nitrate wastes by aerobic granular sludge under anoxic conditions. Chemosphere 85(4): 683–688.

Nancharaiah, Y.V., S. Venkata Mohan and P.N.L. Lens. 2016. Recent advances in nutrient removal and recovery in biological and bioelectrochemical systems. Bioresource Technology 215: 173–185.

Nancharaiah, Y.V. and G. Kiran Kumar Reddy. 2018. Aerobic granular sludge technology: mechanisms and biotechnological applications. Bioresource Technology 247: 1128–1143.

Nielsen, P.H. 1996. Adsorption of ammonium to activated sludge. Water Research 30(3): 762–764.

Pronk, M., T.R. Neu, M.C.M. van Loosdrecht and Y.M. Lin. 2017. The acid soluble extracellular polymeric substance of aerobic granular sludge dominated by *Defluviicoccus* sp. Water Research 122: 148–158.

Sarma, S.J. and J.H. Tay. 2018. Aerobic granulation for future wastewater treatment technology: challenges ahead. Environmental Science: Water Research and Technology 4: 9–15.

Seviour, T., M. Pijuan, T. Nicholson, J. Keller and Z. Yuan. 2009. Understanding the properties of aerobic sludge granules as hydrogels. Biotechnology and Bioengineering 102(5): 1483–1493.

Seviour, T., L.K. Lambert, M. Pijuan and Z. Yuan. 2010. Structural determination of a key exopolysaccharide in mixed culture aerobic sludge granules using NMR spectroscopy. Environmental Science and Technology 44(23): 8964–8970.

Seviour, T.W., L.K. Lambert, M. Pijuan and Z. Yuan. 2011. Selectively inducing the synthesis of a key structural exopolysaccharide in aerobic granules by enriching for Candidatus "Competibacter phosphatis". Applied Microbiology and Biotechnology 92(6), 1297–1305.

Seviour, T., Z. Yuan, M.C.M. van Loosdrecht and Y. Lin. 2012. Aerobic sludge granulation: a tale of two polysaccharides. Water Research 46(15): 4803–4813.

Tay, J.H., Q.S. Liu and Y. Liu. 2001. The role of cellular polysaccharides in the formation and stability of aerobic granules. Letters in Applied Microbiology 33(3): 222–226.

van Loosdrecht, M.C.M. and D. Brdjanovic. 2014. Water treatment. Anticipating the next century of wastewater treatment. Science 344(6191): 1452–1453.

Wang, L., X. Liu, D.J. Lee, J.H. Tay, Y. Zhang, C.L. Wan and X.F. Chen. 2018. Recent advances in biosorption by aerobic granular sludge. Journal of Hazardous Materials 357: 253–270.

Wang, R., Y. Peng, Z. Cheng and N. Ren. 2014. Understanding the role of extracellular polymeric substances in an enhanced biological phosphorus removal granular sludge system. Bioresource Technology 169: 307–312.

Wang, Z.W., Y. Liu and J.H. Tay. 2005. Distribution of EPS and cell surface hydrophobicity in aerobic granules. Applied Microbiology and Biotechnology 69(4): 469–473.

Wagner, M., N.P. Ivleva, C. Haisch, R. Niessner and H. Horn. 2009. Combined use of confocal laser scanning microscopy (CLSM) and Raman microscopy (RM): investigation on EPS-matrix. Water Research 43(1): 63–76.

Wei, D., B. Wang, H.H. Ngo, W. Guo, F. Han, X. Wang, B. Du and Q. Wei. 2015. Role of extracellular polymeric substances in biosorption of dye wastewater using aerobic granular sludge. Bioresource Technology 185: 14–20.

Wilfret, P., P.S. Kumar, L. Korving, G.J. Witkamp and M.C.M. van Loosdrecht. 2015. The relevance of phosphorus and iron chemistry to the recovery of phosphorus from wastewater: a review. Environmental Science and Technology 49(16): 9400–9414.

Winkler, M.K.H., R. Kleerebezem, W.O. Khunjar, B. de Bruin and M.C.M. van Loosdrecht. 2012. Evaluating the solid retention time of bacteria in flocculent and granular sludge. Water Research 46(16): 4973–4980.

Xiong, Y. and Y. Liu. 2013. Importance of extracellular proteins in maintaining structural integrity of aerobic granules. Colloids and Surfaces B: Biointerfaces 112: 435–440.

Yan, L., Y. Liu, Y. Wen, Y. Ren, G. Hao and Y. Zhang. 2015. Role and significance of extracellular polymeric substances from granular sludge for simultaneous removal of organic matter and ammonia nitrogen. Bioresource Technology 179: 460–466.

Yilmaz, G., E. Cetin, U. Bozkurt and K. Aleksanyan Magden. 2017. Effects of ferrous iron on the performance and microbial community in aerobic granular sludge in relation to nutrient removal. Biotechnology Progress 33(3): 716–725.

Yuan, Z., S. Pratt and D.J. Batstone. 2012. Phosphorus recovery from wastewater through microbial processes. Current Opinion in Biotechnology 23(6): 878–883.

Zhang, L., X. Feng, N. Zhu and J. Chen. 2007. Role of extracellular protein in the formation and stability of aerobic granules. Enzyme and Microbial Technology 41: 551–557.

Zhang, H.L., W. Fang, Y.P. Wang, G.P. Sheng, C.W. Xia, R.J. Zeng and H.Q. Yu. 2013. Species of phosphorus in the extracellular polymeric substances of EPBR sludge. Bioresource Technology 142: 714–718.

Zhu, L., H. Qi, M. Lv, Y. Kong, Y. Yu and X. Xu. 2012. Component analysis of extracellular polymeric substances (EPS) during aerobic sludge granulation using FTIR and 3D-EEM technologies. Bioresource Technology 124: 455–459.

CHAPTER 4

Simultaneous Nitrification, Denitrification, and Phosphorus Removal by Aerobic Granular Sludge

Sheng Chang,[1,][*] *Tao Tao*[1] *and* *Ping Wu*[2]

1. Introduction

Since Heijnen and van Loosdrecht published the first aerobic granular process patent in 1998 (Heijnen and van Loosdrecht 1998), aerobic granular sludge (AGS) wastewater treatment technology has attracted great attention in the last two decades. The AGS is self-immobilized microbial aggregates developed without any carrier material. The successful cultivation of AGS in a sequencing batch reactor (SBR) was reported in 1997 by Morgenroth et al. (Morgenroth et al. 1997). In the last 20 years, extensive studies have been conducted on the cultivation, formation mechanisms and applications of AGS (Beun et al. 1999, de Kreuk et al. 2005a, b). Nowadays, the AGS process is considered as one of the promising biological nutrient removal technologies. In contrast to the conventional activated sludge, the AGS features some excellent attributes (Tay et al. 2001a, b, Su and Yu 2005), including

a. compact microbial structure and excellent settleability;
b. high solid separation efficiency;
c. simultaneous nitrification, denitrification, and phosphorus removal;
d. tolerance to high organic loading rates.

Numerous laboratory-scale studies have illustrated the successful cultivation of AGS in sequencing batch reactors (SBR) with various wastewater characteristics and operation conditions. The early stage studies on AGS mainly focused on the mechanisms of sludge granulation and the factors affecting the formation of AGS in SBRs that only contained the aerobic reaction period (Morgenroth

---

[1] School of Engineering, University of Guelph, Guelph, Ontario, Canada N1G 2W1.
[2] Ontario Ministry of Agriculture, Food and Rural Affairs, Guelph, Ontario, Canada N1G 4Y2.
[*] Corresponding author: Schang01@uoguelph.ca

et al. 1997, Beun et al. 1999). It was reported that the content of the readily biodegradable COD (rbCOD) in wastewater, length of the aerobic reaction duration, hydrodynamic shear force, sludge settling time, reactor configuration, and organic loading rate are important selection pressures affecting the formation of AGS (Beun et al. 1999, de Kreuk et al. 2005, Morgenroth et al. 1997, Adav et al. 2007a, Su and Yu 2005, Tay et al. 2002a).

The interests in the AGS technology were initially focused on the excellent settleability of the granular sludge because it can largely improve the liquid-solid separation efficiency of the biological wastewater treatment. However, this technology has recently received increased attention due to its capacity to achieve simultaneous nitrification, denitrification, and phosphorus removal. The nitrogen and phosphorus constituents present in the wastewater stimulate the growth of algae and other aquatic vegetation in receiving water, leading to eutrophication. It is increasingly required for the wastewater treatment processes to remove nitrogen and phosphorus from industrial and municipal wastewater to protect the environment. The conventional enhanced biological nutrient removal (EBNR) processes are designed to promote the growth of different groups of microorganisms specifically for the removal of COD, nitrogen, or phosphorus. While the EBNR is currently the standard design for the biological nitrogen and phosphorus removal, there is great interest in improving the EBNR processes by reducing the process energy and carbon source consumption. Recent research has demonstrated that the same groups of microorganisms, named as the denitrifying phosphate accumulating organisms (DPAOs), can simultaneously remove nitrogen and phosphate (P) using the same carbon source.

Thus, the AGS performing simultaneous nitrification, denitrification and phosphorus removal (SNDNPR) is a promising technology for the wastewater treatment. This chapter introduces the concept of the SNDNPR, the formation mechanisms of AGS, the performance and cycle behavior of SNDNPR in AGS reactors, and the role of DPAOs in the SNDNPR.

2. Principle of Simultaneous Nitrification, Denitrification and Phosphorus Removal (SNÐNPR)

2.1 Nitrification and denitrification

Aerobic nitrification is a two-step process, involving the biological conversion of ammonia to nitrite and nitrite to nitrate, by two functionally defined bacterial groups, named as Ammonia-Oxidizing Bacteria (AOB) and Nitrite-Oxidizing Bacteria (NOB), respectively. The autotrophic AOB and NOB utilize inorganic carbon (carbonates and bicarbonates) or carbon dioxide as the carbon source for cell synthesis, while for the energy metabolism, they use ammonia and nitrite nitrogen as the electron donors and oxygen as the electron acceptor. The stoichiometric relations of nitrification can be described by the following equations (Metcalf and Eddy 2014):

AOB for NH_4-N oxidation $\qquad\qquad 2NH_4^+ + 3O_2 \rightarrow 4H^+ + 2H_2O + 2NO_2^-$ $\qquad$ (1)

NOB for NO_2^--N oxidation $\qquad\qquad 2NO_2^- + O_2 \rightarrow 2NO_3^-$ $\qquad$ (2)

With the estimated cell yields of 0.12 gVSS/gNH$_4$-N oxidized and 0.04 gVSS/gNO$_2$-N oxidized for the AOB and NOB, respectively, the overall nitrification reaction for the energy metabolism (Eq. 3) and cell synthesis (Eq. 4) can be described by (Metcalf and Eddy 2014):

$NH_4(HCO_3) + 0.9852Na(HCO_3) + 0.0991CO_2 + 1.8675O_2 \rightarrow$ $\qquad$ (3)

$0.01982C_5H_7NO_2 + 0.9852NaNO_3 + 2.9232H_2O + 1.9852CO_2$

As shown in Eq. 3, the complete oxidation of 1 g of NH_4-N to NO_3-N consumes 4.25 g of O_2 and 7.09 g of alkalinity as $CaCO_3$, and produce 0.16 g of biomass. The stoichiometric numbers of the oxygen and alkalinity consumption for the nitrification determined from Eq. 3 are slightly lower than those calculated from Eqs. 1 and 2 because of the consideration of the utilization of ammonia for cell synthesis.

Denitrification is the biological reduction of NO_3-N to N_2 by facultative heterotrophic bacteria that use organic carbon as the carbon source and electron donors and NO_3-N as the electron acceptor. Different groups of microorganisms may be involved in the reduction of NO_3-N to NO_2-N and NO_2 to N_2. Anoxic denitrification occurs in the absence of DO or under a limited DO concentration with the presence of rbCOD. The following equations have been developed from bioenergetics to describe the stoichiometric relations involved in the denitrification ($NO_3 \rightarrow N_2$) and denitritation ($NO_2 \rightarrow N_2$) with the assumption that ammonia is used as the nitrogen source for the cell synthesis (Metcalf and Eddy 2014).

Denitrification:

$$NO_3^- + H^+ + 0.33NH_3 + 1.45CH_3COO^- \tag{4}$$
$$\rightarrow 0.5N_2 + 0.33C_5H_7O_2N + 1.60H_2O + 1.12HCO_3^- + 0.12CO_2$$

Denitritation

$$NO_2^- + H^+ + 0.24NH_3 + 0.98CH_3COO^- \tag{5}$$
$$\rightarrow 0.5N_2 + 0.24C_5H_7O_2N + 1.24H_2O + 0.74HCO_3^- + 0.008CO_2$$

Denitrification consumes organic substrate and acidity. It can be derived from the stoichiometric equations that the direct reduction of NO_2-N to N_2 consumes around 32% less acetate than the complete denitrification ($NO_3 \rightarrow N_2$). When using an exogenous carbon source for the denitrification, normally around 4 grams of BOD is needed for the removal of one gram of NO_3-N. The theoretical alkalinity recovery from the reduction of 1 gram of NO_3^--N is 3.57 grams alkalinity (as $CaCO_3$) so that the denitrification can recover around 50% of alkalinity consumed by the nitrification.

2.2 Oxic and anoxic phosphorus removal

Enhanced biological phosphorus removal (EBPR) involves uptake of phosphorus from wastewater by phosphorus accumulating organisms (PAO) and subsequent removal of phosphorus via sludge wasting. Some PAOs have been identified to belong to the *Rhodocyclus genus* of *Betaprotebacteria* and were named as "*Candidatus Accumulibacter Phosphatis*" or in short as *Accumulibacter* (Bond et al. 1995). *Accumulibacter* can be classified into denitrifying PAOs (DPAO) that use nitrate and nitrite as electron acceptors and PAOs that use oxygen as an electron acceptor (Oehmen et al. 2010, Wang et al. 2013a, Lv et al. 2016). The main characteristics of PAOs and DPAOs are their capacity to accumulate phosphate as intracellular poly-P. They have a phosphorus content between 0.2 to 0.3 g P/g VSS, which is much higher than the typical 0.015 g P/g VSS contained in the ordinary heterotrophic BOD removal bacterial cells (Metcalf and Eddy 2014).

It is well-known that PAOs can be enriched in a sequential anaerobic-aerobic process. Under the anaerobic condition, PAOs assimilate soluble volatile fatty acids (VFA) like acetate to form intracellular poly-β-hydroxyalkanoates (PHA) and release orthophosphate from breaking intracellular polyphosphate to gain energy for the acetate transport and PHA synthesis. Typically, around 0.5 gram of phosphorus is released by PAOs for the uptake of 1 gram of acetate (Gu et al. 2008). In the subsequent aerobic condition, PAOs consume the stored PHA for the energy metabolism and cell synthesis, while take up the external orthophosphate from the liquid to make-up intracellular poly-P reserve for the PHA generation during the anaerobic period. To support the growth of PAOs, wastewater having a molar ratio of P/Mg/K/Ca of 1/0.28/0.26/0.09 is recommended (Sedlak 1991,

Metcalf and Eddy 2014). Metal ions, like Mg^{2+} and K^+, may also accompany the release and uptake of phosphorus of PAOs for the charge balance.

2.3 Simultaneous nitrification, denitrification and phosphorus removal

The conventional EBNR processes consist of the anaerobic, anoxic, and aerobic reactors. Figure 1(a) and (b) are the schematics of the typical EBNR anaerobic-anoxic-aerobic (A^2O) and the University of Cape Town (UCT) processes. The A^2O process consists of the anaerobic, anoxic, and aerobic reaction zones which are designed for the anaerobic VFA uptake and P release by PAOs, anoxic denitrification, and oxic BOD degradation, nitrification, and P uptake. The NO_x recycle line of the A^2O returns the nitrate produced in the oxic zone to the anoxic zone for denitrification, while the settled sludge in the secondary clarifier is returned to the anaerobic zone where it is mixed with the influent to allow the PAOs to take up the VFA for PHA synthesis and P release. Such an integrated anaerobic, anoxic, and aerobic process promotes the growth of PAOs, nitrifiers, and denitrifiers as well as ordinary BOD removing heterotrophic bacteria. The UCT process is distinguished from the A^2O process by returning the settled sludge to the anoxic zone, instead of to the anaerobic zone, and recycling the mixed sludge from the anoxic zone to the anaerobic zone. The main purpose of these modifications is to minimize the introduction of nitrate into the anaerobic zone to avoid denitrifiers competing with PAOs for the VFA.

In addition to the continuous EBNR processes, anaerobic/anoxic/aerobic SBRs are also widely used for nitrogen and phosphorus removal. Figure 2 shows the typical operation cycle of an EBNR SBR, which include the feeding, pre-anoxic, anaerobic, aerobic, post-anoxic, aerobic, settling and decanting periods. Since the mixed liquor in an EBNR SBR is a mixture of the ordinary heterotrophs, nitrifiers, denitrifiers, and PAOs, a pre-anoxic period is inevitably established during or after the

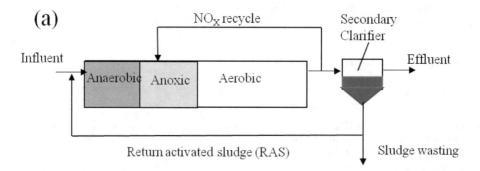

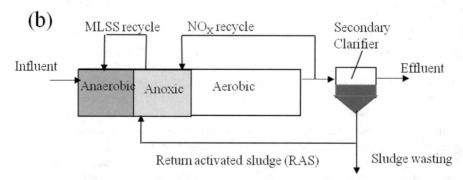

Figure 1. Schematics of the anaerobic, anoxic, and aerobic (A^2O) (a) and the University of Cape Town (UCT) (b) processes.

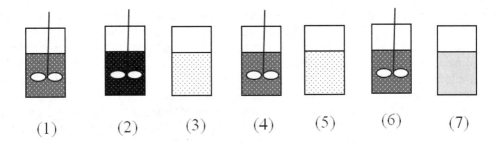

(1) Anoxic & filling; (2) Anaerobic & filling; (3) Aerobic; (4) Anoxic;
(5) Aerobic; (6) Anoxic; (7) settling and decant

Figure 2. Schematics of cyclic SBR operation for nitrogen and phosphorus removal.

feeding period due to the residual of nitrate from the previous cycle. Following the pre-anoxic period, PAOs in the mixed liquor will take up VFA and release PO_4-P during the anaerobic period. Subsequently, BOD degradation, nitrification, and P uptake will occur during the aerobic period. The post-anoxic period will further reduce the nitrogen concentration in the mixed liquor via the denitrification reaction but at a much lower rate than that with the pre-anoxic denitrification due to the low organic carbon concentration in the post-anoxic period. A second aerobic period is normally necessary to further reduce the BOD concentration, enhanced P uptake, and avoid the secondary release of phosphorus by PAOs prior to the settling and decanting period. The ratio of the decanting volume to the feeding volume, the cycle design, and the sludge SRT are key process parameters for SBRs to achieve effective EBNR.

The conventional EBNR processes are characterized by the intergradation of the anaerobic, anoxic, and aerobic reactions to promote the growth of ordinary heterotrophs, nitrifiers, denitrifies, and PAOs. In these systems, the denitrification and phosphorus removal were accomplished by different groups of bacteria. The denitrifiers use the external VFAs as carbon sources and electron donors and nitrate and/or nitrite as the elector acceptors, while the PAOs use the stored intracellular PHB as the electron donors and oxygen as the electron acceptors. The reaction zone composition and the internal circulation of mixed liquor are believed to be important to minimize the competition of the denitrifiers and PAOs for VFAs to obtain a balanced microbial community structure for the enhancement of nitrogen and phosphorus removal.

While the conventional EBNR processes are prevailingly used in the full-scale wastewater treatment, the recent development has demonstrated that the denitrification and phosphorus removal can be simultaneously accomplished by DPAOs that can use nitrate or nitrite as the electron acceptor for the uptake of phosphate. Different from the conventional EBNR processes, a DPAO enriched biological wastewater treatment system can achieve simultaneous denitrification and phosphorus removal with the same carbon source. Numerous studies have demonstrated that the anaerobic-aerobic SBR can effectively cultivate DPAOs enriched granular sludge. The operation cycle of the SNDNPR AGS SBR is normally around 4 to 8 hours, consisting of a short feeding period and a prolonged aerobic stage for the DPAOs to take up phosphate with the consumption of intracellular PHA. The anaerobic feeding period will provide the DPAOs sufficient carbon source for the VFA uptake as well as P release. Also, a short settling time of fewer than 10 minutes is usually used in the AGS SBR to select fast settling sludge.

The SNDNPR AGS can exhibit a strong phosphate release behavior during the anaerobic period and show the simultaneous nitrogen and phosphate removal during the aerobic period. The structure of the microbial community within the SNDNPR AGS will depend on the distribution of oxygen and substrate inside the AGS. Although many studies have shown that the anaerobic-aerobic AGS SBR achieved simultaneous nitrification, denitrification, and phosphorus removal, it is still inconclusive about the end products of denitrification by AGS and the organisms responsible for the AGS denitrification and phosphate uptake. As discussed in Section 7, the microbial community in AGS could include nitrifiers that oxidize ammonium nitrogen to nitrate, denitrifiers and/or glycogen accumulating organisms (GAOs) that reduce nitrate to nitrite, and DPAOs that take up phosphate using nitrate or nitrite as electron acceptors. The distributions of these microorganisms in the granules can be affected by the characteristics and structure of AGS, wastewater characteristics, and SBR operation conditions, such as cycle design and SRT.

SNDNPR AGS processes have been tested for the treatment of synthetic and industrial wastewater (Schwarzenbeck et al. 2005, Othman et al. 2013, Rosman et al. 2014, 2013, Liu et al. 2015, Long et al. 2015). Studies showed that the AGS had a high tolerance to the high COD, TN and TP loading rates. The AGS processes were also used to treat wastewater containing phenol (Adav et al. 2007, Tiong-Lee Tay et al. 2005, Chou and Huang 2005), heavy metals and dyes (Wei et al. 2015, Liu et al. 2015, Wei et al. 2016, Marques et al. 2013). Due to the layered structure of AGS, the mass transfer barrier can limit the concentration of toxic compounds within AGS, which increased the tolerance of AGS to toxic compounds (Liu and Tay 2004). However, despite the success in the cultivation and application of AGS in laboratory-scale studies, the full-scale applications of AGS technology in wastewater treatment are still limited by the long-term stability of the processes, which is inherently affected by the dynamic changes in the physical properties and microbial community structure of AGS over the operation period. Further understanding of the formation and the SNDNPR mechanisms of AGS is critical for the optimization of the operational conditions and performance of AGS SBR.

3. Factors affecting AGS formation

3.1 Reactor height to diameter ratio (H/D)

Most of the granular SBRs (GSBR) tested had a cylindrical configuration. The height to diameter (H/D) ratio of the cylindrical GSBR was once thought to be an important reactor configuration parameter. It was suggested that using a larger H to D ratio in the range of 20–30 (Li et al. 2006, Liu and Tay 2007, Wang 2007) would be important to select the fast settling AGS. However, Kong et al. (2009) assessed the effect of H to D ratios ranged from 20 to 4 using the parallel experiments. They found that the H/D ratios in the tested range had no evident effect on the microbial community structure and physical properties of AGS formed in the reactors, suggesting that the H to D ratio might not be a critical parameter to the formation of AGS.

3.2 Cycle length

Applying certain periods of feasting and fasting during the operation cycles of SBRs is widely accepted to be an important selection pressure for proliferating the populations of nutrient removal and floc formation organisms. The feasting period will provide an opportunity for the PAOs to take up VFA and synthesize intracellular PHA, while the starvation period will cause PAOs to use the stored PHA and take up phosphorus from the bulk liquid. Thus, the periodic feasting and fasting can facilitate the growth of nutrient removal organisms. However, a long starvation period means a long aeration period and hydraulic retention time (HRT), which will result in a high-energy consumption and a reduction in the treatment capacity of the reactor. Several studies were conducted to examine the effect of the cycle time on the aerobic granule formation and properties. Tay et al. (2002) studied the effect of cycle times ranged from 3 to 24 hours on the formation of nitrifying granules. In their study, the SBRs were operated at the same influent filling time of 4 min, settling time of 30 min, and effluent withdrawal of

4 min but with different aeration times of 24, 12, 6, and 3 hours. They found that the nitrifying granules with a mean diameter of 0.25 mm and specific density of 1.014 were developed at the cycle times of 12 and 6 hours, while the granulation failed with the cycle times of 24 and 3 hours. It showed that both the extended and insignificant starvation period might be detrimental to the formation of AGS. However, contradictory to the results reported by Tay et al. (2002), Liu and Tay (2007) reported that large granules were formed at a shorter cycle time of 1.5 hours, suggesting the influence of the cycle time on the sludge granulation might be different under different operating conditions. Liu and Tay (2007) also reported that the observed growth rate of granular sludge increased from 0.063 to 0.316 gVSS/gCOD/day when the SBR's cycle time was reduced from 8.0 hours to 1.5 hours. Their study suggested that a shorter cycle time might proliferate the population of the fast-growing heterotrophs but suppress the population of the slow-growing nutrient removal bacteria. Also, several studies showed that the granule size decreased and the density of granules increased with the increase in the cycle times (Liu and Tay 2007, Pan et al. 2004, Lin et al. 2001).

On the other hand, the AGS formed under different cycle times may exhibit different biological activities. Pan et al. (2004) reported average specific volumetric oxygen consumption rates (SOUR$_{vol}$) of 265.3, 365.7, 316.1, 186.4 mgO$_2$/L/hr for AGS formed at the cycle times of 1, 3, 6, and 12 hr, respectively. They indicated an optimal cycle time might exist for maximizing the COD removal capacity of AGS. Pan et al. (2004) reported that the ratio of polysaccharides to protein (PS/PN) and the hydrophobicity of the extracellular polymeric substance (EPS) extracted from AGS increased with the decrease in the cycle time. The sludge aggregation or flocculation depends on the direct cell-cell interactions via the EPS layers surrounding the cell walls. The high EPS hydrophobicity will facilitate the cell-cell aggregation and the formation of granular sludge, leading to the increased strength of granules. Therefore, a cycle time in the range of 4 to 12 hours could be important for the formation of AGS with a high stability and biological activity.

3.3 Hydrodynamic shear force

Hydrodynamic shear force plays an important role in the formation of aerobic granular sludge because a relatively high shear force can strengthen the cell collision and aggregation, stimulate the extracellular polysaccharide production, and increase the respiration activity of granules (Liu and Tay 2002). In the SBRs, the shear force is generated by the turbulence and particle-particle collisions induced by injecting air at the reactor bottom. Tay et al. (2001) assessed the aeration intensity on the formation of granular sludge. After testing the superficial air velocities of 0.003, 0.012, 0.024, and 0.036 m/s, it was found that the compact and dense granules were only formed at the air superficial velocities higher than 0.012 m/s. Beun et al. (2002a) reported that a superficial air velocity higher than 0.014 m/s was essential for the formation of compact aerobic granular sludge. Adav et al. (2007a) reported that the floc-type sludge, large (3–3.5 mm) granules and compact granules (1–1.5 mm) were formed at superficial air velocities of 0.006, 0.012, and 0.018 m/s, respectively. Thus, as shown by those studies, a critical superficial air velocity between 0.012 and 0.014 may be required for the formation of the stable granular sludge in an SBR.

3.4 Settling time

For the SBR operation, a settling period prior to the decantation is necessary to separate the sludge from the treated wastewater. To gain a clear effluent, the height of the settled sludge should be below the discharge port at the end of the settling period. The distance between the operating liquid level of an SBR and the discharging port can be defined as the separation distance of the SBR. For a given settling distance, the settling time required to achieve clean effluent will depend on the sludge settling velocity. The settling period for the conventional SBRs is in the range of 30 minutes to a few hours. However, for the AGS SBR, it is necessary to apply a much shorter settling time to select the granular sludge with a good settleability while washing out the less dense flocs. A long settling time will prevent less dense flocs from washout, and leave them to compete with granular sludge for the uptake of nutrients and oxygen, which can subsequently lead to the failure of AGS cultivation. Qin et al. (2004) reported that a settling time less than 5 min should be applied for selecting granular sludge. In general, a settling time between 3 to 5 min will be sufficient for granular sludge to settle during the settling period.

3.5 Organic loading rate

Organic loading rate (OLR) is defined as the unit mass of COD fed into the unit volume of reactor per unit of time. Successful cultivation of AGS was reported with OLRs in the range of 2.5 to 15 kg COD/m^3/day (Liu et al. 2005). Moy et al. (2002) found that the glucose-fed granules were able to tolerate a higher organic loading rate than the acetate-fed granules because the former had a loose filamentous or irregular microstructure that facilitated the mass transfer in the interior region of the granules. It was reported that the diameter of the glucose-fed granules increased from 2.70 ± 1.00 mm to 3.30 ± 1.30 mm when the OLR increased from 6 to 15 kg COD/m^3/d, while for the acetate-fed granules, the diameter increased from 1.96 ± 0.92 mm to 4.2 ± 0.1 mm when the OLR increased from 6 to 9 kg COD/m^3/d (Moy et al. 2002). Liu et al. (2003) reported that the granules sludge was successfully developed at the acetate substrate concentrations ranging from 500 to 3000 mg/L COD with the slightly increased granular size observed at the higher OLRs. Liu et al. (2003) also found that the surface hydrophobicity and cell polysaccharide content of the aerobic granules were significantly higher than the seed sludge, indicating that the high surface hydrophobicity and polysaccharide content might play an important role in the formation of granular sludge. Yang et al. (2003) reported that the simultaneous organic and nitrogen removing granules were developed with the wastewater N/COD ratios ranging from 5/100 to 30/100 in an SBR that was operated at an operation cycle consisting of 4 min of feeding, 230 min of aeration, 2 min of settling, and 4 min of effluent withdraw. The developed granules contained heterotrophic, nitrifying, and denitrifying organisms in spite that the operation cycle had no anoxic and anaerobic periods. It was found that the populations of nitrifying and denitrifying organisms increased and the population of heterotrophic organisms decreased with the increase in the wastewater N/COD ratios (Yang et al. 2003). Tsuneda et al. (2003) reported that the nitrifying granules that mainly contained *Nitrosomonas-like* ammonia-oxidizing species were formed in an aerobic up-flow fluidized bed reactor at an NH$_4$-N loading rate of 1.5 kg N/m^3/d. Lin et al. (2003) reported that PAO-enriched aerobic granules were cultivated at the wastewater P/COD ratios ranging from 1/100 to 10/100 in an SBR that was operated with a cycle consisting of 5 min of feeding, 129 min of anaerobic reaction, 226 min of aerobic reaction, 5 min of sludge settling and 4 min of sludge discharging. The PAO granules became smaller and more compact as the wastewater P/COD ratio increased. In general, it seems that loading rates may have an insignificant impact on the formation of granular sludge but the properties, biological activity, and microbial communities of AGS may vary with the change in the COD, nitrogen, and P loading rates.

4. Aerobic granular sludge for SNDNPR

4.1 SNDNPR by aerobic granular sludge

Granular sludge was initially observed in the up-flow anaerobic sludge blanket (UASB) reactors (Lettinga et al. 1980). In 1997, Morgenroth et al. (1997) reported that the granular sludge was able to be cultivated in the aerobic SBR operated with the cycle consisting of the filling, aerobic reaction, and decanting periods. Since then, numerous studies have been conducted to investigate the

Table 1. SNDNPR AGS SBR operation conditions and treatment performance.

Wastewater content (mg/L)	Operation cycle	Conditions	Effluent (mg/L)	References
Synthetic COD (as Glucose); 1800 NH_4-N: 200, PO_4-P: 15; Ca:14.4, Mg: 3.1 Alk: 952.4 (as $CaCO_3$)	Cycle time 360 Filling: 5 min Anoxic: 25 min Aerobic: 300 min Settling: 2 min Decanting and idle: 28 min Exchange ratio: 0.5	3.6 gCOD/L/d 0.6 gN/L/d 0.045 gP/L/d HRT: 16 hr SRT: 15 d MLSS: 14.52 g/L pH: ~ 6.8 to 7.8	COD: n/a NH_4-N: 3.05 ± 0.65 NO_2-N: 0.67 ± 0.13 NO_3-N: 19.03 ± 1.13 TP: 3.61 ± 0.15	Wei et al. 2014
Abattoir wastewater TCOD: 1480 ± 138 SCOD: 1072 (Inc. 750 Ac) TKN: 237.3 ± 6.6 NH_4-N: 221.7 ± 6.1 TP: 34.3 ± 1.2 PO_4-P: 33.6 ± 0.8 TSS: 205 ± 17	Cycle time 480 Filling: 18 min Anaerobic: 60 min Aerobic: 315 min Post anoxic: 80 Settling: 2 min Decanting and idle: 5 min Exchange ratio: 0.45	2.7 gCOD/L/d 0.7 gN/L/d 0.06 gP/L/d HRT: 13 hr SRT: 15–20 d MLSS: 20 g/L pH: 7.0–8.4	TCOD: 467 ± 83 SCOD: 162 ± 33 TKN: 25.3 ± 5.2 NH_4-N: 7.6 ± 0.7 TP: 9.0 ± 2.8 PO_4-P: 0.6 ± 0.3 TSS: 306 ± 88	Yilmaz et al. 2008
Synthetic wastewater TCOD: 600 (acetate) NH_3-N: 60 PO_4^{-3}-P: 10 Ca: 3.8, Mg: 8.9	Cycle time 360 min Filling: 20 min Anaerobic: 90 min Aerobic: 120 min Post anoxic: 120 Settling: 0.5 min Decanting and idle: 9.5 min Exchange ratio: 0.33	0.8 gCOD/L/d 0.08 gN/L/d 0.01 gP/L/d HRT: 18.2 hr SRT: 15 d MLSS 4.75 g/L pH: 7.5–8.1	COD: n/a NH_3-N: < 0.1 NOx-N: < 0.1 TP: 0.3	Kishida et al. 2006
Abattoir wastewater TCOD: 7685 ± 646 SCOD: 5163 ± 470 TKN: 1057 ± 63 NH_3-N: 50 ± 12 TP: 217 ± 38 Alk: 872 ± 42 As $CaCO_3$	Cycle time 357 min Filling anaerobic: 120 min Aerobic: 220 min Settling: 2 min Decanting: 15 min Exchange ratio: 0.083	2.5 gCOD/L/d 0.35 gTKN/L/d 0.025 gP/L/d HRT: 72 SRT: 20 d MLVSS: 8 g/L pH 7.0 to 7.5	TCOD: 106 ± 18 SCOD: 95 ± 17 TKN: 2 ± 2 NH_4-N: 0 NO_3-N: 26 NO_2-N: 0 TP: 4 ± 2 VSS: 42 ± 8	Cassidy and Belia (2005)

formation mechanism and treatment performance of the aerobic granules. Yang et al. (2003) reported that heterotrophic, nitrifying, and denitrifying bacteria had co-existed in the aerobic granules and the AGS was able to achieve effective nitrification and denitrification at the DO concentration of below 0.5 mg/L. Lin et al. (2003) reported that the aerobic granules-enriched with PAOs were able to be developed in an SBR operated with the cycle consisting of both anaerobic and aerobic periods. Meanwhile, numerous studies demonstrated that the anaerobic-aerobic SBR achieved the simultaneous nitrification, denitrification, and phosphorus removal (Wei et al. 2014, Yilmaz et al. 2008, Kishida et al. 2006, Cassidy and Belia 2005).

Table 1 summarizes some important process conditions of the SNDNPR SBR, including the wastewater characteristics, SBR cycle designs, operating conditions, and effluent quality. The VFA content of the wastewater is critical for the formation of the granular sludge because PAOs need to take up VFA to synthesize cellular PHA during the anaerobic period. Most of the studies on the SNDNPR AGS SBRs used the synthetic wastewater with acetate as the sole carbon source at concentrations ranging from 300 mg/L to 1800 mg/L (Wang et al. 2015). For the industrial wastewater treatment, acetate may be needed to be added into the wastewater, depending on the sCOD content of the wastewater. Yilmaz et al. (2008) reported that it is necessary to add 750 mg/L of acetate COD to anaerobically pre-treated abattoir wastewater to achieve the efficient nitrogen and phosphorus removal. Cassidy and Belia (2005) reported that the AGS SBR achieved successful SNDNPR with abattoir wastewater of sCOD content of 5163 mg/L without addition of extra acetate. The recommended maximum P/sCOD ratio for the enhanced biological wastewater treatment processes to achieve a soluble P effluent concentration below 0.50 mg/L was around 0.17 (Metcalf and Eddy 2014). However, for the SNDNPR AGS SBRs, relatively high TN/sCOD and TP/sCOD ratios may be required to promote the growth of nitrifying, denitrifying, and phosphate accumulating organisms (Lin et al. 2003, Yang et al. 2003). For the studies shown in Table 1, the ratios of TN/sCOD and TP/sCOD of the wastewater varied between 0.10 to 0.22 and 0.0083 to 0.042, respectively (Table 1). The molar ratio of (Ca + Mg)/P of the synthetic wastewater used in the AGS SNDNPR studies varied from 0.92 to 1.46 with the molar ratio of Ca to Mg to be around 0.26 (Kishida et al. 2006, Wang et al. 2015, Yuan et al. 2007).

The duration of the operation cycle of the SNDNPR AGS SBR, which consisted of the filling, anaerobic, aerobic, settling, and decanting/idle periods, were normally between 6 to 10 hours (Table 1). A short anaerobic feeding time around 5 to 30 min was commonly used to achieve anaerobic feasting. The length of the anaerobic period could be between 60 to 120 min to ensure a sufficient reaction time for the PAOs to take up VFA and synthesize intracellular PHA. Yuan et al. (2007) showed that having nitrate remaining over the whole anaerobic period will be detrimental to the activities of PAOs because the heterotrophic denitrifiers will compete with PAOs for the VFA carbon source, which could suppress the population of PAOs in the system. Also, with the incomplete depletion of nitrate during the anaerobic period, DPAOs could consume PHA for denitrification instead of uptake of VFA to make PHA during the anaerobic period, which will gradually reduce the PHA contents in the DPAO cells. Thus, the establishment of a real anaerobic phase is vital for developing AGS for achieving SNDNPR. The lengths of the aerobic period of the AGS SBRs could vary between 2 to 8 hours. Yuan et al. (2007) showed that a short oxic period could proliferate the growth of DPAOs but it is essential to have sufficient oxic reaction time for the nitrification and P uptake. While many studies showed that a cycle with the periodic anaerobic and aerobic reactions could achieve the SNDNPR, a post-anoxic period between 1 to 2 hours that followed the aerobic period can also be introduced to enhance the process of denitrification (Yilmaz et al. 2007). The settling times ranging from 0.5 min to 20 min were normally used in the SNDNPR AGS SBRs to select the granular sludge with the good settleability.

The hydraulic retention time (*HRT*) and loading rates (*L_r*) of the AGS SBR can be determined by Eqs. 6 and 7:

$$HRT = \frac{T_c}{\varphi_v} \tag{6}$$

$$L_r = \frac{C_0}{HRT} \tag{7}$$

where T_c is the cycle time, C_0 is the substrate or nutrient concentration in the influent, and φ_v is the SBR volume exchange ratio which is defined as the ratio of the discharged effluent volume at the end of the operation cycle to the reactor working volume.

The main operating parameters of AGS SBR include the hydraulic retention time (HRT), the organic, nitrogen, and phosphorus loading rates, and the solid retention time (SRT). The typical HRTs of AGS SBR were reported in the range of 10 to 20 h (Table 1) but an extended HRT could also be applied, depending on the characteristics of wastewater. The organic, nitrogen, and phosphorus loading rates of the AGS SBR could vary in the range of 2.5 to 3.6 g COD/L/d, 0.35 to 0.7 g N/L/d, and 0.025 to 0.06 g P/L/d, respectively. As discussed in Section 3.5, the formation of the granular sludge is insensitive to the loading rate while the microstructure and microbial community of granules may vary with the SBR loading rates.

The SRT can be approximately defined as the ratio of the total biomass mass in the reactor to the biomass wasted per day. At steady state, the wasted sludge mass is balanced by the sludge mass grown in the reactor, so the SRT represents the reciprocal of the specific growth rate of biomass. The SRT is normally used as an independent design parameter of SBRs because it exerts a strong selection pressure on the bacterial population in the reactors. The SRT of SNDNPR AGS SBR is typically in the range of 10 to 20 days. Compared to the conventional activated sludge process, the effect of the SRT on the microbial community structure of AGS could be less profound because of the aggregation feature of microorganisms in the AGS SBR.

Many studies have shown that the anaerobic-aerobic AGS SBR can achieve excellent simultaneous nitrification, denitrification, and phosphorus removal. However, some studies also demonstrated the challenges to achieve the SNDNPR with AGS SBR (Nancharaiah and Kiran Kumar Reddy 2018). Simultaneous nitrogen and phosphorus removal could be limited by various process factors, including the SBR cycle design, reactor DO concentration, granular size and structure, and microbial community structure of granular sludge. High DO in the bulk liquid during the aerobic period could limit denitrification. Beun et al. (2002) reported that DO levels could have a profound impact on the location of autotrophic bacteria within granules and a 40% saturation of DO was optimal for simultaneous nitrification and denitrification. A process named as the Single reactor High activity Ammonia Removal over Nitrite (SHARON) demonstrated simultaneous nitrification and denitrification with intermittent aeration (Hellinga et al. 1998). Adav et al. (2009b) showed that introduction of an anoxic phase after the SBR aerobic reaction could enhance denitrification. However, the efficiency of the post-anoxic process may be limited by the non-availability of COD (Nancharaiah and Kiran Kumar Reddy 2018). Moreover, there was evidence that low DO operation could result in the emission of N_2O and NO (Hellinga et al. 1998). For phosphorus removal, having an anaerobic period is essential to culture PAO and DPAO although granular sludge can provide an internal microbial anaerobic zones. Our study showed that the phosphorus removal capacity of AGS could be seriously affected by the change in microbial community composition with AGS aging. Also, as shown in some studies, the total suspended solid in the effluent of AGS SBR could be as high as a few hundreds of milligram per liter (Yilmaz et al. 2008) due to the very short settling time used in the operation. A post-settling process or membrane filtration can be coupled with the AGS SBRs to reduce the suspended solids in the effluent.

4.2 SNDNPR AGS cycle behavior

The SNDNPR AGS can exhibit a strong anaerobic phosphorus release, aerobic P uptake, and the simultaneous aerobic nitrification and denitrification throughout the operation cycle. Figure 3a and b show the typical sCOD, PO_4-P, NH_4-N, NO_2-N, and NO_3-N cycle profiles of the SNDNPR AGS observed in our study. In this example, the AGS was formed in an anaerobic/aerobic SBR treating the acetate-amended sewage. The wastewater contained a sCOD of 800 mg/L, NH_4-N of 100 mg/L, and PO_4-P of 15 mg/L. The bioreactor was operated in a sequential batch mode with an operating cycle of 7 h and 13 min which included 10-minute feeding, 2-hour anaerobic reaction, 5-hour aerobic reaction, followed by 3 minutes of settling. In this case, the PO_4-P concentration increased from 8.0 mg/L to 44.8 mg/L during the two-hour anaerobic operation period; continued to increase in the first half-hour of the aerobic period; and then decreased from 46.2 mg/L to 3.4 mg/L during the remaining 4.5-hour aerobic period (Figure 3a). The P release observed during the first half hour of the aerobic period was likely caused by the slow development of the aerobic zone within the granular

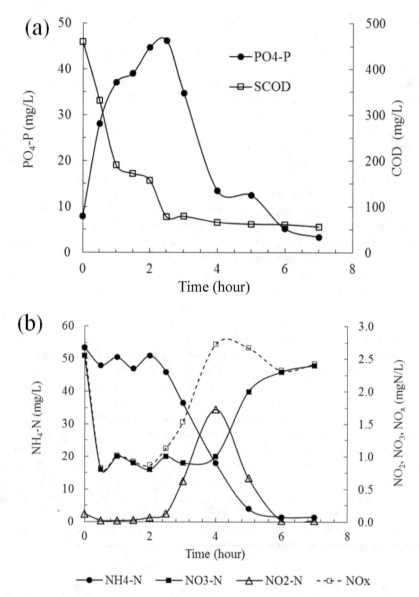

Figure 3. The change of the COD, phosphorus, and nitrogen concentrations during the operation cycle of an anaerobic–aerobic AGS SBR(a) COD and PO_4-P concentration profiles; (b) NH_4-N, NO_3-N, NO_2-N, and NOx-N profiles.

sludge due to the mass transfer limitation. Yilmaz et al. (2008) reported that it took around one hour for oxygen to penetrate the active dense granules after the aeration was initialized in the reactor. The time taken for the development of oxic zone in granular sludge will depend on the microstructure of the granules, and an evident delay of the aerobic condition in granules could be observed only with the dense and compact granules. In the case shown in Figure 3, the sCOD was reduced drastically from 459.5 mg/L to 157 mg/L during the anaerobic period, and then reduced to 78 mg/L in the first half hour of the aerobic period before reaching a slow sCOD degradation stage (Figure 3a). A low COD aerobic period could be important to stimulate PAOs to use the stored PHA for P uptake. The length of such an aerobic starvation period will affect the distribution of PAO and DPAO in the granules. Figure 3b shows the NH_4-N, NO_3-N, NO_2-N, and total NOx-N profiles over the cycle duration. In this example, the total NOx-N accumulated during the aerobic period was significantly lower than the amount of ammonia nitrogen oxidized via nitrification, suggesting the simultaneous nitrification and denitrification under the aerobic condition. The concentration of NH_4-N was reduced from around 50 mg/L to 20 mg/L during the first two hours of the aerobic period, while NO_3-N was maintained at around 1 mg/L and the NO_2-N concentration was increased from nil to a maximum value of 1.7 mg/L. Since the end product of the granular sludge nitrification is nitrate (Zeng et al. 2003), the denitrification in the granules will include steps of nitrate to nitrite and nitrite to nitrogen gas. The slight accumulation of nitrite during the aerobic period could suggest that the control step of the denitrification was the nitrite reduction step.

4.3 GAO denitrification

Although studies have shown that DPAOs can accomplish the SNDNPR, Zeng et al. (2003) demonstrated that the simultaneous nitrification and denitrification observed with an SBR treating synthetic wastewater was accomplished by denitrifying glycogen-accumulating organisms (DGAO). GAOs consume glycogen, take up VFA, and synthesize PHA under anaerobic conditions, while consume PHA for glycogen synthesis under aerobic conditions. Thus, there is no P uptake, and poly-P accumulation involved in the GAO metabolism. Zeng et al. (2003) demonstrated that GAOs cultivated in an anaerobic-aerobic SBR reduced nitrite to N_2O when the concentrations of NO_2-N were below 1 mg/L. Lemaire et al. (2008) also observed that it was the DGAOs rather than DPAOs that were responsible for the simultaneous nitrification and denitrification in GAOs and PAOs enriched SBR system. According to these studies, it can be speculated that there are two different types of SNDNPR AGS: DPAO granules and PAO-GAO granules. The DPAO granules have a surface zone that is dominated by nitrifiers and an anoxic core where the DPAOs are dominant. The DPAO granules achieve SNDNPR by DPAO using the same carbon source. The PAO-GAO granules may have outer layers where nitrifiers and PAO are dominant and an anoxic core where the DGAOs are dominant. The SNDNPR in a PAO-GAO granule could be accomplished by two different groups of microorganisms. The PAOs in the outer layers take up P using O_2 as the electron acceptors, while the GAOs in the anoxic core reduce NO_2-N to N_2O. The PAO-GAO AGS process has no advantages over the conventional EBNR regarding carbon source saving because the PAOs and GAOs in the granules use the different electron acceptors and carbon sources. Moreover, the heat-trapping capacity of N_2O in the atmosphere is 298 times larger than CO_2, so the reduction of NO_2-N to N_2O causes serious environmental concerns. However, Lemaire et al. (2008) found that the N_2O production observed in their study could be attributed to the low diversity of the denitrifiers associated with the sole carbon source feeding. They observed that the N_2O production in the SNDNPR SBR was immediately reduced after adding the sludge from the conventional nutrient removal process, suggesting that the diversity of denitrifiers in the reactor could exert a critical impact on the denitrification and phosphate uptake pathways.

4.4 Effect of chemical precipitation

The SNDNPR behavior of an AGS SBR can also be affected by biologically induced chemical precipitation in the treatment of high nitrogen content wastewater. The oversaturation of Ca^{2+}, Mg^{2+}, NH_4^+, and PO_4^{3-}, could induce the formation of various metal precipitates, e.g., apatites ($Ca_2HPO_4(OH)_2$, $CaHPO_4$, hydroxydicalcium phosphate (HDP), newberyite ($MgHPO_4.3H_2O$), struvite ($MgNH_4PO_4$), etc. (Maurer et al. 1999). Yilmaz et al. (2008) suggested that a significant decrease in the concentration of NH_4-N during the anaerobic period might indicate the anaerobic precipitation of ammonium-containing minerals, such as struvite. Regarding the effect of chemical precipitation on P removal, Yilmaz et al. (2003) found that an insignificant amount of the precipitated P was detected at the end of the aerobic periods although significant P precipitation was observed during the anaerobic period. They suggested that most of the P metal precipitates were dissolved during the aerobic period and the dissolved P was taken up immediately by PAOs for poly-P synthesis. Therefore, it appeared that the overall impact of the chemical precipitation on the P removal in the SNDNPR SBR is insignificant (Yilmaz et al. 2008).

5. Properties of SNDNPR granules

Figure 4 shows the microscopy pictures of an anaerobic granule, fresh aerobic granule, and matured aerobic granule related to the SNDNPR cycle behavior shown in Figure 3a and 3b. The appearance and physical properties of granules may change through a long-term operation period. For the granules shown in Figure 4, the fresh aerobic granule looked a little transparent under the microscope, while the matured aerobic granules were dark and compact with a fluffy outer surface. In general, the aerobic granules had a regular round shape with size distribution in a relatively narrow range of 1 to 2 mm and a specific gravity around 1.03. The 30-min Sludge Volume Index (SVI) of granular sludge could be smaller than 30 mL/g, and the settling velocity of AGS can be greater than 20 m/hr (Table 2).

Liu et al. (2003) studied the elemental compositions of nitrifying aerobic granules cultivated in an aerobic SBR with a 4-hour operation cycle using the ethanol-type synthetic wastewater that had the N/C ratios ranging from 5/100 to 30/100. The granules in the SBR was determined to contain around 42% of carbon, 8.5 to 9.5% of N, 0.75 to 0.85% of P, 0.23 to 0.54% of Ca, 0.07 to 0.14% of Mg, and 0.11 to 0.48% of Na. For the SNDNPR granules, the phosphorus and calcium contents could be much higher than the nitrifying granules. The chemical analysis showed that the granules associated with the cycle behaviors shown in Figure 3 contained around 32.3% of organic carbon, 10.7% of TKN, 5% of P, 5% of Ca, and 0.445% of Mg. Moreover, the analysis revealed that the granule P and TKN contents decreased and the TOC content of the granular increased after the P release. The calcium

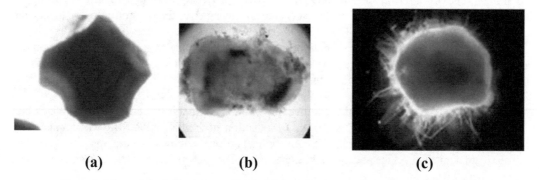

| (a) | (b) | (c) |

Figure 4. Microscopic pictures of anaerobic (a), fresh aerobic (b), and matured aerobic granules (c).

Color version at the end of the book

Table 2. Physical properties of aerobic granules.

	Mean diameter (mm)	Specific gravity	ZSV (m/h)	SVI (L/mg)
Cassidy and Belia 2005	1.7 ± 0.3	1.035 ± 0.001	51 ± 9	22 ± 8
Wei et al. 2014	1–3	1.033	36	n/a
Kishida et al. 2006	~ 1	n/a	24	20
Wang et al. 2015	1.92 ± 0.78	1.171 ± 0.001	241 ± 29	13.7 ± 1.2

content of the granules also increased after the P release, likely due to the Ca precipitation within the granular sludge which could be induced by the P and Ca release during the anaerobic period as suggested by Yilmaz et al. (2008). Thus, the compositions of the SNDNPR granules in the AGS SBRs vary throughout the operation cycle due to the biological and chemical reactions occurring during the aerobic and anaerobic period.

6. DO and microbial profiles of SNDNPR granules

Yilmaz et al. (2008) measured the oxygen and microbial distributions in the granules using the oxygen microsensors and revealed that the depth of the aerobic zone in the granules was gradually extended and reached a maximum depth of 300 to 400 µm during the aeration period. Using the similar technique, Kishida et al. (2006) found the DO penetration depth was around 100 µm when the DO concentration in the reactor was around 5.5 mg/L. They also revealed by using Fluorescence *In Situ* Hybridization (FISH) technique the co-existence of PAOs and GAOs in granules with the GAOs near the granule surface and the PAOs in both the outer and inner surface of the granules. de Kreuk et al. (2005a, b) suggested a layered granule structure that was characterized by the outer oxic layers containing a mixture of heterotrophic PAOs and autotrophic nitrifier and the inner anoxic core containing DPAOs. Lemaire et al. (2008) studied the DO and microbial distributions in granules and determined that the depth of the aerobic penetration was around 250 µm for the granules with a size larger than 500 µm. By using the FISH technique, they found *Accumulibacter* spp. and *Competibacter* spp. were only PAOs and GAOs detected in the granules. *Accumulibacter* spp. were dominated in the region between 0 to 200 µm from the granule surface and *Competibacter* spp. were dominated in the region from 200 µm inwards. The distributions of microorganisms were highly correlated with the oxygen profiles in the granules. In general, it is reasonable to see different DO distributions and microbial community structures in granules cultured under different conditions because of the difference in the physical and biological properties of granules.

7. Role of DPAOs in AGS SNDNPR

Anoxic P removal in lab-scale and full-scale systems by DPAOs have been reported by numerous researchers (Akin and Ugurlu 2004, Bortone et al. 1999, Carvalho et al. 2007, Hu et al. 2003, Kuba et al. 1997, Kuba et al. 1997, Kerrn-Jespersen and Henze 1993). It was reported that the cell yield of DPAOs is 20 ~ 30% lower than that of PAOs (Murnleitner et al. 1997). The SNDNPR with DPAOs can save 50% of carbon sources and 30% of energy due to the utilization of the same carbon sources for both denitrification and phosphorus removal (Lochmatter et al. 2009, Kuba et al. 1997, Kuba et al. 1997, Bassin et al. 2012, Wang 2007, Lv et al. 2016). The occurrence of the AGS SNDNPR under the aerobic condition suggested that DPAOs play an important role in the AGS denitrification and P uptake. DPAOs can remove phosphorus using nitrate/nitrite as the electron acceptors in the absence of oxygen. The limited DO, scarcity in external carbon source, and rich in nitrate or nitrite will proliferate the population of DPAOs in the interior region of aerobic granules.

Studies have identified three types of PAOs by using anaerobic-anoxic-aerobic batch tests, including non-DPAOs that use oxygen as electron acceptors only; DPAOs that use oxygen or nitrate or nitrite as electron acceptors; and DPAOs that use nitrate or nitrite as elector acceptors only (Carvalho et al. 2007, Freitas et al. 2005, Kerrn-Jespersen and Henze 1993, Meinhold et al. 1999). Although PAOs have not been isolated from enhanced biological nutrient removal processes, it is widely accepted that the primary PAO, named as *Candidatus Accumulibacter phosphatis* (*Accumulibacter*), is affiliated with *Rhodocyclus* genus within the *Betaproteobacteria.* Kong et al. (2004) studied the behavior of Rhodocyclus-related polyphosphate-accumulating (RPAO) bacteria in the full-scale EBPR plants and found that the RPAOs were able to take up orthophosphate and accumulate polyphosphate by using

Table 3. Stoichiometric values of anaerobic conversion with PAO, combined PAO-GAO, and GAO metabolisms.

Metabolisms	$P_{release}/HAC_{uptake}$ (P-mol/C-mol)	PHB_{stroed}/HAC_{utake} (C-mol/C-mol)	$Gly_{consumed}/HACc_{uptake}$ (C-mol/C-mol)	PHV/PHB (C-mol/C-mol)	References
PAO clade I	0.64	1.27	0.29	0.07	Welles et al. (2015)
PAO clade II (PAO-GAO)	0.22	1.24	0.96	0.19	Welles et al. (2015)
GAO	0.13	1.34	1.06	0.32	Welles et al. (2015)
GAO	0.01	1.28	1.2	0.34	Lopez-Vazques et al. (2007), Welles et al. (2015)

*PHB: Polyhydroxybutyrate; PHV: polyhydroxyvaterate; Gly: Glycogen

oxygen, nitrate, and nitrite as electron acceptors, suggesting RPAOs can perform denitrification. They also found that RPAOs did not take up phosphorus when acetate was available together with oxygen, nitrate, and nitrite as electron acceptors, implying the importance of starvation for the SNDNPR. In addition to Rhodocyclus-related PAOs, Kong et al. (2005) also identified that two morphotypes of Actinobacteria, coccus and short rod morphotypes can take up orthophosphate with oxygen and nitrate as electron acceptors. Zeng et al. (2003) determined that the primary PAOs cultured in both anaerobic-aerobic and anaerobic-anoxic systems were the same Accumulibacter organisms but P uptake could be retarded when switching the process from the anaerobic-aerobic condition to the anaerobic-anoxic condition. Carvalho et al. (2007) investigated P uptake and denitrification behaviors of DPAOs in the SBRs fed with acetate and propionate. They found that DPAOs cultured with propionate were able to maintain a stable denitrification and P removal performance when the SBR was gradually switched from the anaerobic-aerobic to anaerobic-anoxic conditions. They reported that the P removal and denitrification of the acetate-fed SBR collapsed after the aerobic phase was eliminated, suggesting that the PAOs in the acetate-fed SBR became inactive under the anaerobic-anoxic condition. The FISH characterization revealed that both reactors had *Accumulibacter* as dominant microorganisms, but the cell morphotypes in each reactor were different. The *Accumulibacter* in the acetate-fed SBR had a dominant *coccus* morphotype while most of those in the propionate-fed SBR showed a *rod* morphotype. Based on these findings, Carvalho et al. (2007) suggested that the *coccus Accumulibacter strains* might only be able to use oxygen and nitrite as the electron acceptors and the *rod Accumulibacter strains* were able to use nitrate, nitrite, and oxygen as electron acceptors to perform SNDNPR (Carvalho et al. 2007).

McMahon et al. (2002) identified two types of poly-P kinase (*ppk*), Type I and Type II, that were related to the *Rhodocyclus-like PAOs* and found that a *ppk* related to Type I *ppk* (*ppk 1*) is transcribed in the enhanced biological phosphorus removal processes. He et al. (2007) investigated the population structures of the *Accumulibacter* lineage using *ppk 1* as a genetic marker and determined the relative distributions of five *Accumulibacter* clades. Welles et al. (2015) investigated the metabolism characteristics of *Accumulibacter* clades I (PAO I) and *Accumulibacter* clade II (PAO II) and found that PAO I performed the typical PAO metabolism (showing a higher P/HAC ratio of 0.64), while PAO II exhibited a mixed PAO-GAO metabolism (showing a lower P/HAC ratio of 0.22). Table 3 shows the stoichiometric values of anaerobic P release, PHB synthesis, and glycogen depletion of PAO I, PAO II, and GAO. For the typical PAO metabolism, the relatively high P/HAC and low PHV/PHB and Gly/HAC ratios can be observed compared to those with the GAO and PAO-GAO metabolisms. Also, Welles et al. (2015) reported both PAO I and PAO II gradually shifted their metabolism to GAO metabolism when the intracellular poly-content decreased. Their results indicated that maintaining the proper conditions for the anaerobic P release and aerobic P uptake will be crucial to achieve a stable long-term P removal.

8. Conclusion

Simultaneous nitrification, denitrification, and phosphorus removal (SNDNPR) is an advanced biological wastewater treatment technology. The aerobic granular technology was proven to be able to achieve the SNDNPR in a simple anaerobic-aerobic SBR reactor. The formation of aerobic granules can be affected by the reactor configuration, SBR cycle design, hydrodynamic condition, organic loading rate, and settling time. The SNDNPR capacity of aerobic granules mainly depends on the DPAO composition and population in granules, which can be significantly affected by the microstructure of the granules and the substrate characteristics.

Acknowledgment

The authors thank Ontario Ministry of Agriculture, Food and Rural Affairs (OMAFRA) for the support.

References

Adav, S.S., D.J. Lee and J.Y. Lai. 2007a. Effects of aeration intensity on formation of phenol-fed aerobic granules and extracellular polymeric substances. Applied Microbiology and Biotechnology 77(1): 175–182.

Adav, S.S., M.Y. Chen, D.J. Lee and N.Q. Ren. 2007b. Degradation of phenol by aerobic granules and isolated yeast *Candida tropicalis*. Biotechnology and Bioengineering 96(5): 844–852.

Akin, B.S. and A. Ugurlu. 2004. The effect of an anoxic zone on biological phosphorus removal by a sequential batch reactor. Bioresource Technology 94(1): 1–7.

Bassin, J.P., R. Kleerebezem, M. Dezotti and M.C.M. van Loosdrecht. 2012. Simultaneous nitrogen and phosphate removal in aerobic granular sludge reactors operated at different temperatures. Water Research 46(12): 3805–3816.

Bernet, N., P. Dangcong, J.P. Delegès and R. Moletta. 2001. Nitrification at low oxygen concentration in biofilm reactor. Journal of Environmental Engineering 127(3): 266–271.

Beun, J.J., A. Hendriks, M.C.M. van Loosdrecht, E. Morgenroth, P.A. Wilderer and J.J. Heijene. 1999. Aerobic granulation in a sequencing batch reactor. Water Research 33(10): 2283–2290.

Beun, J.J., M.C.M. van Loosdrecht and J.J. Heijnen. 2002a. Aerobic granulation in a sequencing batch airlift reactor. Water Research 36(3): 702–712.

Beun, J.J., J.J. Heijnen and M.C.M. van Loosdrecht. 2002b. Nitrogen removal in a granular sludge sequencing batch airlift rector. Biotechnol Bioengineering 75(1): 82–92.

Blackburne, R., Z. Yuan and J. Keller. 2008. Demonstration of nitrogen removal via nitrite in a sequencing batch reactor treating domestic wastewater. Water Research 42(8-9): 2166–2176.

Bond, P.L., P. Hugenholtz, J. Keller and L.L. Blackall. 1995. Bacterial community structures of phosphate-removing and non-phosphate-removing activated sludges from sequencing batch reactors. Applied and Environmental Microbiology 61(5): 1910–6.

Bortone, G., S.M. Libelli, A. Tilche and J. Wanner. 1999. Anoxic phosphate uptake in the dephanox process. Water Science and Technology 40(4): 177–185.

Carvalho, G., P.C. Lemos, A. Oehmen and M.A.M. Reis. 2007. Denitrifying phosphorus removal: Linking the process performance with the microbial community structure. Water Research 41(19): 4383–4396.

Cassidy, D.P. and E. Belia. 2005. Nitrogen and phosphorus removal from an abattoir wastewater in a SBR with aerobic granular sludge. Water Research 39(19): 4817–4823.

Chou, H.H. and J.S. Huang. 2005. Comparative granule characteristics and biokinetics of sucrose-fed and phenol-fed UASB reactors. Chemosphere 59(1): 107–116.

Freitas, F., M. Temudo and M.A.M. Reis. 2005. Microbial population response to changes of the operating conditions in a dynamic nutrient-removal sequencing batch reactor. Bioprocess and Biosystems Engineering 28(3): 199–209.

Gu, A.Z., A. Saunders, J.B. Neethling, H.D. Stensel and L.L. Blackall. 2008. Functionally relevant microorganisms to enhanced biological phosphorus removal performance at full-scale wastewater treatment plants in the United States. Water Environment Research 80(8): 688–98.

He, S., D.L. Gall and K.D. McMahon. 2007. "*Candidatus accumulibacter*" population structure in enhanced biological phosphorus removal sludges as revealed by polyphosphate kinase genes. Applied and Environmental Microbiology 73(18): 5865–5874.

Heijnen, J.J. and M.C.M. van Loosdrecht. 1998. Method for acquiring grain-shaped growth of a microorganism in a reactor. Biofutur 1998(183): 50.

Hellinga, C., A.A.J.C. Schellen, J.W. Mulder, M.C.M. van Loosdrecht and J.J. Heijnen. 1998. The SHARON process: An innovative method for nitrogen removal from ammonium-rich waste water. Water Science and Technology 37(9): 135–142.

Kerrn-Jespersen, J.P. and M. Henze. 1993. Biological phosphorus uptake under anoxic and aerobic conditions. Water Research 27(4): 617–624.

Kishida, N., J. Kim, S. Tsuneda and R. Sudo. 2006. Anaerobic/oxic/anoxic granular sludge process as an effective nutrient removal process utilizing denitrifying polyphosphate-accumulating organisms. Water Research 40(12): 2303–2310.

Kong, Y., J.L. Nielsen and P.H. Nielsen. 2004. Microautoradiographic study of Rhodocyclus-related polyphosphate-accumulating bacteria in full-scale enhanced biological phosphorus removal plants. Applied and Environmental Microbiology 70(9): 5383–90.

Kong, Y., J.L. Nielsen and P.H. Nielsen. 2005. Identity and ecophysiology of uncultured actinobacterial polyphosphate-accumulating organisms in full-scale enhanced biological phosphorus removal plants. Applied and Environmental Microbiology 71(7): 4076–4085.

Kong, Y., Y.Q. Liu, J.H. Tay, F.S. Wong and J. Zhu. 2009. Aerobic granulation in sequencing batch reactors with different reactor height/diameter ratios. Enzyme and Microbial Technology 45(5): 379–383.

de Kreuk, M.K., J.J. Heijnen and M.C.M. van Loosdrecht. 2005a. Simultaneous COD, nitrogen, and phosphate removal by aerobic granular sludge. Biotechnology and Bioengineering 90(6): 761–9.

de Kreuk, M.K., M. Pronk and M.C.M. van Loosdrecht. 2005b. Formation of aerobic granules and conversion processes in an aerobic granular sludge reactor at moderate and low temperatures. Water Research 39(18): 4476–4484.

Kuba, T., M.C.M. van Loosdrecht, F.A. Brandse and J.J. Heijnen. 1997. Occurrence of denitrifying phosphorus removing bacteria in modified UCT-type wastewater treatment plants. Water Research 31(4): 777–786.

Kuba, T., M.C.M. van Loosdrecht and J.J. Heijnen. 1997. Biological dephosphatation by activated sludge under denitrifying conditions pH influence and occurrence of denitrifying dephosphatation in a full-scale waste water treatment plant. Water Science and Technology 36(12): 75–82.

Lemaire, R., Z. Yuan, L.L. Blackall and G.R. Crocetti. 2008. Microbial distribution of *Accumulibacter* spp. and *Competibacter* spp. in aerobic granules from a lab-scale biological nutrient removal system. Environmental Microbiology 10(2): 354–363.

Li, Z.H., T. Kuba and T. Kusuda. 2006. Selective force and mature phase affect the stability of aerobic granule: An experimental study by applying different removal methods of sludge. Enzyme and Microbial Technology 39(5): 976–981.

Lin, C.-Y., F.-Y. Chang and C.-H. Chang. 2001. Treatment of septage using an upflow anaerobic sludge blanket reactor. Water Environment Research 73(4): 404–408.

Liu, Q.S., J.H. Tay and Y. Liu. 2003. Substrate concentration independent aerobic granulation in sequential aerobic sludge blanket reactor. Environmental Technology 24(10): 1235–1242.

Liu, W., J. Zhang, Y. Jin, X. Zhao and Z. Cai. 2015. Adsorption of Pb(II), Cd(II) and Zn(II) by extracellular polymeric substances extracted from aerobic granular sludge: Efficiency of protein. Journal of Environmental Chemical Engineering 3(2): 1223–1232.

Liu, Y.Q. and J.H. Tay. 2007. Influence of cycle time on kinetic behaviors of steady-state aerobic granules in sequencing batch reactors. Enzyme and Microbial Technology 41: 516–522.

Liu, Y. and J.H. Tay. 2002. The essential role of hydrodynamic shear force in the formation of biofilm and granular sludge. Water Research 36(7): 1653–1665.

Liu, Y. and J.H. Tay. 2004. State of the art of biogranulation technology for wastewater treatment. Biotechnology Advances 22(7): 533–563.

Lochmatter, S., D. Weissbrodt, G. Gonzalez-Gil and C. Holliger. 2009. Denitrifying PAO and GAO in aerobic granular biofilm cultivated with acetate and propionate. SSM Annual Assembly. https://infoscience.epfl.ch/record/136739?ln=en.

Lv, X., L.J. Li, F. Sun, C. Li, M. Shao and W. Dong. 2016. Denitrifying phosphorus removal for simultaneous nitrogen and phosphorus removal from wastewater with low C/N ratios and microbial community structure analysis. Journal of Chemical Technology and Biotechnology 57(4): 1890–1899.

Marques, A.P., A.F. Duque, V.S. Bessa, R.B. Mesquita, A.O. Rangel and P.M. Castro. 2013. Performance of an aerobic granular sequencing batch reactor fed with wastewaters contaminated with Zn^{2+}. Journal of Environmental Management 128: 877–882.

Maurer, M., D. Abramovich, H. Siegrist and W. Gujer. 1999. Kinetics of biologically induced phosphorus precipitation in waste-water treatment. Water Research 33(2): 484–493.

McMahon, K.D., M.A. Dojka, N.R. Pace, D. Jenkins and J.D. Keasling. 2002. Polyphosphate kinase from activated sludge performing enhanced biological phosphorus removal. Applied and Environmental Microbiology 68(10): 4971–8.

Meinhold, J., C.D.M. Filipe, G.T. Daigger and S. Isaacs. 1999. Characterization of the denitrifying fraction of phosphate accumulating organisms in biological phosphate removal. Water Science and Technology 39(1): 31–42.

Metcalf and Eddy. 2014. Wastewater Engineering: Treatment and Resource Recovery, 5th ed., McGraw-Hill, New York.

Morgenroth, E., T. Sherden, M.C.M. van Loosdrecht, J.J. Heijnen and P.A. Wilderer. 1997. Aerobic granular sludge in a sequencing batch reactor. Water Research 31(12): 3191–3194.

Murnleitner, E., T. Kuba, M.C.M. van Loosdrecht and J.J. Heijnen. 1997. An integrated metabolic model for the aerobic and denitrifying biological phosphorus removal. Biotechnology and Bioengineering 54(5): 434–450.

Nancharaiah, Y.V. and G. Kiran Kumar Reddy. 2018. Aerobic granular sludge technology: Mechanisms of granulation and biotechnological applications. Bioresource Technology 247: 1128–1143.

Oehmen, A., G. Carvalho, F. Freitas and M.A. Reis. 2010. Assessing the abundance and activity of denitrifying polyphosphate accumulating organisms through molecular and chemical techniques. Water Science and Technology 61(8): 2061–2068.

Othman, I., A. Nor-Anuar, Z. Ujang, N.H. Rosman, H. Harun and S. Chelliapan. 2013. Livestock wastewater treatment using aerobic granular sludge. Bioresource Technology 133: 630–634.

Pan, S., J.H. Tay, Y.X. He and S.T. Tay. 2004. The effect of hydraulic retention time on the stability of aerobically grown microbial granules. Letters in Applied Microbiology 38(2): 158–163.

Qin, L., J.H. Tay and Y. Liu. 2004. Selection pressure is a driving force of aerobic granulation in sequencing batch reactors. Process Biochemistry 39(5): 579–584.

Schwarzenbeck, N., J.M. Borge and P.A. Wilderer. 2005. Treatment of dairy effluents in an aerobic granular sludge sequencing batch reactor. Applied Microbiology and Biotechnology 66(6): 711–718.

Sedlak, R. 1991. Phosphorus and nitrogen removal from municipal wastewater: Principles and practice. Lewis Publishers.

Su, K.-Z. and H.-Q. Yu. 2005. Formation and characterization of aerobic granules in a sequencing batch reactor treating soybean-processing wastewater. Environmental Science and Technology 39(8): 2818–2827.

Tay, J.H., Q.S. Liu and Y. Liu. 2001. The effects of shear force on the formation, structure and metabolism of aerobic granules. Applied Microbiology and Biotechnology 57(1-2): 227–233.

Tay, J.H., S.F. Yang and Y. Liu. 2002a. Hydraulic selection pressure-induced nitrifying granulation in sequencing batch reactors. Applied Microbiology and Biotechnology 59(2-3): 332–337.

Tay, J.H., Q.S. Liu and Y. Liu. 2002b. Characteristics of aerobic granules grown on glucose and acetate in sequential aerobic sludge blanket reactors. Environmental Technology 23(8): 931–936.

Tiong-Lee Tay, S., B.Y. Moy, H.L. Jiang and J.H. Tay. 2005. Rapid cultivation of stable aerobic phenol-degrading granules using acetate-fed granules as microbial seed. Journal of Biotechnology 115: 387–395.

Tokutomi, T. 2004. Operation of a nitrite-type airlift reactor at low DO concentration. Water Science and Technology 49(5-6): 81–88.

Wang, Y., J. Geng, Z. Ren, G. Guo, C. Wang and H. Wang. 2013. Effect of COD/N and COD/P ratios on the PHA transformation and dynamics of microbial community structure in a denitrifying phosphorus removal process. Journal of Chemical Technology and Biotechnology 88(7): 1228–1236.

Wang, Y., X. Jiang, H. Wang, G. Guo, J. Guo, J. Qin and S. Zhou. 2015. Comparison of performance, microorganism populations, and bio-physiochemical properties of granular and flocculent sludge from denitrifying phosphorus removal reactors. Chemical Engineering Journal 262: 49–58.

Wang, Z. 2007. Insights into mechanism of aerobic granulation in sequencing batch reactor. PhD. thesis. https://pdfs.semanticscholar.org/d2e0/9acdf355577618f50abb5d477fa84c78039b.pdf.

Wei, D., B. Wang, H.H. Ngo, W. Guo, F. Han, X. Wang, B. Du and Q. Wei. 2015. Role of extracellular polymeric substances in biosorption of dye wastewater using aerobic granular sludge. Bioresource Technology 185: 14–20.

Wei, D., M. Li, X. Wang, F. Han, L. Li, J. Guo, L. Ai, L. Fang, L. Liu, B. Du and Q. Wei. 2016. Extracellular polymeric substances for Zn (II) binding during its sorption process onto aerobic granular sludge. Journal of Hazardous Materials 301: 407–415.

Welles, L., W.D. Tian, S. Saad, B. Abbas, C.M. Lopez-Vazuez, C.M. Hooijmans, M.C.M. van Loosdrecht and D. Brdjanovic. 2015. Accumulibacter clades Type I and II performing kinetically different glycogen-accumulating organisms metabolisms for anaerobic substrate uptake. Water Research 83: 354–366.

Yilmaz, G., R. Lemaire, J. Keller and Z. Yuan. 2007. Effectiveness of an alternating aerobic, anoxic/anaerobic strategy for maintaining biomass activity of BNR sludge during long-term starvation. Water Research 41(12): 2590–2598.

Yilmaz, G., R. Lemaire, J. Keller and Z. Yuaz. 2008. Simultaneous nitrification, denitrification, and phosphorus removal from nutrient-rich industrial wastewater using granular sludge. Biotechnology and Bioengineering 100(3): 529–541.

Yuan, L., W. Han, L. Wang, Y. Yang and Z. Wang. 2007. Simultaneous denitrifying phosphorus accumulation in a sequencing batch reactor. Frontiers of Environmental Science & Engineering in China 1(1): 23–27.

Zeng, R.J., A.M. Saunders, Z. Yuan, L.L. Blackall and J. Keller. 2003. Identification and comparison of aerobic and denitrifying polyphosphate-accumulating organisms. Biotechnology and Bioengineering 83(2): 140–148.

CHAPTER 5

Biofilm and Granular Sludge Bioreactors for Textile Wastewater Treatment

G. Kiran Kumar Reddy[1,2] *and Y.V. Nancharaiah*[1,2,]*

1. Introduction

The world's first commercially successful synthetic dye was invented in 1856 by William Henry Perkin. By the end of the 19th century, nearly ten thousand new synthetic dyes were developed for various applications (Wesenberg et al. 2003). Dyes are the chemical substances which impart colour to the substrates by different mechanisms, including adsorption, entrapment, covalent bonding and complexation (Bafana et al. 2011). Dyes impart colour to the medium by forming solutions. Textile manufacturers are the largest consumers of dyes. More than 80% of the commercially produced dyes are used in textile industry. In addition, dyes are used in paper industry, food industry, agricultural research, and in making hair colorants, light harvesting arrays and photo-electrochemical cells (Forgacs et al. 2004). In countries like China, European Union and India, textile manufacturing is one of the largest job-creating industries. In fact, textile industry is an important contributor of economic growth (Ghaly et al. 2013). This industry is expected to grow further and provide more job opportunities especially in developing countries.

Dyes are classified based on their origin, chemical/physical properties and application type (Bafana et al. 2011, Wesenberg et al. 2003). Based on chemical composition, dyes are classified into azo, nitro, nitroso, diarylmethane, triarylmethane, xanthene, anthraquinoid, acridine, cyanine, quinone-imine, pthalocyanine, and thiazole dyes. Based on the application type, dyes are classified as acid, azoic, basic, direct, disperse, mordant, reactive, sulphur and vat dyes. The colour of the dye depends on absorption of specific wavelength of visible light, which in turn depends on the chemical

[1] Biofouling and Biofilm Processes Section, Water and Steam Chemistry Division, Bhabha Atomic Research Centre, Kalpakkam-603102, Tamil Nadu, India.
[2] Homi Bhabha National Institute, Anushakti Nagar, Mumbai 400094.
* Corresponding author: yvn@igcar.gov.in, venkatany@gmail.com

structure of the dye molecules. The chromophore groups such as azo, anthraquinone, methine, nitro, arylmethane and carbonyl are responsible for imparting colour to dye molecules. The auxochrome groups such as amine, carboxyl, sulfonate and hydroxyl groups are responsible for deepening the colour intensity. The auxochrome groups are often bonded to the aromatic or heterocyclic rings of dye molecules.

Azo dyes alone constitute nearly 70% of the organic dyes produced in the world. Azo dyes are the largest group of dyes both by the number and the volume of production. These dyes are characterised by the presence of at least one azo group, that is, "-N=N-". This azo bond is flanked by a minimum of one or usually two aromatic groups. The chemical structures of azo group, methyl orange (single azo bond), amido black (two azo bonds) and direct black (three azo bonds) are shown in Figure 1.

Major textile operations include preparation of fibres, converting fibres to yarn and fabric manufacturing. Several wet processing steps such as sizing and desizing, scouring, bleaching, dyeing, printing and final finishing steps are used while converting yarn to fabric. Each one of these steps employs different chemical constituents. The type of chemical constituents used and the characteristics of wastewater generated in different stages of textile operations are shown in Figure 2. Large quantities of effluents are generated during various steps of textile manufacturing. It has been estimated that processing about 12–20 tonnes of fabric generates nearly 1,000–3,000 m^3 of liquid effluent (Ghaly et al. 2013). The textile processes such as dyeing, bleaching and washing steps produce large quantities of dye containing effluents. During dyeing, only a part of the dye is incorporated onto fabric and much of it enters into the effluent stream. About 10–50% of the total applied dye finally enters into the effluent streams depending on the process, fabric and dye (Forgacs et al. 2004, Chequer et al. 2013). Thus, quality of textile effluents varies in terms of chemical oxygen demand (COD), biochemical oxygen demand (BOD), colour, colour intensity, surfactants, metals, nitrates, sulphates, pH and alkalinity. A residual dye as low as 1 mg L^{-1} can impart colour to the effluent and its intensity increases with an increase dye concentration (Pereira and Alves 2012).

Complex synthetic dyes present in the textile effluents are toxic and pose threat to health and environment. Various toxicological studies indicated toxicity and mutagenicity of synthetic dyes (Jager et al. 2004, Mathur et al. 2012). The release of these effluents into natural environments causes environmental pollution (Pereira and Alves 2012). Due to the adverse effects of dye effluents on environment and health, strict discharge limits are imposed by various legislations across the world. Many industries are required to follow zero liquid discharge norms set out by the pollution control boards for preventing pollution of natural water bodies. Therefore, effective treatment of textile effluents is necessary before their environmental discharge. Many physical, chemical, biological or combination approaches are available for the treatment of textile effluents. The advantages and

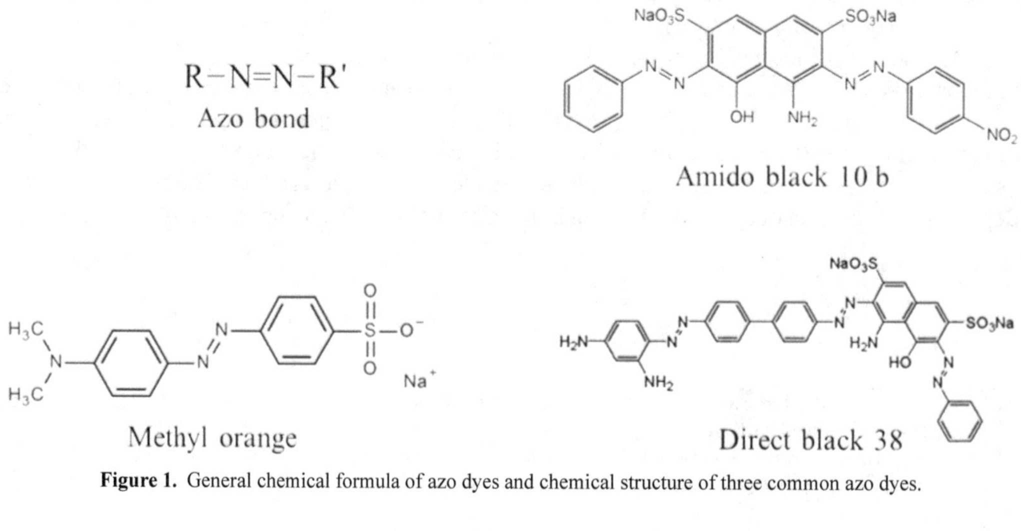

$$R-N=N-R'$$
Azo bond

Amido black 10 b

Methyl orange

Direct black 38

Figure 1. General chemical formula of azo dyes and chemical structure of three common azo dyes.

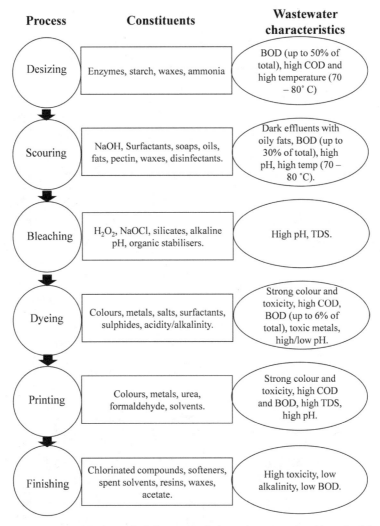

Figure 2. Important processes used in the textile industry, chemical constituents employed in each of the process and the characteristics of wastewater generated in each step.

disadvantages of various treatment methods are presented in Table 1. Biological treatment methods are preferred due to their low cost operation and eco-friendly nature. Due to complex nature of textile effluents, biological methods are often combined with physical-chemical pre-treatment methods.

Biological treatment of textile wastewater is widely studied using various organisms. Besides algae, fungi and plants, treatment by bacteria is most commonly applied due to their diversity and metabolic capabilities for transforming or metabolizing persistent dye stuffs. There are many studies on bioremediation of individual or mixture of textile dyes by bacterial isolates. Bioremoval of dyes by different bacterial strains and removal mechanisms were reviewed by Saratale et al. (2011). Dye removal involves biosorption, oxidative enzymes and reductive enzymes or combination of some or all. Compared to planktonic growth of bacteria, biofilms offer some specific advantages like higher tolerance to toxic pollutants and improved degradation by means of high cell densities. In addition, multispecies biofilms are helpful in removing pollutants by multiple mechanisms like sorption, biodegradation and entrapment. Higher diversity, genetic exchange and metabolic co-operation are advantages of biofilms that offer improved biological treatment of wastewater. The biofilm life cycle

Table 1. Various physical, chemical and biological processes for treating textile wastewaters.

Removal method	Type	Process description	Advantages	Disadvantages
Physico-chemical process	Membrane filtration	Physical separation	Effective across all dye types	Generation of concentrated dye sludge
	Adsorption	Sorption onto surface	Can remove large variety of dyes	Dye saturated surface needs regeneration or replacement makes the process expensive
	Ion exchange	Exchange of charged dyes onto counter ions	Resins can be regenerated for reuse	Can't be used for uncharged dyes
	Coagulation	By adding chemical coagulants	Simple and relatively cheap	High sludge production needs further treatment
	Reverse osmosis	Reversing the osmosis process under high pressure	High retention of dyes leaving good quality permeate	Costly equipment, membrane fouling and expensive to maintain
Oxidative processes	Fenton reagent	H_2O_2-Fe(II) based oxidation	Effective against soluble and insoluble dyes	High sludge generation
	Ozonation	Ozone gas is used as oxidising agent	Simple, effective and fast. Also improves biodegradability	Short half-life (20 min). Activity depends on pH, temperature and co-contaminants in waste water
	Photochemical	Reactive radicals from H_2O_2 in presence of UV and catalyst	Effective and no sludge production	Energy intensive and formation of by-products
	NaOCl	Reactive oxidising radicals (OCl^{-1})	No sludge production.	Carcinogenic by-products (aromatic amines) are formed
Biological process	Microbes	Aerobic and/or anaerobic process using bacteria or fungi	Green, cost effective without any chemical sludge formation	Enrichment and selection of microbes is a slow process. Some complex dyes are tough to degrade and may require both anaerobic and aerobic treatments
	Enzymes	Selective immobilised enzymes for catalysing dye degradation	High conversion rates and specificity	Extraction, purification and immobilisation of enzymes makes the process expensive and no complete mineralisation

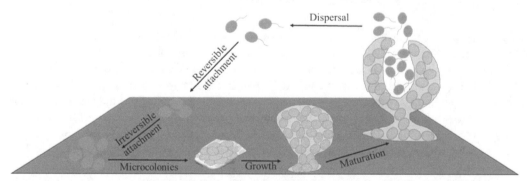

Figure 3. Typical biofilm life cycle.

is shown in Figure 3. Biofilms are microbial communities enmeshed in a self-produced extracellular polymeric substances (EPS) matrix formed on a substratum. In addition, microbial granules formed without a substratum also share biofilm characteristics. Consequently, these microbial granules are referred to as particulate biofilms or granular sludge biofilms in the literature. Both substratum-grown biofilms and self-immobilized granular biofilms are attractive for the treatment of dye wastewaters.

This chapter is focused on the application of biofilm and biogranule based systems for textile wastewater treatment. Dye removal from synthetic and real wastewaters by biofilms under aerobic, anaerobic and sequential anaerobic–aerobic conditions is presented. The research performed on textile wastewaters using biofilms or biogranules is categorized into synthetic and real textile wastewater treatment. Lastly, potential application of bio-electrochemical systems for removing textile dyes is also presented.

2. Treatment of azo dye containing synthetic wastewater using biofilms

Microbial biofilms were mainly developed under aerobic, anaerobic and sequential anaerobic–aerobic biofilm reactors for the treatment of synthetic dye wastewaters containing azo dyes. Most of these studies were performed by spiking different concentrations of various azo dyes either singly or as mixture of dyes in mineral salts media containing essential nutrients. Carbon sources such as starch, molasses, acetate, glucose, lactate or ethanol were also included in the synthetic textile wastewater to support microbial growth. These easily metabolisable organic substrates act as source of carbon, electrons and energy for the development of biofilms and biodecolourization of dyes.

2.1 Dye removal by aerobic biofilms

Previously, most dyes were considered to be non-degradable and non-transformable by microorganisms under aerobic conditions (Zimmermann et al. 1982). This was mainly attributed to the presence of the strong electron withdrawing character of the azo group, whose reduction is not favourable under aerobic conditions. However, reduction of dyes under aerobic conditions was noticed in prolonged continuous cultures of bacteria. This was attributed to spontaneous mutations, resulting in development of bacterial strains with oxygen-insensitive azoreductase enzymes (Coughlin et al. 1997). A bacterial consortium TXB65 was enriched from mixed liquor of a wastewater treatment plant for the biodegradation of two azo dyes, namely AO7 and AO8. This study showed that the azo dyes were removed by initial reduction, followed by biodegradation. Microbial community analysis using 16S rRNA gene sequencing showed that the enriched consortium contained two azo dye degrading bacteria and several other non-degraders. Reduction of azo dyes under aerobic conditions was linked to azoreductase enzyme system of dye degrading strains. Two bacterial strains (MC1 and MI2) belonging to *Sphingomonas* capable of degrading azo dyes were isolated and characterised. MI2 was able to use both the dyes (AO7 and AO8) as the sole carbon, nitrogen and energy source. However, MC1 was able to decolourize both the dyes, i.e., AO7 and AO8 but only in presence of an exogenous carbon and nitrogen source.

Abraham et al. (2003) studied biodegradation of azo dye mixtures using aerobic bacterial consortium in a rotating biological contactor (RBC). A mixture of seven different dyes comprising mono- and di-azo dyes were used at different concentrations of 25, 50 and 100 mg L^{-1} each. RBC with 42 L capacity was seeded with a bacterial consortium comprising of three different azo dye degrading strains for developing biofilm in dye-spiked minimal media. At 100 mg L^{-1} dye concentration, degradation efficiency increased from 53% to 69% with an increase in hydraulic retention time (HRT) from 3 h to 73 h. Among the three strains used for biofilm formation in RBC, one isolate was known for expressing laccase capable of azo dye degradation. Naresh Kumar et al. (2014) reported bioremoval of azo dye, C.I. Acid Black 10 B by biofilms in an aerobic periodic discontinuous batch reactor (PDBR). Dye removal efficiency of biofilm reactor was compared with a suspended growth reactor. PDBR built with Perspex material was operated in 48 h cycles with 47 h aerobic phase. Stone chips were used as carrier material for biofilm development in the biofilm PDBR, while the suspended growth PDBR was operated without carrier material. Initial operation of PDBRs was stabilised with

350 mg L^{-1} dye in the synthetic dye wastewater containing 3 g L^{-1} glucose. Concentration of dye in PDBRs was gradually increased to 500, 750, 1000 and 1250 mg L^{-1}. Shock loading tolerance and removal efficiencies were reported to be high for biofilm PDBR over suspended growth PDBR. At a high dye loading concentration of 1250 mg L^{-1}, biofilm PDBR removed 75% colour. But, the colour removal was only 42% in suspended growth PDBR.

Recently, Sarvajith et al. (2018) evaluated biodecolourization potential of aerobic granular sludge under microaerophilic conditions. A model aerobic granular biofilm reactor is shown in Figure 4A. Biodecolourization of a commercial yellow dye (YD) was determined. For this, a 1 L volume reactor was seeded with aerobic granular sludge cultivated from acetate fed synthetic wastewater and was initially fed with synthetic textile wastewater containing 5 mg L^{-1} YD and 3.74 mM lactate. The reactor was operated in sequencing batch reactor (SBR) mode with 72 h cycle. After 16 cycles of SBR operation, the cycle time was decreased to 24 h. The YD concentration was gradually raised to 10, 20, 30, 40 and 50 mg L^{-1} with 7 cycles of operation at each concentration. Azo dye, total organic carbon and ammoniacal nitrogen removal efficiencies of 89–100%, 79–95% and 92–100%, respectively, were achieved. Dye removal was also stable during 80 days of operation. This study demonstrated efficient removal efficiencies for colour, organic carbon and ammonium by operating granular sludge SBR under microaerophilic conditions.

2.2 Dye removal by anaerobic biofilms

Carliell et al. (1995) and Razo-Flores et al. (1996) have reported biodecolourization of a wide range of dyes under anaerobic conditions. The dye decolourization was found to be more effective and efficient under anaerobic conditions. Two general mechanisms were reported for dye biodecolourization under anaerobic conditions: (i) cleavage of azo bonds by a four electron reduction reactions, and (ii) by unique enzyme cofactors, F430 and vitamin B12 present in methanogenic or acidogenic bacteria (Razo-Flores et al. 1996). The electrons required for the reduction are often provided by adding volatile fatty acids (VFAs). Although colour is removed efficiently, the dyes are not completely degraded under anaerobic conditions. Aromatic amines formed due to reduction of dyes are sometimes more toxic than the original dyes. Moreover, aromatic amines formed via reduction of dyes can be potential carcinogens or mutagens (Carliell et al. 1995). A model anaerobic biofilm reactor operated as upflow anaerobic sludge blanket (UASB) or fluidised bed reactor (FBR) is shown in Figure 4B. Sen and Demirer (2003a) studied anaerobic treatment of reactive azo dye, Remazol Brilliant Violet 5R, in a fluidised bed reactor (FBR). For this, biofilm biomass was developed on pumice carrier in a 4 L capacity FBR. Mixed anaerobic culture from anaerobic digester of a wastewater treatment plant was used as the inoculum for developing biofilms. During the start-up of the FBR, biofilm was developed by feeding synthetic wastewater spiked with glucose, methanol and yeast extract. Subsequently, the FBR was supplied with synthetic textile wastewater containing acetate and starch. FBR was operated in different stages with a gradual increase in Remazol Brilliant Violet 5R concentration. At 300 mg L^{-1} dye concentration, colour and COD removal efficiencies of 94% and 60%, respectively, were achieved. Sulfonated aromatic amines (SAAs) were found to be the chief reduction products during anaerobic treatment of Remazol Brilliant Violet 5R.

Talarposhti et al. (2001) evaluated a two phase upflow anaerobic filter for the treatment of simulated mixed dye wastewater containing 7 different cationic dyes. Anaerobic culture from an anaerobic digester of sewage treatment plant was used as the inoculum for growing biofilm. Cationic dye mixture containing 7 different dyes at a total dye concentration of 1000 mg L^{-1} was added with whey powder and orthophosphate to yield a C:P:N ratio of 200:5:1. Using this approach, almost 90% biodecolourization was achieved at a loading rate of 5 kg COD m^{-3} d^{-1}. A decrease in decolourization efficiency was noticed with an increase in influent dye concentrations. Longer HRTs have improved the colour removal efficiency (Talarposhti et al. 2001).

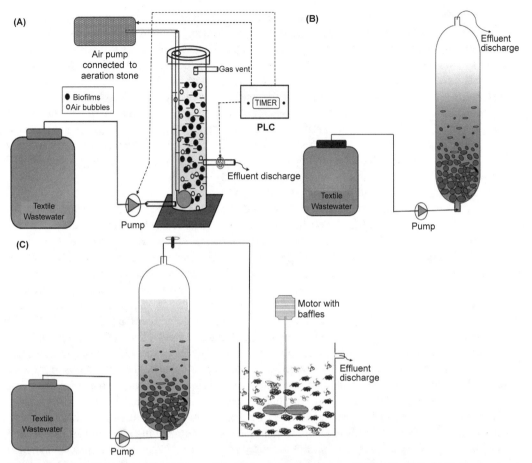

Figure 4. Different bioreactor configurations used for treating textile wastewaters. (A) Aerobic SBR with biofilm developed on carrier or as self-immobilised granules, (B) Anaerobic UASB reactor and (C) Sequential anaerobic UASB coupled to sludge biomass containing aerobic CSTR.

2.3 Dye removal by biofilms under sequential anaerobic–aerobic conditions

Decolourization of dyes under anaerobic condition is initiated by azoreductase-catalysed reduction or by cleavage of azo bonds resulting in formation of aromatic amines (Sponza and Isik 2005a). The efficiency of anaerobic biological colour removal has been reported to be higher than 80% and in some cases even 100% for a variety of azo dyes (Da Silva et al. 2012). The extent and rate of anaerobic decolourization of azo dyes are influenced by various parameters such as dye chemical structure, dye concentration, supplementation with different carbon and nitrogen sources, electron donor and redox mediators (Vanderzee and Villaverde 2005, Da Silva et al. 2012). The presence of alternative electron acceptors such as oxygen, nitrate, sulphate and ferric ion may compete with the azo dye for reducing equivalents, which can lead to inefficient colour removal under anaerobic condition.

Cirik et al. (2013) compared anaerobic colour removal rate constants of Brilliant Violet 5R in the presence and absence of nitrate. The authors noted a decrease in decolourization efficiency up to 63% with increase in concentration of nitrate to > 100 mg L^{-1}. NO_3^-–N. Mohanty et al. (2006) reported reduction in decolourization efficiency from 96% to 19% when the dissolved oxygen concentration increased from 0.5 to 3.0 mg L^{-1}. These studies showed that both oxygen and nitrate compete for the electrons, and thus, slow down azo dye reduction and decolourization. Also, single-step anaerobic processes have limitations in terms of low chemical oxygen demand (COD) removal (Kapdan and Ozturk 2005) and the toxicity of reduced by-products of dyes.

Aerobic conditions are effective for removing dye by-products (aromatic amines) and residual COD from anaerobically treated textile wastewaters. The biodegradation of dye by-products are found to be generally metabolism-dependent and coupled to the oxygen sensitive enzymatic systems (e.g., Catechol 1,2-dioxygenase) (Viliesid and Lilly 1992). In order to determine the efficiency of different stages (aerobic, anaerobic and sequential anaerobic–aerobic) of biological treatment, toxicity tests were employed to assay the toxicity of treated effluent. The toxicity of the effluents was assayed using respiration inhibition and anaeroic toxicity assays (ATA) using *Daphnia magna* and *Vibro fischeri* as test organisms. Toxicity evaluation of effluents suggested that, 2-stage biological treatment process of sequential anaerobic and aerobic treatment is effective for removing dyes and attenuating toxicity of textile wastewater. Studies on the treatment of synthetic textile wastewaters using combination of anaerobic and aerobic systems are summarized in Table 2.

A sequential anaerobic upflow fixed bed column and aerobic agitated tank for the removal of azo dyes such as Orange G, Amido black 10B, Direct red 4BS and Congo red was evaluated (Rajaguru et al. 2000). With glucose as the co-substrate, these dyes were reduced to corresponding aromatic amines in the anaerobic reactor and were subsequently biodegraded in the aerobic reactor. Sponza and Isik (2002) evaluated upflow anaerobic sludge blanket (UASB) reactor-completely stirred tank reactor (CSTR) for decolourization and degradation, respectively, of azodye Reactive Black 5. Schematic representation of a similar setup is shown in Figure 4C. Partially granulated anaerobic sludge from an operating methanogenic reactor and activated sludge from dye wastewater treatment plant were seeded in the UASB reactor and CSTR, respectively. About 100 mg L^{-1} Reactive Black 5 (52 mg COD L^{-1}) was added to synthetic textile wastewater along with glucose to yield a final COD of 3000 mg L^{-1}. Maximum colour removal of up to 98% was observed in the UASB reactor, while the maximum COD removal was achieved in the CSTR. The removal efficiency was dependent on operating parameters such as HRT, organic loading rate and sludge retention time. Removal of monoazo based Reactive Red 195 (Sumifix Supra Br. Red 3BF 150% gran) in a sequential anaerobic–aerobic system was studied (Kapdan et al. 2003). Stainless steel metal wire based carrier was used in anaerobic packed column reactor and was seeded with anaerobic mixed bacterial cultures containing *Alcaligenes faecolis* and *Commomonas acidourans*. Activated sludge was used in aerobic tank and it was connected to a final sedimentation tank. Anaerobic reactor was fed with synthetic textile wastewater containing 100 mg L^{-1} dye and molasses as the co-substrate to give a final COD of 5000 mg L^{-1}. The treated effluent from the UASB reactor was used as the influent for activated sludge reactor. In the aerobic reactor, a DO of 2–3 mg L^{-1} was maintained. In anaerobic reactor, the decolourization efficiency increased from 60 to 85% with the increase in HRT from 12 to 18 h. Decolourization efficiencies up to 100% was noticed at HRTs longer than 60 h. Decolourization efficiencies of about 10 to 20% were seen in the aerobic unit. In the final effluent after the treatment in anaerobic–aerobic system, the decolourization efficiency ranged from 65 to 95% at different HRTs with a contribution of anaerobic and aerobic systems to around 85 and 15%, respectively.

Removal of high concentrations (3200 mg L^{-1}) of Direct Black 38 (DB 38) azo dye from synthetic wastewater was studied using UASB-CSTR system (Isik and Sponza 2004a). The synthetic wastewater had 3200 mg L^{-1} of DB 38 azo dye, 3000 mg L^{-1} COD of glucose and 3000 mg $NaHCO_3$ L^{-1} of alkalinity. After steady-state operation at an OLR of 1.14 kg m^{-3} d^{-1}, an HRT of 3.6 d was maintained in the UASB reactor, while a HRT of CSTR was maintained at 18 d corresponding to 0.5 L d^{-1} flow rate. Under these conditions, the total COD removal was found to be 92% with 77% removal in UASB reactor. The total colour removal was 94% wherein UASB reactor achieved 81% removal in colour. Toxicity levels were determined for the feed, UASB reactor effluent and CSTR effluent using three tests such as ATA, respiration inhibition assay and *Daphnia magna* test. Feed and UASB reactor effluents were found to be more toxic than the CSTR effluent. Higher toxicity of synthetic wastewater and UASB reactor effluent was attributed to the presence, respectively, of dyes and aromatic amines. Biodegradation of aromatic amines during aerobic treatment led to minimal toxicity. Later, similar approach was successfully demonstrated for the treatment of synthetic wastewater containing Direct Red 28 (Sponza and Isik 2005b).

Table 2. Sequential anaerobic–aerobic biofilm reactors for removing azo dyes.

Dye	Type of reactor	Inoculum; reactor operating conditions	Treatment performance	Remarks	References
Orange (OG), Amido black 10B (AB), Direct red 4BS (DR), Congo red (CR)	Anaerobic–Aerobic bioreactor System; Up-flow fixed bed column (anaerobic) and agitated tank (aerobic)	Inoculum: 4 strains of Pseudomonads Feed: SWW—5 g L^{-1} glucose; 0.2 g L^{-1} yeast extract and dyes. Flow rate—30 ml h^{-1} HRT—3 days	Removal rate: OG—60.9 mg L^{-1} per day AB—571.3 mg L^{-1} per day DR—112.5 mg L^{-1} per day CR—134.9 mg L^{-1} per day	Sequential anaerobic and aerobic treatment completely degraded and mineralized all 4 dyes within 3 days	Rajaguru et al. 2000
Reactive Black-5	Anaerobic UASB and Aerobic CSTR in sequence	Inoculum: partially granulated anaerobic granules in UASB and activated sludge in CSTR Feed: MSM with glucose (3000 mg L^{-1} COD), Dye—100 mg L^{-1}; Flow rate—4 L d^{-1}. OLR increased from 4.83 to 12 kg COD m^{-3} d^{-1} after 30 days of start-up	98% dye removal during anaerobic reaction and mineralization of intermediates during aerobic phase. 96% overall COD removal efficiency	Increased OLR had decreased COD removal and methane gas production efficiency by 51.8% and 21% in UASB	Sponza and Isik 2002
Reactive Red 195	Anaerobic packed - column reactor connected to activated sludge unit followed by a sedimentation tank in sequence	Inoculum: facultative anaerobic bacteria consortium-PDW and activated sludge Feed: Nutrient media with molasses (5000 mg L^{-1} COD), Dye—100 mg L^{-1} Varying HRT in Anaerobic unit. COD/N/P = 100/5/1; DO = 2–3 mg L^{-1}	Upto 90% decolourization with 3000 mg L^{-1} COD at 24 h HRT in anaerobic unit; 85% and 15% decolourization in anaerobic and aerobic reactor	Sequential anaerobic and aerobic reactor unit was capable of decolourization with at least 18 h HRT and minimal (2–3 mg L^{-1}) DO	Kapdan et al. 2003
Direct Black 38	Anaerobic UASB and aerobic CSTR in sequence	Inoculum: partially granulated anaerobic granules for UASB and activated sludge for CSTR; Feed: SWW with glucose (3000 mg L^{-1} COD); Dye—100 to 3200 mg L^{-1}; Flow rate –0.5 L d^{-1}; HRT—3.6 d and SRT 86 d	Cumulative COD removal of 92% and colour removal of 94%	Efficient colour and COD removal in sequential reactors; Final effluent had reduced ecotoxicity	Isik and Sponza 2004a
Combination of 50 mg L^{-1} each of Reactive Black 5, Direct Red 28, Direct Black 38, Direct Brown 2 and 250 mg L^{-1} Direct Yellow 12	Continuously fed anaerobic UASB and aerobic CSTR in sequence	186 d of total operation; steady-state achieved in 46 d; Inoculum: Partially granulated anaerobic granules for UASB (29,000 mg L^{-1} MLSS); activated sludge for CSTR (2700 mg L^{-1} MLSS); Feed: SWW with 2062 mg L^{-1} COD; HRT: 1.22, 6.05, 8.5, and 19.7 days	At lower HRT of 1.22 days, COD removal was observed in aerobic phase and significant dye removal in anaerobic phase (UASB—94% colour and 74% COD removal)	Complete biodegradation of mixed dyes. Two HRTs tested did not show any significant difference in total COD and colour removal	Isik and Sponza 2008

Table 2 contd. ...

...Table 2 contd.

Dye	Type of reactor	Inoculum; reactor operating conditions	Treatment performance	Remarks	Reference
Reactive black-5	Anaerobic UASB and Aerobic CSAR	Inoculum: anaerobic granules for UASB and activated sludge for CSAR; Feed: SWW with 3000 mg L⁻¹ glucose COD, 150 mg L⁻¹ dye, Ammonium chloride—400 mg L⁻¹; HRT = 3.2 to 30.1 h	Decrease in HRT from 30.1 to 3.2 h reduced colour removal efficiency from 99.8 to 90.7%	Operational parameters such as HRT, OLR, VFA, F/M ratio, SRT, and SVI affect the reactor performance in colour removal	Karatas et al. 2010
Sumifix Black EXA, Sumifix Navy Blue EXF and Synozol Red K-4B	Single granular sludge SBR with intermittent anaerobic–aerobic phases	Mixed sewage, textile mill sludge and anaerobic granules from UASB were used as inoculum; 4 L SBR in 6 h cycles with 340 m reaction time (40 m anaerobic–130 m aerobic phase and another 40 m second anaerobic–130 m second aerobic phase). 50 mg L⁻¹ mixed dye with organic carbon spiked to 1270 mg COD L⁻¹ and 1020 ADMI	94% COD removal, 95% ammonia removal and 62% colour removal was observed	Granules of an average size 2.3 ± 1.0 mm, settling velocity of 80 ± 8 mh⁻¹ and a biomass concentration of 7.3 ± 0.9 g L⁻¹ were developed in SBR	Muda et al. 2010
Remazol red (RR), Remazol blue (RB), Remazol yellow (RW)	Anaerobic–aerobic sequential reactors	Inoculum: anaerobic sludge with either plastic kaldnes carriers or poraver carriers; Activated sludge from 3 different plants treating sewage, nitroaromatics and textile dye in aerobic reactor; Feed: MSS with 1 g L⁻¹ glucose, vitamins, trace elements; HRT: 0.5 to 3.0 d	100–2000 mg L⁻¹ dye decolourization without acclimatization. 98% decolourization in anaerobic reactor with poraver carriers and slightly lower with kaldness carriers; RW autooxidised in aerobic phase and resisted further degradation	Complete dye decolourization and removal of 3 different dyes using carrier based approach gave promising results. However, autooxidised by-products removal was not achieved	Jonstrup 2011
Congo Red (CR) and Reactive Black 5 (RB5)	Mesophilic UASB and aerobic SBR	Anaerobic sludge from operating UASB and activated sludge in aerobic SBR. 24 h HRT for CR and 12 h HRT for RB5 in UASB. 24 h SBR cycle in aerobic reactor. SWW having CR (400 mg L⁻¹) or RB5 (200 mg L⁻¹) with ethanol (1000 mg COD L⁻¹)	Decolourization efficiencies of 96% and 75% for CR and RB5, respectively, in anaerobic reactor. Overall COD removal of 88% for both dyes	Ecotoxicity tests with *D. magna* for CR effluents indicated toxicity by both anaerobic and aerobic effluents. In case of RB5, aerobic step highly reduced the toxicity	Da Silva et al. 2012

Table 2 contd. ...

Acid Red 18 (AR 18)	Integrated anaerobic/ aerobic fixed-bed sequencing batch biofilm reactor (FB-SBBR)	FB-SBBR1 with pumice stones and FB-SBBR2 with plastic media as biofilm carrier; 100 mg L⁻¹ AR18 with 500 mg L⁻¹ each glucose, lactose in SWW. Reactors in 24 h SBR cycles with 14 h anaerobic phase and 8 h aerobic phase	Removal efficiencies of 93% COD, 97% colour in FB-SBBR1 and 95% COD, 97% colour in FB-SBBR2 were seen	Maximum removal of both colour and COD was seen in anaerobic phase with very small removal in aerobic phase	Koupaie et al. 2013
Methyl orange (MO)	Integrated anaerobic– aerobic biofilm (IAABF) reactor	3 L IAABF reactor with anaerobic bottom portion and aerobic top portion separated by perforated plate. Coconut fibre as biofilm carrier was filled in both portions. Activated sludge seed for biofilm formation. The aerobic phase was 3 h till 100 mg L⁻¹ and was gradually increased to 24 h at 300 mg L⁻¹	With increase in MO concentration from 50, 100, 200, 250 and 300 mg L⁻¹, colour removals were 97%, 96%, 97%, 97%, and 96%; COD removals were 75%, 72%, 63%, 81%, and 73% at respective concentrations	Efficient colour removal (> 95%) at all the studied concentrations. COD removal increased with increase in aeration phase period at 200 and 300 mg L⁻¹	Murali et al. 2013
Acid red 14 (AR 14)	Granular sludge based anaerobic–aerobic phase SBR	1.5 L SBR operated in 6 h cycles with 3 to 3.5 h aerobic phase and 2 h anaerobic phase. MSM with starch based organic carbon to yield 1000 mg COD L⁻¹ along with 20 mg L⁻¹ AR 14 was used as waste water. Activated sludge as the seed for granule cultivation. A control SBR with same feed composition except dye was also operated	Colour and COD removal efficiencies were 85% and 80%, respectively	Surprisingly, presence of dye in the waste water significantly enhanced granulation compared to dye lacking SBR	Mata et al. 2015

Muda et al. (2010) studied treatment of synthetic wastewater in SBR by cultivating aerobic granular sludge. A 4 L working volume SBR was inoculated with mixed sewage and textile mill sludge along with anaerobic granules collected from an operating UASB reactor. The SBR was operated with a 6 h cycle containing anaerobic and aerobic phases. Each 6 h cycle contained 5 min filling, 340 min reaction (40 min anaerobic–130 min aerobic phase and another 40 min second anaerobic–130 min second aerobic phase), 5 min settling, 5 min decanting and 5 min idle periods. The SBR was fed with synthetic textile wastewater containing mixed dyes, i.e., Sumifix Black EXA, Sumifix Navy Blue EXF and Synozol Red K-4B with a total concentration of 50 mg L^{-1} (colour: 1020 ADMI). Carbon sources such as glucose (0.5 g L^{-1}), ethanol (0.125 g L^{-1}) and sodium acetate (0.5 g L^{-1}) were added to synthetic wastewater to obtain a total COD of 1270 mg L^{-1}. Over a period of 66 d, the granules formed in the reactor had an average size of 2.3 ± 1.0 mm, settling velocity of 80 ± 8 m h^{-1} and a biomass concentration of 7.3 ± 0.9 g L^{-1}. This study has demonstrated cultivation of granular sludge while treating synthetic textile wastewater. The removal efficiencies of COD and colour were high at 94% and 62%, respectively.

Jonstrup et al. (2011) studied anaerobic–aerobic treatment system for removing three azo dyes, namely Remazol Red RR, Remazol Blue RR and Remazol Yellow RR. Two types of carrier, namely, plastic Kaldness carriers and glass Poraver carriers, were evaluated for biofilm formation and decolourization in anaerobic reactors. Poraver carriers consist of porous recycled glass material. Both these carrier materials are known to allow attachment and biofilm formation. Anaerobic reactors with biofilms on Poraver carriers showed near complete decolourization of dye concentrations between 100 and 2000 mg L^{-1}. Partial degradation of some of the aromatic amines was achieved using activated sludge in the aerobic reactor, while certain aromatic amines were removed by auto-oxidation processes. Congo Red (CR) (400 mg L^{-1}) and Reactive Black 5 (RB5) (200 mg L^{-1}) removal in UASB reactors and aerobic activated sludge SBRs was studied (Da Silva et al. 2012). Maximum decolourization of 96% for CR and 75% for RB5 was observed in UASB reactors. Significant COD removal in aerobic SBR led to a final removal efficiency of 88% for both the dyes. Ecotoxicity experiments using *D. magna* revealed that anaerobic and aerobic treatments for CR dye did not decrease the toxicity compared to the influent. The toxicity of anaerobic effluent was lower than that in the influent in the case of RB5. The aerobic treatment completely eliminated the toxicity of synthetic textile wastewater containing RB5 towards *D. magna*.

Integrated anaerobic/aerobic fixed-bed sequencing batch biofilm reactor (FB-SBBR) for the decolourization and biodegradation of azo dye Acid Red 18 (AR18) (100 mg L^{-1}) was evaluated (Koupaie et al. 2013). For this, two different FB-SBBRs were installed with different carrier materials. Volcanic pumice stones and polythene were used as the carrier material in FB-SBBR1 and FB-SBBR2, respectively, for biofilm development. These SBRs were operated in 24 h cycles comprising of 14 h anaerobic phase and 8 h aerobic phase. Maximum COD and colour removal was achieved in the anaerobic phase, wherein 93% COD and 97% colour removals were achieved in FB-SBBR1. Similarly, 95% COD and 97% colour removals were achieved in the anaerobic phase of FB-SBBR2. Additionally, 1 to 2% COD was removed in the aerobic phase. Comparatively, FB-SBBR2 containing polyethylene media was more effective in removing COD and colour from synthetic textile wastewater.

Murali et al. (2013) studied the ability of biofilms developed on coconut carrier material in a single reactor system containing perforated plate separating the anaerobic and aerobic portions for methyl orange (MO) removal. Decolourization and COD removal were determined for concentrations ranging from 50 to 300 mg L^{-1} MO. Decolourization efficiencies over 95% were seen across all the dye concentrations. With an increase in MO concentration, the COD removal decreased due to the accumulation of aromatic amines. But, the COD removal could be improved by increasing the length of aeration phase in the reactor. Mata et al. (2015) studied the decolourization and COD removal in Acid red 14 (AR 14) (20 mg L^{-1}) containing synthetic wastewater having 1000 mg L^{-1} COD. The SBR was operated with 3 to 3.5 h anaerobic phase and 2 h aerobic phase. Granulation was found to be improved in presence of dye as compared to control SBR operated without azo dye. Efficient colour (85%) and COD (80%) removal efficiencies were observed. These studies suggested that efficient decolourization followed by effective organic carbon removal in biofilm based sequential anaerobic–aerobic reactors. Most of the studies indicated that sequential anaerobic–aerobic treatment processes decreased the ecotoxicity of textile wastewaters.

3. Treatment of real textile wastewater

Effective treatment of simulated textile wastewater by biofilms under aerobic, anaerobic or sequential anaerobic–aerobic conditions has been reported (Abraham et al. 2003, Talarposhti et al. 2001, Jonstrup et al. 2011). The ultimate objective of these studies was to develop an appropriate system for application in treatment of real textile wastewater. Unlike the treatment of simulated wastewaters, where the composition is defined and concentrations of dyes can be controlled, the treatment of real textile wastewater is more challenging due to complex wastewater characteristics and variable environmental parameters. The composition of textile wastewater depends on the dying process and the individual dye types used. Any alteration in process parameters or process type can greatly change the wastewater characteristics. Besides varying dye mixtures, the wastewater also contains biodegradable and non-biodegradable organic and inorganic contaminants. The biodegradable organic compounds like starch, cellulose, enzymes, fibres, fats, surfactants, waxes, grease and phenols are commonly found in textile effluents. The brightly coloured dyes impart aesthetically unacceptable colour to the effluents. Acids (acetic, sulphuric acid), bases (sodium hydroxide, sodium carbonate), oxidising agents (hydrogen peroxide) and reducing agents (sodium sulphide) used in different processes of textile industry are sources of inorganic contaminants in textile wastewater. Salts of sodium and ammonium, sulphates, nitrates, carbonates and bicarbonates are the various inorganic contaminants. Toxic metals (cadmium, chromium, lead, mercury, copper and zinc) are also present in effluents, depending on the type, and process of dying and can negatively impact the biological treatment performance (Correia et al. 1994, Paul et al. 2012). Typical textile effluents have COD and BOD in the range of about 1000 mg L^{-1} and 300–500 mg L^{-1}, respectively. The typical BOD/COD ratios encountered in textile wastewaters are in the range of 0.3–0.5 (Isik and Sponza 2004b).

Presence of a wide variety of organic, inorganic and heavy metal contaminants at different proportions makes the biological treatment of real textile wastewater more challenging. In spite of complex nature of wastewater, several biofilm based aerobic and anaerobic treatment systems capable of treating real textile effluents are developed. Among the available options, sequential anaerobic–aerobic systems are found to be more effective in treating real textile effluents. Often, the biological methods are combined with chemical methods for effective treatment of dye contaminated wastewater.

3.1 Treatment of real textile effluents using aerobic biofilms

Treatment of real textile wastewater was evaluated in a biological aerated filter (BAF) by Chang et al. (2002). In a BAF, wastewater flows through the reactor filled with carrier media for biofilm formation. Microbial biofilm developed on carrier media is mainly responsible for removing pollutants from the wastewater. In addition, BAF also act as a filter for removing contaminants. Unlike conventional activated sludge process, BAF does not require a separate clarifier tank, thus occupies smaller foot print. Additionally, lower power consumption and improved treatment are the benefits of BAFs in wastewater treatment. Different types of carrier media are available for use in BAF. For example, the treatment performance of a zeolite containing lab-scale BAF treating textile effluent was assessed for a period of 7 months and compared with that of a sand media BAF. At an OLR of 1.2 to 3.3 kg COD m^{-3} day, the COD removal efficiencies in zeolite BAF and sand BAF averaged at 88 and 75%, respectively. Also, total nitrogen (TN) removal was efficient in zeolite BAF due to the natural ion exchange capacity of zeolite and presence of large number of nitrifying bacteria in the biofilm developed on zeolite. Based on these results, a 12 m^3 d^{-1} capacity pilot-scale zeolite media BAF was installed for treating textile effluent containing 2150 mg L^{-1} COD, 1630 mg L^{-1} BOD,

72 mg L^{-1} TN and 740 colorimetry units (CU). The pilot scale BAF was successfully operated for 5 months with removal efficiencies of 99% BOD, 92% COD, 92% TN and 78% colour (Chang et al. 2002). The biofilm formed on zeolite was effective in removing BOD, COD and TN from real textile wastewater. The removal of colour, COD and BOD was proposed to occur through adsorption on the zeolite-biofilm and subsequent microbial degradation processes.

Ibrahim et al. (2009) studied treatment of textile wastewater in an upflow aerobic biofilm reactor (ABR). In this study, bacteria present in real textile wastewater were enriched using azo dyes and raw textile wastewater. The microbial consortium enriched with dye decolourizing bacteria was used as the inoculum for biofilm formation in the ABR. For evaluating textile wastewater treatment, two ABRs with a capacity of 1 m^3 and 2 m^3 were connected in series, filled with inert HDPE support and inoculated with enriched microbial consortium. Schematic of the ABR is shown in Figure 4A. In ABR, wastewater is directly fed from the bottom. Real textile wastewater was fed into ABRs at a flow rate of 2 m^3 d^{-1} to maintain a HRT of 12 and 24 h in ABR1 and ABR2, respectively. Mixing and aeration in ABRs was achieved by recirculation in ABR1 (DO < 2 mg L^{-1}) and by aerator in the ABR2 (DO ~ 4 mg L^{-1}). Real textile wastewater fed into ABRs had a pH around 9.1 to 9.5 with a COD of 200–5000 mg L^{-1} and a colour intensity of 280–980 ADMI (American Dye Manufacturers Institute). The ABR2 was connected to a clarifier tank after which the treated water was discharged. During the period of 90–100 days of ABR operation, flocs of size 0.5 to 1 mm were collected in the clarifier. When independent batch tests were performed using these flocs, nearly 45 to 55% COD removal (initial COD –548 mg L^{-1}) and 40 to 70% colour removal were noticed. Molecular analysis indicated occurrence of *Bacillus*, *Paenibacillus*, and *Achromobacter* along with other genera. Elemental analysis of flocs showed the presence of divalent cations such as Ca^{+2}, Fe^{+2} and Mg^{+2}. Presence of such ions in the wastewater would have aided in floc formation (Ibrahim et al. 2009).

Aerobic fluidised bed biofilm reactor (FABR) of 1000 L volume was installed in a textile unit (Tirupur, India) for effluent treatment (Kumar and Saravanan 2009). The FABR was inoculated with activated sludge (collected from common effluent treatment plant) and polyurethane (PU) cubes were added as the carrier media and the reactor was operated for 30 days for biofilm development. The textile wastewater (pH: 8, COD: 1600 mg L^{-1}, TDS: 15000 mg L^{-1}) was drawn from a storage tank and fed into the FABR. Mixing and aeration were provided by air diffusers (DO: 3.5 ± 1.5 mg L^{-1}) located at the bottom of the reactor. The FABR was evaluated for treating textile wastewater at 4 different HRTs ranging from 3 to 8 h. COD removal efficiency increased from 69 to 94%, when the HRT was increased from 3 to 4.5 h and remained same up to 8 h. FABR was connected to a coagulant (Fenton's reagent) aided flocculator unit and a final neutralisation unit for improving the treatment performance. Chemical coagulation step significantly reduced TDS and after the final treatment step, COD and TDS removal efficiencies reached 94.2 and 93.9%, respectively.

Lotito et al. (2012) evaluated aerobic granular sludge and biofilms for treating dyeing and finishing wastewater in a sequencing batch biofilter granular reactor (SBBGR). The reactor had bottom and top compartments, respectively, for developing biofilms and aerobic granular sludge. The bottom compartment was filled with plastic support material for developing biofilms, while the top compartment had the liquid phase and fitted with bubble aeration for developing aerobic granular sludge. Due to the presence of carrier materials and aeration induced shear force, the SBBGR system contained both biofilms (formed on plastic carrier) and aerobic granules (formed without a carrier material). The textile wastewater was fed from the base of the reactor through bottom compartment and distributed uniformly in both compartments by re-circulation with a peristaltic pump. The reactor was operated in SBR mode with filling, aerobic degradation and final drawing phases in the cycle. In aerobic degradation phase, the water on top compartment was continuously aerated and re-circulated through bottom bed compartment using a peristaltic pump. Textile wastewater had an average composition of 126 ± 63 mg L^{-1} TSS, 688 ± 280 mg L^{-1} COD, 21 ± 6 mg L^{-1} TKN and 13 ± 7 mg L^{-1} total surfactants. The wastewater composition was highly variable and BOD/COD ratio was around 0.1 indicating poor biodegradability. The SBBGR was operated with real textile wastewater for a period of 200 days. Throughout the period of operation, the COD removal efficiencies varied between 36 to 80%, depending on the OLR applied and also due to fluctuations in influent wastewater quality. But, COD in the treated water was above 160 mg L^{-1} in most of the cycles, thus, treatment did not meet the discharge limit of 125 mg L^{-1} COD in Italy. However, the effluent quality met the requirement for discharging into direct sewerage systems. Efficient TSS removal with at least 50% (sometimes up to 90%) was noticed in every cycle up to OLR of 2.4 to 2.6 kg COD m^{-3} d^{-1}. The TSS values in the effluent were very well within the discharge limits. Increase in TSS at OLR 3.0–3.4 kg COD m^{-3} d^{-1} was attributed to the development of less compact biomass at high OLR and detachment of biomass at high shear forces by continuous recirculation. TKN removal decreased with an increase in OLR. Near complete TKN removal was noticed at OLR of 0.4–0.6 kg COD m^{-3} d^{-1}, whereas it decreased to 29% at 3.0–3.4 kg COD m^{-3} d^{-1}. This was due to the competition between heterotrophic and autotrophic organisms for the available oxygen at high OLR. Colour removal efficiency fluctuated widely between 0 and 60%. Presence of colour in the treated effluent, though not suitable for direct discharge, was well within the limits for discharging into sewer systems.

Later, same authors evaluated SBBGR for the treatment of sewage mixed textile wastewater collected from centralized wastewater treatment plant located in the textile district of Como, Fino Mornasco, Italy (Lotito et al. 2014). The sewage mixed textile wastewater had a relatively low COD (249 ± 65 mg L^{-1}), TSS (87 ± 36 mg L^{-1}) and TKN (34.2 ± 5.1 mg L^{-1}). Due to the mixing of municipal wastewater, even the colour intensity was low due to dilution. During the steady state operation, 82% COD, 95% TSS and 87.5% TKN were removed with in a HRT of 11 h. Colour of the treated effluent was sufficiently low to allow direct discharge. For achieving the similar effluent quality as that of SBBGR, longer HRT of 30 h was required with conventional activated sludge system. Additionally, SBBGR is a simple system with less sludge production and sludge had better dewatering and filterability properties. These studies suggested the possibility of upgradation of existing conventional activated sludge process system with granular biofilm based SBBGRs for improved treatment of sewage mixed with textile wastewater.

3.2 Treatment of real textile effluents using anaerobic biofilms

Aerobic treatment is very effective for removing COD from textile wastewaters. But colour removal by biological processes is not that efficient under aerobic conditions. In contrast, colour removal by microorganisms is more efficient under anaerobic conditions. For that reason, UASB reactors with anaerobic granular sludge have been explored for removing colour of textile wastewaters. Anaerobic treatability of reactive dye bath effluents of colours black, red and blue were determined in two phase UASB reactor system (Chinwetkitvanich et al. 2000). General schematic of the model UASB reactor is shown in Figure 4B. Three separate UASB reactors were set up for treating dye bath effluents of black, red and blue dyeing units. The USAB reactors were connected to independent acidification tanks (3 L capacity). The pH of black, red and blue dye bath effluents was 9.8, 9.4 and 9.95, respectively. The colour intensity of the dye batch effluents was represented as space units (SU). The colour of the black, red and blue dye bath effluents were 3300, 1000, and 2550 SU, respectively. Even the COD of the dyeing effluents were high at 2250, 2700 and 1650 mg L^{-1} in black, red and blue effluents, respectively. Tapioca (starch extracted from cassava root) was used as the low-cost carbon source to facilitate microbial growth and biodecolourization of textile dyes. The natural fermentation of tapioca at room temperature yields volatile fatty acids in the acidification tank, and is thus responsible for reducing the alkaline pH of dye bath effluents. The wastewater was diluted to get a final colour of 150 SU before feeding to two-phase UASB reactors. Anaerobic granular sludge collected from an operating UASB reactor treating synthetic sewage and fruit canning wastewater was used as the inoculum at a mixed liquor suspended solids (MLSS) of 40 to 50 g L^{-1}. In case of black dye bath effluents, with increase in tapioca concentration from 500 to 1500 mg L^{-1}, COD removal increased and removal efficiency ranged from 74 to 90%. Colour removal was around 70%. In case of red dye bath effluents, with increase in tapioca concentration from 0 to 500 mg L^{-1}, colour removal varied between 39 and 57% and the COD removal was between 27 and 45%. When the tapioca concentration was increased from 0 to 500 mg L^{-1} in blue dye bath effluents, the colour removal was between 48 and 56% and the COD removal was between 13 and 84%. Increase in tapioca concentration had a significant effect on decreasing the pH and improving colour and COD removal efficiencies.

Sen and Demirer (2003b) operated fluidised bed reactor (FBR) under anaerobic conditions for treating real cotton textile wastewater. A 4 L FBR was filled with pumice as carrier material for biofilm development. The FBR was inoculated with anaerobic sludge from anaerobic digester of a wastewater treatment plant and fed with synthetic wastewater containing methanol, glucose and yeast extract for biofilm formation. After 128 d of start-up period, the FBR was fed with real textile dye bath effluent. Reactive dyes such as Remazol, Everzol and Levafix types of dyes are commonly used in the dyeing. The FBR was operated with real effluent in 6 different stages over a period of 118 d. In stage I, which lasted from 1 to 46 d, raw effluent with an average COD of 1100 mg L^{-1} was fed at a HRT of 24 h to obtain an OLR of 1 kg COD $m^{-3}d$. By the addition of real wastewater, the COD removal efficiency got reduced to 27% by 18 d, possibly due to toxic effects of the influent on the organisms. This was followed by the slow acclimatisation and steady increase in removal performance. On day 33, COD removal efficiency reached 68%. No colour removal was noticed in the initial 24 d, but it slowly increased to 37% by day 41. In stage II (47 to 62 d), the HRT was increased to 50 h to bring the OLR to 0.5 kg COD $m^{-3}d$ to check the improvement in colour removal. During this stage, COD and colour removal efficiencies decreased to 35% and 19%, respectively. During stage III (63 to 77 d), the textile effluent was mixed with synthetic sewage in 3:1 ratio. COD of the synthetic sewage mixed textile wastewater was 800–900 mg L^{-1}. HRT was maintained at 50 h in this stage and the OLR applied was around 0.38 kg COD $m^{-3}d$. COD removal remained stable at 40%. Colour removal was negatively affected by reaching 0%, but slowly increased to a maximum of 12%. The composition of the influent wastewater significantly affected the decolourization performance of the FBR. In stage IV (78 to 91 d), 500 mg L^{-1}, glucose was added to raw effluent and the HRT was reduced to 24 h to yield an OLR of 1.3 kg COD $m^{-3}d$. Glucose addition led to increase in COD removal (62–66%) and colour removal (40–44%). When the glucose concentration was increased to 2000 mg L^{-1} (OLR of 3 kg COD $m^{-3}d$) in stage V (92 to 105 d), COD removal got enhanced to 78–82% and the colour removal also increased to 54–59%. During the final stage from 106–118 d, the glucose concentration was increased to 5000 mg L^{-1} at the same HRT to obtain an OLR of 5 kg COD $m^{-3}d$. This further improved the COD removal to 89% and colour removal to 62%. These results suggested the effect of effluent quality and presence of easily degrading organics (benign carbon sources) on COD and colour removal. An increase in glucose concentration in the influent improved the FBR performance in terms of COD and colour removal. Recalcitrant organic compounds namely, para-nitrophenol, tributyl phosphate, dibutyl phosphite, nitrilotriacetic acid were rapidly metabolized when acetate was included in the influent wastewater (Kiran Kumar Reddy et al. 2014, Nancharaiah et al. 2006, 2015a, 2016, Suja et al. 2012). Therefore, use of benign carbon sources is a strategy considered for improving the biological treatment of biodegradation of recalcitrant compounds (Nancharaiah and Kiran Kumar Reddy 2018).

Several studies have shown that the presence of easily degradable carbon source enhances COD and colour removal from textile wastewater. But, the treatment process becomes feasible only if the added carbon source is cost effective. In order to provide a cheap carbon source, Senthilkumar et al. (2011) evaluated tapioca sago wastewater (pH: 4.5; COD: 6000 mg L^{-1}) as the co-substrate for treating textile wastewater (pH: 12.8; COD: 1600 mg L^{-1}). For this, a two phase pilot scale UASB reactor with 56 L acidogenic and 230 L methanogenic reactors were operated with HRTs of 6 h and 24 h, respectively. Sago waste water and textile wastewaters were mixed in different ratios of 90:10, 80:20, 75:25, 70:30 and 65:35 to give an OLR of 6.32, 6.16, 6.0, 5.6 and 5.2 kg COD $m^{-3}d$, respectively. For these ratios, the final colour intensities (measured in terms of absorbance at 600 nm) were 0.222, 0.258, 0.272, 0.295 and 0.318. For initial mixing ratios, the colour removal efficiency ranged from 83.4 to 89.4%. A maximum of 91.8% was achieved at the mixing ratio of 70:30. At these varied ratios, the COD values were ranging from 6320 to 5200 mg L^{-1}. The COD removal in the acidogenic reactor decreased from 53 to 23% with increase in the proportion of textile wastewater. However, 81 to 88.5% COD removal was noticed in the methanogenic reactor. Higher COD removal of 88.5% was achieved at a mixing ratio of 70:30. Also, maximum biogas production of 312 L d^{-1} was observed at 70:30 ratio. This study demonstrated that sago wastewater can be a cheap supplementary carbon source for improving the treatment of textile wastewater. However, to make the scheme cost-effective, the availability as well as transportation of these wastewaters to the site of textile industry should be evaluated while contemplating their use in textile wastewater treatment.

Somasiri et al. (2008) examined treatment of real textile water using UASB reactor. For this, anaerobic granular sludge was cultivated in a UASB reactor by feeding glucose containing synthetic wastewater. Subsequently, anaerobic granular sludge was gradually adapted to real textile wastewater by increasing the proportion of real textile wastewater in synthetic wastewater. Finally, the UASB reactor was fed directly with real textile wastewater. The wastewater collected from textile factory had a COD of 832 ± 52 mg L^{-1} and a colour of $\lambda_{580\,nm}$ = 0.18 to 0.35 OD. COD removal efficiencies ranged from 90 to 97%, while, colour removal efficiency of over 92% was achieved. This study showed that gradual adaptation of anaerobic granular sludge was necessary for achieving efficient removal of COD and colour from textile wastewater.

3.3 Treatment of real textile effluents using biofilms under sequential anaerobic–aerobic conditions

Biological treatment of textile wastewater is not complete using aerobic or anaerobic processes solely. Bio-decolourization of azo dyes is very effective in anaerobic bioreactors. But, removal of azo dye compounds (removal of aromatic amines) and COD removal is incomplete. On the other hand, aerobic treatment is more effective in removing aromatic amines or the bio-transformed dyes. Therefore, sequential anaerobic–aerobic systems are proposed for effective biodecolourization and biodegradation of dyes in textile wastewaters. In these systems, biodecolourization by reduction is achieved in anaerobic reactors, whereas the reduced dye by-products (i.e., aromatic amines) are completely biodegraded in aerobic reactors. A model sequential anaerobic (UASB reactor)–aerobic (SBR) system is shown in Figure 4C, wherein the effluent from the UASB reactor is treated in the aerobic SBR.

Sequential anaerobic (UASB reactor) and aerobic (SCAS–Semi Continuous Activated Sludge reactor) system was evaluated for the treatment of carpet dyeing and printing factory wastewater (Kuai et al. 1998). The UASB reactor was inoculated with anaerobic granular sludge from an operating UASB reactor treating carbohydrate wastewater. The SCAS was inoculated with activated sludge. Additionally, granular activated carbon (GAC) was added to the UASB reactor to minimize the toxicity of wastewater (COD: 5000 mg L^{-1}, TKN: 100 mg L^{-1}, Sulphate-S: 100 mg L^{-1}, Colour λ400 nm = 1, λ500 nm = 0.8) on anaerobic granules. The GAC acted as an initial adsorption layer for the toxic pollutants (dyes) present in the real textile wastewater. UASB-SCAS system was operated for 16 weeks by dividing the whole operation into 4 periods. During period 1 (adsorption period, week 1–5), the removal of COD and colour were over 96%, mainly due to the adsorption of wastewater components (dyes) onto GAC. Methane production was poor and corresponded to only 15 to 30% of total COD removed. During period 2 (growth period, week 6–10), GAC saturation with adsorbed organics and biofilm growth on GAC was observed. During this period, the effluent COD values consistently increased from 200 to 1000 mg L^{-1}. Effluent colour also increased consistently and methane production was also poor. During period 3 (toxic shock period, week 11), highly concentrated wastewater (COD of 8760 mg L^{-1}) was fed into the reactor system, which led to complete inhibition of methane production. During period 4 (equilibrium period, week 12–16), the reactor was fed with regular textile wastewater. Methane production, COD removal, colour removal increased gradually in the bioreactors. At the end of the equilibrium period, consistent removal of COD (80%) and colour (75%) was observed. When the operating performance of UASB-SCAS reactor was compared with

SCAS alone, efficient COD and colour removal was observed in the sequential system. SCAS reactor alone was very poor in removing colour from the wastewater.

In another study, UASB reactor seeded with methanogenic granular sludge was coupled to an activated sludge process (ASP) system for the treatment of mixed dyeing industrial complex wastewater (Oh et al. 2004). Both the reactors were initially acclimatised with synthetic textile waste water and gradually shifted to real waste water. During steady-state operation, removal of around 92% COD and 75% colour was reported. Colour removal was mostly seen in the anaerobic reactor, whereas both the UASB and AS reactors were responsible for achieving COD removal (Oh et al. 2004). A sequential anaerobic packed column reactor coupled to activated sludge reactor was studied for treating real textile wastewater. The wastewater was alkaline (pH 9.27) in nature with an average COD of 900 mg L^{-1} and average TCU (true colour units) of 0.363. Nutrient and glucose spiked wastewater having a C:N:P ratio of 100:5:1 was fed into a 6 L capacity anaerobic reactor filled with carrier metal sponges. Facultative anaerobic bacterial consortium consisting of *Alcaligenes faecalis* and *Commomonas acidourans* was immobilised on the carrier media in the form of biofilms. When fed with real textile wastewater in pilot plant, the final decolourization efficiency after anaerobic and aerobic treatment was 85%. Similarly, COD in the treated effluent after final treatment was around 90 mg L^{-1}, resulting in > 85% removal efficiency (Kapdan and Alparslan 2005). Efficient removal of colour and COD in a full scale anaerobic–aerobic system for the treatment of wastewater originated from bleaching, scouring and dyeing processes reported by Frijters et al. (2006). Anaerobic granular sludge in a fluidised bed reactor coupled to the aerobic basin was effective in decolourization and the treated effluent was non-toxic to the indicator organism, *Vibrio fischeri*. The effluents had an average COD of 4500 mg L^{-1} and average dye concentration of 40 mg L^{-1} and the process was efficient in removing > 90% COD and colour.

A sequential anaerobic fixed film reactor-microaerophilic fixed film reactor was operated for the treatment of textile wastewater rich in COD (10,000 mg L^{-1}) and colour (3340 Pt-Co) (Balapure et al. 2016). In anaerobic fixed film reactor, anaerobic biofilm was developed from cattle dung slurry on pumice stone carrier by incubating the slurry for 35 to 40 d. Mixed microaerophilic bacterial consortium was enriched by feeding Bushnell Hass Medium (BHM) amended with 100 mg L^{-1} model dye (Reactive Blue 160) and 0.5% yeast extract. The consortium contained organisms such as *Alcaligenes* sp., *Bacillus* sp., *Escherichia* sp., *Pseudomonas* sp., *Providencia* sp., *Acinetobacter* sp. and *Bacillus* sp. This consortium was incubated for 15 d in microaerophilic fixed film reactor for biofilm development. After establishing the biofilm, the anaerobic reactor was fed with industrial textile wastewater. For the first 3 cycles, a HRT of 10 days was maintained. Then HRT was gradually reduced to 1 d with a gradual increase in OLR. The anaerobic reactor effluent was fed into microaerophilic reactor at varying HRTs ranging from 4 to 24 h. At steady state operation, the combined system was able to efficiently remove 97% of both COD and colour. However, maximum colour removal (80%) was noticed in the anaerobic reactor. The bioreduced dyes were oxidatively degraded in the microaerophilic reactor. Oxido-reductive enzymes such as lignin peroxidase and azo reductase were detected in the cell free extracts of the microaerophilic reactor. The phytotoxicity evaluation of final effluent indicated the non-toxicity of the treated water.

Bae et al. (2016) have evaluated treatment of dyeing wastewater using UASB reactor coupled to the biological aerated filter (BAF) reactor. The dyeing wastewater had a soluble COD of 720 ± 68 mg L^{-1} and colour of 1288 ± 196 ADMI (American Dye Manufacturers Institute) units. The UASB reactor was seeded with anaerobic granular sludge from an operating brewery wastewater treatment facility. Bacteria isolated from textile wastewater treatment facility were cultured as biofilms on polyethylene glycol (PEG) pre polymer additive carrier and used in BAF. Colour removal (70%) was mainly confined to the UASB reactor, while around 32% COD was also removed in the UASB with an HRT of 8 h. Majority of the refractory compounds left over in the UASB effluent were degraded in the BAF reactor. A total of 76% soluble COD was removed in the BAF. Thus, the integrated UASB-BAF system was effective in treating dyeing wastewater. Different studies on sequential anaerobic–aerobic biofilm reactors for the treatment of real textile effluents are summarised in Table 3.

Table 3. Sequential anaerobic–aerobic biofilm reactors for removing azo dyes in real waste water.

Waste water	Waste water quality (in mg L^{-1})	Reactor type and operation	Biomass	Reactor performance	Remarks	References
Carpet dyeing and printing factory	Average conc. CODs = 5000, TKN = 100, SO$_4^{-2}$-S = 100, pH = 5.5, $\lambda_{400\,nm}$ = 1.0 OD, $\lambda_{500\,nm}$ = 0.8 OD	GAC amended UASB-SCAS reactor. 2.3 L UASB with 0.3 L head space. HRT in UASB was 1–2 d depending on strength. HRT in SCAS was 1 d	Anaerobic granular sludge in UASB and activated sludge in SCAS	Around 16 weeks of steady-state operation, consistent removal of COD (80%) and colour (75%) was observed	Performance of sequential GAC amended UASB-SCAS was efficient over SCAS alone system	Kuai et al. 1998
Dyeing industrial complex waste water from approximately 50 textile-dyeing companies in the complex	COD = 440–928; BOD$_5$ = 289–489. Colour = 500–1400 ADMI; TSS = 24–67; pH = 7.2–11.5 and n-hexane = 45–91	5 L UASB reactor fed with SWW having 2000 mg L^{-1} glucose (for 85 d) was gradually raised from 10% to 100% textile waste water (85 to 138 d). Aerobic AS reactor (8 L) fed with SWW having 1000 mg L^{-1} glucose (for 46 d) was gradually increased with 50 to 100% UASB effluent for next 50 d	Methanogenic granular sludge from paper-mill waste water in UASB and activated sludge from sewage treatment plant	92% COD removal and around 75% colour removal was noticed	Although there was not much reduction in colour in UASB effluent, AS reactor helped in improving COD removal	Oh et al. 2004
Textile industry waste water	Average COD = 900, True Colour Units (TCU) = 0.363	6 L anaerobic reactor filled with metal carrier sponges was connected to aerobic activated sludge reactor	*Alcaligenes faecolis* and *Commomonas acidourans* mixed culture biofilm in anaerobic reactor and activated sludge in aerobic reactor	85% colour and > 85% COD removal was observed	Anaerobic reactor was effective in colour removal, whereas effective COD removal was observed in activated sludge reactor	Kapdan and Alparslan 2004
Bleaching and dyeing processes in the textile factory	COD = 2500–7500, Dyes = 10–150	A combination of an anaerobic fluidized bed reactor (70 m³) and a aerobic basin (450 m³) with integrated, tilted plate settlers were constructed	Anaerobic granular sludge from a operating paper mill treatment plant and aerobic sludge was used in aerobic basin	Overall efficiency of anaerobic/aerobic system was 80–90% for COD and 80–95% for colour removal was observed	With the combined anaerobic-aerobic system, the treated effluent was non-toxic to indicator organisms	Frijters et al. 2006

Table 3 contd. ...

...Table 3 contd.

Waste water	Waste water quality (in mg L⁻¹)	Reactor type and operation	Biomass	Reactor performance	Remarks	References
Textile industry waste water	COD = 10,000 ± 34.3, Colour = 3340 ± 25.7 Pt-Co., Phosphate = 650 ± 22.5, Sulphate = 930 ± 29.5	Sequential anaerobic fixed film reactor with 1.5 L working volume and microaerophilic fixed film reactor with 0.75 L working volume. Reactors were filled with pumice stone carrier material	Cattle dung slurry as inoculum for biofilm development in anaerobic reactor. Enriched bacterial consortium developed from a model dye, Reactive blue 160, in microaerophilic reactor	The combined system could efficiently remove 97% of COD and colour at a combined HRT of 60 h	Cleavage of dyes in anaerobic condition and mineralisation in microaerophilic reactor was noticed. The final effluent was not phytotoxic	Balapure et al. 2016
Dyeing industry waste water	Soluble COD = 720 ± 68, colour = 1288 ± 196 ADMI units	10 L working volume UASB reactor was connected to PEG media filled BAF	Anaerobic granular sludge from operating reactor was seeded into UASB. Bacteria isolated from textile treatment plant was immobilised on PEG and seeded in BAF	70% colour removal in UASB with 8 h HRT. 76% COD$_s$ removal with a 24 h contact time	Most of the colour was removed in UASB. BAF was responsible for the maximum removal of refractory compounds	Bae et al. 2016

4. Azo dye decolourization in bioelectrochemical systems

In bioelectrochemical systems (BES), microbial oxidation of organic matter available in the wastewater is either coupled to the electricity generation (microbial fuel cells (MFCs)) or hydrogen generation (microbial electrolysis cells (MECs)). The operation of BESs is also coupled to transformation or removal of various organic/inorganic pollutants from wastewater (Mu et al. 2009). Compared to the energy intensive aerobic industrial/domestic wastewater treatment, the BES process has the advantages of resource generation (energy recovery) coupled to wastewater treatment. The metabolic activity of microorganisms from the oxidation of organic matter in wastewater generates electrons and these released electrons (exoelectrons) are either directly or indirectly transferred to the anode. Certain microorganisms possess the ability to directly transfer the electrons to extracellular electron acceptor (anode) without using mediators. Indirect transfer of electrons to the anode is mediated by redox agents (mediators or electron shuttles) such as humic acids, quinones, phenazines, thionine, methyl viologen, methyl blue and neutral red (Rozendal et al. 2008, Solanki et al. 2013). The microorganisms grow on the electrodes in a self-produced extracellular polymeric substances matrix called electroactive biofilms.

Bacteria belonging to *Geobacter* and *Shewanella* are commonly noticed in the biofilms on the anode and can mediate electron transfer to the anode using direct electron transfer mechanisms (Borole et al. 2011). These bacteria can use membrane proteins such as cytochromes or nanowires for transferring the electrons directly to the anode. The electrons transferred to the anode are transferred to the cathode via external circuit for generating electricity in MFCs or H_2 production in MECs. The electron transfer in BES has been coupled to remove various contaminants like dyes, nutrients and metal ions. Typically, BESs are built either with single chamber or two chambers, as shown in Figure 5. The dual chamber BES will have anode and cathode compartments separated by a proton exchange membrane (PEM or cation exchange membrane (CEM)). The anaerobic oxidation of waste organic matter present in the wastewater by electroactive biofilms generates electrons and protons in the anode chamber. These electrons are directly transferred to the anode. Electrons travelling to the cathode through an external circuit are ultimately consumed by the final electron acceptors in the cathode chamber. The electron acceptors in the cathode chamber are oxygen (MFC) or protons (MEC). The terminal electron acceptors may also include wastewater containing oxidised contaminants such as nitrates, sulphates, redox metals or dyes (Nancharaiah et al. 2015b). In the case of single chamber BES, there is no definitive cathode compartment and PEM may or may not exist. Oxygen from atmosphere directly flows through the porous cathode from side of the wall and it also allows the protons to pass through.

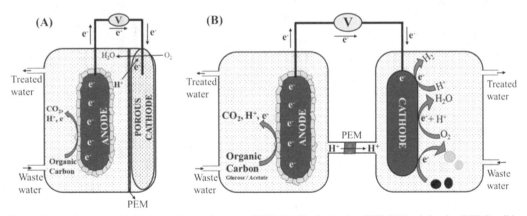

Figure 5. Azo dye removal in bioelectrochemical systems (BES). (A) Single chamber BES. (B) Dual chamber BES. Possible biochemical reactions along with dye decolourization in the cathode chamber are shown in B.

Table 4. Summary of azo dye removal in bioelectrochemical systems.

Dye	BES configuration	Electron donor in anode	Biomass inoculum	Dye removal efficiency	Current generation	Remarks	References
Active brilliant red X-3B (ABR X-3B) (300 to 1500 mg L^{-1})	1 L single chamber MFC with microfiltration membrane. Anode—porous carbon papers, Cathode—membrane-less air cathode with a carbon paper containing 0.5 mg cm^{-2} of Pt catalyst on water facing side and a polytetrafluoroethylene diffusion layer on the air-facing side	Glucose, acetate, sucrose, and confectionery wastewater (500 mg COD L^{-1}) were tested	Mixture of aerobic and anaerobic sludge from STP in 1:1 ratio to a final concentration of 2 g VSS L^{-1}	~ 90% removal up to 900 mg L^{-1} and 77% at 1500 mg L^{-1} with in 48 h	274 mW/m^2 without dye, 234 mW/m^2 with 300 mg L^{-1}, 150 mW/m^2 with 900 mg L^{-1} and 110 mW/m^2 with 1500 mg L^{-1} dye	Glucose was the optimal co-substrate for decolourization and acetate was poor	Sun et al. 2009
Methyl Orange (MO) (0.05 mM)	Dual chamber MFC with 75 mL working volume in each chamber. Ti wire inserted carbon felt were used as the electrodes in both anaerobic anode and AO fed cathode, separated by PEM. Strict anaerobic condition in cathode chamber was maintained by N$_2$ pumping	Glucose (3 g L^{-1}) in mineral salt medium	*Klebsiella pneumoniae* strain L17 in anode chamber	Near complete removal at a rate of 0.298 μmol min^{-1} at pH 3. With increase in pH, dye removal efficiency decreased and at pH 7, poor dye removal was observed	At an external resistance of 2000 Ω, the initial operation of MFC about 2.7 h produced a voltage of 250 $\pm$ 15 mV. Beyond this, a sharp drop of voltage to 75 $\pm$ 10 mV was found and was attributed to the lack of MO in the cathode	Sulfanilic acid and N,N-dimethyl-p-phenylenediamine were the degradation products of methyl orange	Liu et al. 2009
Acid Orange 7 (AO7) (0.19 mM)	Dual chamber BES made from Perspex frames separated by cation exchange membrane was constructed. Working volume of BES was 336 mL. 5 mm diameter graphite rods were used as electrodes to connect to the external circuit	320 mg L^{-1} sodium acetate as electron donor in mineral salt medium	Microbial consortium previously enriched in MFCs with acetate as energy source	79% removal with 0.19 mM influent concentration at a rate of 2.64 mol m^{-3} d^{-1} and was decreased to 35% at 0.7 mM	0.31 to 0.60 W m^{-3} NCC when the influent AO7 concentration increased from 0.19 to 0.70 mM at an external resistance of 8.5 Ω	When BES was supplied with an external power of 0.012 kWh mol^{-1} AO7, the decolourization rate was increased to 13.18 mol m^{-3} NCC d^{-1}	Mu et al. 2009

Amaranth (75 mg L⁻¹)	MFC—electrochemical Fenton system. 80 mL each dual chamber MFC with granular graphite as anode and spectrographically pure graphite (SPG) rods as cathode were constructed. O₂ purging in cathode chamber for two electron transfer to oxygen to generate H₂O₂. After the H₂O₂ generation, Ferrous iron was added to generate free radicals for the oxidative degradation of dye through generated free radicals	Glucose (700 mg L⁻¹ COD) was used as electron donor	Anaerobic sludge from sewage treatment plant	100% removal at 25 mg L⁻¹ was reduced to 76% at 75 mg L⁻¹ in presence of 1.0 mmol L⁻¹ iron catalyst at pH 3	The maximum power density was up to 28.3 W m³	Power density in MFC—electrochemical Fenton system is larger than that of MFC with K₃Fe(CN)₆ as electron acceptor	Fu et al. 2010
Reactive blue 160 (RBu 160) (450 to 1350 mg L⁻¹)	Transparent polymethyl methacrylate (PMMA) made single chamber air cathode; membrane less MFC with 200 mL working volume was constructed. Activated carbon cloth (without catalyst) was used as anode. Hydrophobic carbon cloth with a polytetrafluoroethylene (PTFE) diffusion layer on the air-facing side was used as air cathode	Luria-Bertani medium	*Proteus hauseri* isolated from natural environments	97% removal at a rate of 14.62 mg L⁻¹ ODU⁻¹ h⁻¹. With increase in dye concentrations, longer incubation time was required for complete decolourization	Progressive increase in RBu160 concentration from 0, 450, 900, and 1350 mg L⁻¹ led to decrease in voltage output to 245, 197, 152, and 109 mV, respectively	Simultaneous electricity generation and dye decolourization was observed in BES although with increase in dye concentration, electricity generation is decreased due to competition for reductive decolourization	Chen et al. 2010

Table 4 contd. ...

...Table 4 contd.

Dye	BES configuration	Electron donor in anode	Biomass inoculum	Dye removal efficiency	Current generation	Remarks	References
decolourization liquid (DL) of Active brilliant red X-3B (ABRX3)	Aerobic bio cathode MFC for reducing COD in decolourization liquid of ABRX3. 900 mL volume each dual-chamber MFC made of polycarbonate separated by proton exchange membrane was constructed. Porous carbon papers were used as electrodes and cathode does not contain catalyst	Mineral media with glucose contributing to a final COD of 500 mg L⁻¹	Aerobic and anaerobic sludge from waste water treatment plant and wetland sediment was used for richer biofilm diversity at anode and cathode	Decolourized ABRX3 water from single chamber air cathode MFC was fed to aerobic bio cathode MFC. Within 12 h, 25% COD was removed. The aerobic degradation of aromatic amines present in DL is responsible for this	Addition of DL to aerobic bio cathode resulted in 300% increase in maximum power density from 50.74 to 213.93 mW m⁻²	The redox nature of by-products generated during ABRX3 decolourization is responsible for enhanced electron transfer to oxygen resulting in increased power output	Sun et al. 2011
Congo red (300 mg L⁻¹)	Comparison studies using two MFCs. (1) Dual chamber aerobic bio cathode MFC with 400 mL each capacity. Porous carbon papers without any catalyst were used as electrodes. (2) Single chamber air cathode MFC. 400 mL working volume Plexiglas containing porous carbon papers as electrodes was constructed. Cathode was coated with Pt (0.5 mg cm⁻²) catalyst	Mineral media with glucose contributing to a final COD of 500 mg L⁻¹ was added to anode chamber. Same medium was also used for initial biofilm development on aerobic bio cathode chamber	Aerobic and anaerobic sludge (1:1, v:v) collected from municipal wastewater treatment plant at a concentration of 2 g VSS L⁻¹	96.4% decolourization within 29 h in aerobic bio cathode and 107 h with the air-cathode	The maximum power density in bio cathode MFC is 122 mW m⁻². In air cathode MFC, it was 324 mW m⁻² which was about 166% higher than bio cathode MFC	The decolourization was faster in bio cathode MFC. Community analysis revealed the presence of δ-Proteobacteria and they could be responsible for the decolourization	Hou et al. 2012

Table 4 contd. ...

Dye	Reactor	Electron donor/substrate	Inoculum	Results	Power	Remarks	Reference
Acid Orange 7 (AO7) (0.14 to 2.00 mM)	Sleeve-type BES 10 cm height. Perspex tubes with inner anode chamber and outer cathode chamber of 200 mL each were separated by PEM. Electrodes were made of carbon fibre and were connected at an external resistance of 10 Ω	Sodium acetate was used as the electron donor	Adapted microorganism consortium from cow ruminal contents	Complete decolourization (> 98%) was observed	–	Sleeve-type BES decreases the distance between electrodes, thereby decreasing the internal resistance and enhancing the decolourization efficiency	Kong et al. 2013
Synthetic waste water containing 18 azo dyes of 20 mg L⁻¹ each to give a final concentration of 360 mg L⁻¹	Dual chamber fed-batch type BES. Two glass chamber with a 200 mL working volume. Carbon cloth electrodes with cathode having a Pt catalyst coated on it	Molasses (4 g L⁻¹) was used as the co-substrate along with 20 mg L⁻¹ each of 18 azo dyes in anode compartment	Anaerobic consortium from sewage treatment plant was acclimatised with mixed dye (18 dyes) solution. Microcrystalline cellulose was used as carbon source	Decolourization and COD removal was over 90% in repeated cycles of operation	Peak power density of 25 mW m⁻² at 50°C and 27 mW m⁻² at 2% W/V salt content was found	Studies with elevated temperatures and high salinity revealed the possibility using MFC for dye waste water of industrial relevance such as high temperature and salinity	Ferrando et al. 2013
Alizarin Yellow R (AYR) 100 mg L⁻¹	Up-flow bio-electrocatalyzed electrolysis reactor (UBER) containing bottom cathode and top anode zones without membranes. 250 mL volume UBER with 3–5 mm graphite granules as cathode and 4.5 cm diameter carbon brush as anode and were fixed at a distance of 2 cm. UBER was connected to aerobic bio-contact oxidation reactor (ABOR) and the effluent from UBER is transferred to ABOR for complete mineralisation	Sodium acetate (1.0 g L⁻¹) in mineral salt media was used as electron donor in anode	Biofilm acclimatised and developed as anode respiring bacteria (ARB) on carbon brush electrode in MFC was directly placed in UBER. ABOR was inoculated with activated sludge from Sewage treatment plant	94% colour and 93% COD removal was estimated in a combined process of UBER and ABOR in 6 h overall HRT (2.5 h in UBER + 3.5 h in ABOR)	–	Increased decolourization efficiency with increase in cathode size was observed. UBER reduction by-products such as p-phenylenediamine and 5-aminosalicylic acid were effectively mineralised in ABOR	Cui et al. 2014

...Table 4 contd.

Dye	BES configuration	Electron donor in anode	Biomass inoculum	Dye removal efficiency	Current generation	Remarks	References
Direct Red 80 (DR 380) (200 mg L⁻¹)	Dual chamber MFC. Graphite felt was used as anode and platinum-coated graphite cloth was used as air cathode and was separated by PEM	Glucose (1000 mg L⁻¹) and DR 80 (200 mg L⁻¹) in mineral salt media. Decolourization was also studied with acetic acid, propionic acid, lactic acid electron donors	Anaerobic digestion sludge from STP. VSS of the sludge was 11.46 g L⁻¹	85% colour and 75% COD removal in presence of glucose and dye	Maximum power density of 477 and 455 mW m⁻² with glucose alone and glucose + dye, respectively	Addition of dye to the glucose acclimatised MFC did not impact power generation. Glucose was effective co-substrate followed by acetic, propionic, and lactic acid in terms of decolourization and power production	Waheed et al. 2015
Acid orange 7 (AO 7) (200 mg L⁻¹). By varying HRT, loading rates varied from 200 to 800 g m⁻³ d⁻¹	1.25 L working volume BES with two anodes and two cathodes made from granular graphite made as 4 cm height rods. Three similar BES were constructed for evaluating decolourization in presence of electron donors such as domestic waste water, glucose and acetate	Filtered domestic waste water with a COD of 309 ± 18 mg l⁻¹ was mixed with 200 mg L⁻¹ AO 7. Control BES with acetate and glucose were adjusted to get similar COD and were also added with 200 mg L⁻¹ AO 7	Anaerobic activated sludge to give a final VSS of 17.11 g l⁻¹ was used as inoculum	Across the tested loading rate, the decolourization efficiencies were over 98% in presence of domestic waste water, acetate or glucose fed BES	Current generation was nearly similar in all the three BES and was increased with increase in loading rate (at low HRT). ~ 4.5 mA current was generated at a loading rate of 800 g m⁻³ d⁻¹	Sulfanilic acid was one of the AO 7 reduced products identified in effluent	Cui et al. 2016a
Alizarin Yellow R (AYR) (60–200 mg L⁻¹)	Continuous stirred tank reactor with built-in bioelectrochemical system (CSTR-BES) 0.75 L cylindrical glass BES with graphite fibre brushes as the electrodes were constructed. Mixing was achieved by placing the BES on magnetic stirrer. BES were operated at an external resistance of 10 Ω and applied voltage of 0.5 V	Mineral medium added with 500–2000 mg L⁻¹ sodium acetate and 60–200 mg L⁻¹ AYR	Anode was inoculated with mixed inoculation consisting of effluent from a long-term operated BES and activated sludge from STP. To this established BES, 3 g L⁻¹ anaerobic sludge was added to make CSTR-BES	At 100 mg L⁻¹ AYR, decolourization efficiency of CSTR-BES was 97% in 7 h period with 3 g L⁻¹ sludge concentration	–	AYR removal efficiencies were increased with increased electron donor (Sodium acetate) and sludge (Biomass) concentration	Cui et al. 2016b

Methyl orange (MO) (0.92 mM)	MFC–MEC coupled system Two Perspex chambers each with anode and cathode compartment with a working volume of 0.12 L were added with graphite granules as the electrodes. Cathode chamber of MFC was fed with $K_3Fe(CN)_6$ and MEC with MO. Both the chambers were deaerated with nitrogen gas	Phosphate buffered mineral salt medium containing 320 mg l^{-1} sodium acetate was used as electron donor	Electroactive anodic biofilm was developed from anaerobic sludge	At an acetate concentration of 2.56 mM in anodes of MFC and MEC, the decolourization efficiency reached upto 98.5%	With increase in acetate concentration from 0.75 to 3.64 mM, the current generation was increased from 6.5 to 20.8 mA	Coupled MFC–MEC was more efficient compared to the single MFC. Under similar operating conditions, DE in MFC-MEC was 92%, where as in MFC alone it was 65%	Li et al. 2016
Mono azo dyes (New Coccine, NC and Acid Orange 7, AO 7) and di azo dyes (Reactive Red 120, RR 120 and Reactive Green 19, RG 19). (25 mg L^{-1} each)	Dual Chamber MFC. Two identical Perspex chambers with 1.5 L each working volume. Carbon rod inserted into the carbon felt was used as the electrodes. Bioanode and abiotic cathode were connected through a copper wire at an external resistance of 1000 Ω	Nutrient medium with 440 mg L^{-1} sodium acetate as electron source	Mixed sludge collected from rubber gloves production waste water treatment plant was used	Decolourization efficiency of NC–95%, AO 7–95%, RG 19–81% and RR 120–78%	Voltage output (mV) with NC–359, AO 7–334, RR 120–314 and RG 19–309	Decolourization efficiency and electricity generation was dependent on the type of azo dye used as electron acceptor in cathode chamber and was high for mono azo dyes over di azo dyes	Oon et al. 2017

MFC—Microbial fuel cell, MEC—Microbial electrolysis cell, HRT—Hydraulic retention time, DE—Decolourization efficiency, NCC—Net cathodic compartment, BES—Bio electrochemical system, PEM—Proton exchange membrane, VSS—Volatile suspended solids, COD—Chemical oxygen demand, Pt—Platinum

Sun et al. (2009) reported simultaneous electricity production and azo dye [active brilliant red X-3B (ABRX3)] decolourization in MFCs for the first time. A 1 L volume single-chamber, air cathode MFC with microfiltration membrane and porous carbon papers as the anode was constructed. Cathode was a membrane-less air cathode with a carbon paper containing 0.5 mg cm^{-2} of Pt catalyst on water facing side and a polytetrafluoroethylene diffusion layer on the air-facing side. It was covered with a porous Plexiglas for allowing oxygen to reach the cathode. The MFC was inoculated with aerobic and anaerobic sludge collected from municipal sewage treatment plant in 1:1 ratio to get a final sludge concentration of 2 g volatile suspended solids (VSS) L^{-1}. Glucose, acetate, sucrose or confectionary wastewater (CW) was used as the source of carbon and electrons in the anode chamber. Dye removal and electricity generation were studied at different concentrations of dye (300–1500 mg L^{-1}), using different co-substrates and at different resistance loads. Glucose was found to be an efficient co-substrate for dye decolourization in MFC. CW supported efficient dye decolourization and a cheap substrate compared to other carbon sources. Decolourization was higher at lower resistance (50 Ω) and decolourization decreased with an increase in resistance. Electricity generation was not impacted up to 300 mg L^{-1} dye, whereas addition of 1500 mg L^{-1} dye lowered the electricity production by 60% to 110 mW m^2.

Liu et al. (2009) determined the potential of a dual chamber MFC for decolourization of azo dyes using *Klebsiella pneumoniae* in the anode chamber. Glucose was used as the electron donor in the anode compartment. The cathode chamber was made anaerobic and irrigated with methyl orange (MO), Orange I or Orange II solution. The current generation was dependent on the availability of azo dye in the cathode compartment, to which the electrons are finally transferred. The dye with high redox potential was reduced first and the rate of reduction was in the order of methyl orange > Orange I > Orange II. Inverse relation between rate of dye removal and the medium pH was evident with acidic pH (3.0) favouring the dye reduction. Decolourization of acridine orange 7 (AO7) was also determined in a dual chamber MFC constructed with graphite rods using microbial consortium in the anode chamber (Mu et al. 2009). The decolorization of AO7 in the cathode chamber was achieved under closed circuit conditions along with power generation. However, the decolourization efficiency increased with the supply of external power.

Dye degradation was also coupled to the MFC based advanced oxidation process wherein H$_2$O$_2$ generated in the MFC was used for oxidative degradation of dyes. Amaranath, an azo dye, was subjected to oxidative degradation in the cathode chamber of a MFC-electrochemical Fenton reaction which generated free radicals (Fu et al. 2010). Excess oxygen purging in the cathode leads to two electron transfer to oxygen, thus producing H$_2$O$_2$. By the addition of iron catalyst, reactive free radicals are produced and used for dye degradation. Chen et al. (2010) isolated dye decolourizing strains, *Proteus hauseri*, from natural environments and acclimatised them in MFC for biofilm formation and steady-state power generation using nutrient rich Luria Bertani medium in the anode chamber. The dye decolourization was largely dependent on azo dye structure and concentration. Simultaneous decolourization of the azo dye reactive blue 160 (RBu 160) and electricity generation was studied by varying dye concentration from 0 to 1350 mg L^{-1}. BES was able to decolourize such high concentrations of dye with longer incubation periods. However, current generation decreased with an increase in dye concentration due to the competition of dye for reductive transformation (Chen et al. 2010).

Aromatic amines are the primary by-products of azo dye reduction. These amines are not biodegraded further in the BES due to the prevailing anaerobic conditions. The azo dye ABRX3 decolourization liquid (DL) from single chamber air cathode MFC (Sun et al. 2009) was subjected to further treatment in a dual chamber MFC with aerobic biocathode. Around 25% COD was removed in the cathodic compartment of aerobic biocathode MFC. Due to the presence of ABRX3 by-products in DL, around 300% increase in power density was observed (Sun et al. 2011). Comparative study of dual chamber aerobic biocathode MFC and single chamber air cathode MFC for simultaneous Congo red decolourization and electricity generation was studied (Hou et al. 2012). Decolourization was faster in biocathode MFC, whereas the power density was high (166%) in Pt catalyst coated single chamber air cathode MFC. The reduced internal resistance and presence of *Geobacter* sp., which can directly transfer the electrons to electrodes, are responsible for the increased power generation in single chamber air cathode MFC. In air cathode MFC, large numbers of aerobic or facultative organisms were present due to relatively high dissolved oxygen. Aerobic or facultative microbes were not detected in biocathode MFC. Sulphate reducing bacteria belonging to δ-*Proteobacteria* observed in the MFC were responsible for dye decolourization.

Azo dye decolourization in BES is mainly studied in dual chamber and single chamber systems. The efficiency of dual chamber systems is dependent on the overall internal resistance, which is relatively high due to large size (two different compartments) which hampers the efficiency. This can be overcome by the compact single chamber BES. But, these systems are limited by the unavailability of cathode for dye removal.

A modified sleeve-type BES with low resistance and effective usage of cathode was constructed for efficient AO7 decolourization (Kong et al. 2013). Anode chamber was placed inside the cathode chamber and both the chambers were separated by PEM. In order to simulate real textile wastewater, synthetic wastewater containing 18 different azo dyes was prepared for determining decolourization, COD removal and electricity generation in dual chamber BES (Fernando et al. 2013). A dual chamber BES with 20 mg L^{-1} of each dye (total 360 mg L^{-1} dye) and molasses (4 g L^{-1}) in the anode compartment and 50 mM phosphate buffer in the cathode compartment was used. Influence of operating parameters such as temperature and salinity on BES performance was evaluated. Increase in temperature from 20 to 50°C led to increase in decolourization efficiency from 0.04 to 0.27 K-decolourization h^{-1}, 0.0114 to 0.04 K-COD removal h^{-1} and a power density of 13.22 to 25.6 mW m^{-2}. Increase in salinity to 1% did not negatively impact BES performance, whereas further increase in salinity to 2.5% led to a 4-fold decrease in decolourization efficiency.

A single chamber, membrane less upflow bio-electrocatalyzed electrolysis reactor (UBER) was connected to an aerobic bio-contact oxidation reactor (ABOR) for determining decolourization and COD removal (Cui et al. 2014). Alizarin Yellow R (AYR) decolourization was tested in UBER using acetate as electron donor. The decolourized liquid was passed into ABOR for further treatment (Cui et al. 2014). Biofilm developed on carbon brush anode electrodes was connected to granular graphite cathode and fed with mineral media containing acetate and azo dye. The decolourized liquid was transferred to the ABOR for complete mineralisation. This approach achieved > 93% decolourization and COD removal, indicating efficient treatment of dyes. Bacterial strain *Pseudomonas* sp. WYZ-2 isolated from the electroactive biocathode biofilm was shown to utilise electrode as the electron donor for decolourizing acid black 1 (AB1) (Wang et al. 2014). In the decolourization studies of Direct Red 80 (DR 80), glucose acclimatised anaerobic sludge was able to decolourize DR 80 in presence of glucose co-substrate in the anode compartment along with simultaneous electricity production. Other carbon sources such as acetic, propionic or lactic acid were able to support decolourization of DR 80, but at a slower rate. Addition of dye did not impact the power production (Miran et al. 2015).

In most of the studies, addition of labile carbon sources such as glucose or organic acids as the electron donor in BES for azo dye decolourization was reported. Recently, Cui et al. (2016a) used domestic wastewater as the electron donor for decolourizing AO7. This method accomplished the effective decolourization of azo dye besides sustainable wastewater treatment. The decolourization efficiency using wastewater was comparable to the glucose or acetate fed BES (Cui et al. 2016a). In a modified coupled system, a CSTR was coupled to a BES (CSTR-BES) for decolourization of Alizarin Yellow R (AYR) (Cui et al. 2016b). The decolourization efficiencies of 55 and 91%, respectively, were achieved in CSTR and BES. The colour removal efficiency was higher at 97% in the integrated CSTR-BES system. Such modifications can be implemented for converting the existing anaerobic treatment systems to efficient combined treatment systems. Various methods such as use of modified electrodes, changing operating conditions, cost-effective electron donors, efficient organisms which can form electroactive biofilms were evaluated for efficient azo dye decolourization with/without coupling to electricity generation.

In a different approach, MFC and MEC were coupled for efficient degradation of methyl orange (MO). It has been shown in several studies that dye decolourization is efficient in MECs. But, the energy required for powering MECs is a drawback of these systems. This can be overcome by the coupling of power generating MFC to a MEC. The power generated in the MFC is directly used to power a MEC, where azo dye decolourization is achieved. For these studies, two different rectangular Perspex chambers were connected with graphite granules as the electrodes and each chamber had anode and cathode compartments. Bioanodes were developed in both the chambers from anaerobic sludge using acetate as carbon and energy source. At steady-state operation, complete removal of MO was noticed in the MFC-MEC system. The decolourization potential of the coupled system was nearly 2-fold higher as compared to single MFC system (Li et al. 2016). Yang et al. (2016) determined the effect of various parameters such as cathode potential, catholyte DO concentration and cathode biofilm on decolourization of MO. A decrease in cathode potential from –0.2 to –0.8 V relative to reference electrode led to increase in MO decolourization efficiency from 0 to 95%. The increase in DO from 0 to 5.8 mg L^{-1} led to a decrease in decolourization efficiency from 87 to 28%. Formation of biofilm on the cathode decreased the decolourization efficiency from 92% to 59% at a cathode potential of –0.6 V. However, the decolourization efficiency could be improved to 85% by adding glucose to the cathode chamber. Electrons generated from the glucose metabolism by microbial cells in the biofilm are responsible for the MO decolourization. Recently, Oon et al. (2017) showed that decolourization of mono azo dyes, i.e., New Coccine and Acid Orange 7, is more efficient over di azo dyes, i.e., Reactive Red 120 and Reactive Green 19, in MFCs. Moreover, simultaneous electricity generation during the decolourization was also dependent on azo dye in cathode compartment (Oon et al. 2017).

5. Conclusions

Various studies indicated the ecotoxicity and carcinogenicity of various azo dyes and their by-products which necessitates proper treatment of dye wastewater generated in textile manufacturing and dying. Among the various treatment approaches, biological treatment is more economical and do not produce secondary waste. Biofilms as well as self-immobilised granular sludge can tolerate shock loadings and fluctuation in dye wastewater composition, thereby suitable for treatment of textile wastewaters. By and large, aerobic or anaerobic treatment systems singly cannot completely decolourize or degrade the azo dyes. Completely anaerobic treatment systems can even produce amine forms of dyes which are more toxic than the original dye compounds. Sequential anaerobic–aerobic treatment systems in various reactor configurations such as UASB-agitated tank or UASB-CSTR or single granular sludge system with intermittent anaerobic–aerobic phases are found to be suitable for efficient decolourization and degradation of azo dyes in synthetic or real wastewater.

Biofilms developed on electrodes of BES are found to be efficient in azo dye removal. Electrons generated from the oxidation of organic matter in the anodic compartment are coupled to the direct or indirect reduction of azo dyes in the cathode chamber. Various configurations of BES such as single chamber or dual chamber or integrated systems were studied for azo dye removal. Acetate, glucose, sucrose, propionic acid and lactic acid are used as the carbon and electron source for developing electrogenic biofilms on the anode of BES. There are few studies on the utilization of electrons generated from oxidation of domestic or industrial waste water for reduction of azo dyes with/without simultaneous electricity generation. In indirect approaches, electrons generated in anode are coupled to the generation of reactive free radicals which are subjected to the advanced oxidation process for the degradation of azo dyes. However, biofilm based approaches are efficient and tolerant to complex azo dye removal over other biological approaches; complete removal of azo dyes is still a big issue to meet the zero discharge limits for dye waste water. Enrichment of efficient biofilm consortia for degrading mixture of various azo dyes or integrated biological approaches with physico-chemical methods could meet the strict discharge limits imposed on dye waste water.

References

Abraham, E.T., R.C. Senan, T.S. Shaffiqu, J.J. Roy, T.P. Poulose and P.P. Thomas. 2003. Bioremediation of textile azo dyes by an aerobic bacterial consortium using a rotating biological contactor. Biotechnology Progress 19: 1372–1376.

Bae, W., D. Han, E. Kim, R.A. de Toledo, K. Kwon and H. Shim. 2016. Enhanced bioremoval of refractory compounds from dyeing wastewater using optimized sequential anaerobic/aerobic process. International Journal of Environmental Science and Technology 13(7): 1675–1684.

Bafana, A., S.S. Devi and T. Chakrabarti. 2011. Azo dyes: past, present and the future. Environmental Reviews 19: 350–371.

Balapure, K., K. Jain, N. Bhatt and D. Madamwar. 2016. Exploring bioremediation strategies to enhance the mineralization of textile industrial wastewater through sequential anaerobic-microaerophilic process. International Biodeterioration and Biodegradation 106: 97–105.

Borole, A.P., G. Reguera, B. Ringeisen, Z.W. Wang, Y. Feng and B.H. Kim. 2011. Electroactive biofilms: Current status and future research needs. Energy and Environmental Science 4(12): 4813.

Carliell, C.M., S.J. Barclay, C.A. Buckley, D.A. Mulholland and E. Senior. 1995. Microbial decolorization of a reactive azo dye under anaerobic conditions. Water South Africa 21: 61–69.

Chang, W.S., S.W. Hong and J. Park. 2002. Effect of zeolite media for the treatment of textile wastewater in a biological aerated filter. Process Biochemistry 37(7): 693–698.

Chen, B.Y., M.M. Zhang, C.T. Chang, Y. Ding, K.L. Lin, C.S. Chiou, C.C. Hsueh and H. Xu. 2010. Assessment upon azo dye decolorization and bioelectricity generation by *Proteus hauseri*. Bioresource Technology 101(12): 4737–4741.

Chequer, F.M.D., G.A.R. de Oliveira, E.R.A. Ferraz, J.C. Cardoso, M.V.B. Zanoni and D.P. de Oliveira. 2013. Textile Dyes: Dyeing Process and Environmental Impact. Rijeka: InTech.

Chinwetkitvanich, S., M. Tuntoolvest and T. Panswad. 2000. Anaerobic decolorization of reactive dye bath effluents by a two-stage UASB system with tapioca as a co-substrate. Water Research 34(8): 2223–2232.

Cirik, K., M. Kitiş and Ö. Çinar. 2013. Effect of nitrate on anaerobic azo dye reduction. Bioprocess and Biosystems Engineering 36: 69–79.

Correia, V.M., T. Stephenson and S.J. Judd. 1994. Characterisation of textile wastewaters—a review. Environmental Technology 15(10): 917–929.

Coughlin, M.F., B.K. Kinkle, A. Tepper and P.L. Bishop. 1997. Characterization of aerobic azo dye-degrading bacteria and their activity in biofilms. Water Science and Technology 36: 215–220.

Cui, D., Y.Q. Guo, H.S. Lee, H.Y. Cheng, B. Liang, F.Y. Kong, Y.Z. Wang, L.P. Huang, M.Y. Xu and A.J. Wang. 2014. Efficient azo dye removal in bioelectrochemical system and post-aerobic bioreactor: Optimization and characterization. Chemical Engineering Journal 243: 355–363.

Cui, M.H., D. Cui, L. Gao, A.-J. Wang and H.-Y. Cheng. 2016a. Azo dye decolorization in an up-flow bioelectrochemical reactor with domestic wastewater as a cost-effective yet highly efficient electron donor source. Water Research 105: 520–526.

Cui, M.H., D. Cui, L. Gao, H.-Y. Cheng and A.-J. Wang. 2016b. Efficient azo dye decolorization in a continuous stirred tank reactor (CSTR) with built-in bioelectrochemical system. Bioresource Technology 218: 1307–1311.

Da Silva, M., P. Firmino, M. De Sousa and A. Dos Santos. 2012. Sequential anaerobic/aerobic treatment of dye-containing wastewaters: colour and COD removals and ecotoxicity tests. Applied Biochemistry and Biotechnology 166: 1057–1069.

Fernando, E., T. Keshavarz and G. Kyazze. 2013. Simultaneous co-metabolic decolourisation of azo dye mixtures and bio-electricity generation under thermophillic (50°C) and saline conditions by an adapted anaerobic mixed culture in microbial fuel cells. Bioresource Technology 127: 1–8.

Forgacs, E., T. Cserháti and G. Oros. 2004. Removal of synthetic dyes from wastewaters: A review. Environment International 30: 953–971.

Frijters, C.T.M.J., R.H. Vos, G. Scheffer and R. Mulder. 2006. Decolorizing and detoxifying textile wastewater, containing both soluble and insoluble dyes, in a full scale combined anaerobic/aerobic system. Water Research 40(6): 1249–1257.

Fu, L., S.-J. You, G.-q. Zhang, F.-L. Yang and X.-h. Fang. 2010. Degradation of azo dyes using *in-situ* Fenton reaction incorporated into H_2O_2-producing microbial fuel cell. Chemical Engineering Journal 160(1): 164–169.

Ghaly, A., R. Ananthashankar, M. Alhattab and V. Ramakrishnan. 2013. Production, characterization and treatment of textile effluents: A critical review. Journal of Chemical Engineering and Process Technology 5: 1–19.

Hou, B., Y. Hu and J. Sun. 2012. Performance and microbial diversity of microbial fuel cells coupled with different cathode types during simultaneous azo dye decolorization and electricity generation. Bioresource Technology 111: 105–110.

Ibrahim, Z., M.F.M. Amin, A. Yahya, A. Aris, N.A. Umor, K. Muda and N.S. Sofian. 2009. Characterisation of microbial flocs formed from raw textile wastewater in aerobic biofilm reactor (ABR). Water Science and Technology 60(3): 683–688.

Isık, M. and D.T. Sponza. 2004a. Monitoring of toxicity and intermediates of C.I. Direct Black 38 azo dye through decolourization in an anaerobic/aerobic sequential reactor system. Journal of Hazardous Materials 114: 29–39.

Isik, M. and D.T. Sponza. 2004b. Anaerobic/aerobic sequential treatment of a cotton textile mill wastewater. Journal of Chemical Technology and Biotechnology 79(11): 1268–1274.

Isık, M. and D.T. Sponza. 2008. Anaerobic/aerobic treatment of a simulated textile wastewater. Separation and Purification Technology 60: 64–72.

Jäger, I., C. Hafner and K. Schneider. 2004. Mutagenicity of different textile dye products in *Salmonella typhimurium* and mouse lymphoma cells. Mutation Research–Genetic Toxicology and Environmental Mutagenesis 561: 35–44.

Jonstrup, M., N. Kumar, M. Murto and B. Mattiasson. 2011. Sequential anaerobic–aerobic treatment of azo dyes: Decolourisation and amine degradability. Desalination 280: 339–34.

Kapdan, I.K., M. Meryem Tekol and F. Sengul. 2003. Decolorization of simulated textile wastewater in an anaerobic/aerobic sequential treatment system. Process Biochemistry 38: 1031–037.

Kapdan, I.K. and R. Ozturk. 2005. Effect of parameters on colour and COD removal performance of SBR: sludge age and initial dyestuff concentration. Journal of Hazardous Materials 123: 217–222.

Kapdan, I.K. and S. Alparslan. 2005. Application of anaerobic–aerobic sequential treatment system to real textile wastewater for color and COD removal. Enzyme and Microbial Technology 36(2-3): 273–279.

Karatas, M., S. Dursun and M.E. Argun. 2010. The decolorization of azo dye reactive black 5 in a sequential anaerobic–aerobic system. Ekoloji 19: 15–23.

Kiran Kumar Reddy, G., Y.V. Nancharaiah and V.P. Venugopalan. 2014. Aerobic granular sludge mediated biodegradation of an organophosphorus ester, dibutyl phosphite. FEMS Microbiology Letters 359(1): 110–115.

Kong, F., A. Wang, B. Liang, W. Liu and H. Cheng. 2013. Improved azo dye decolorization in a modified sleeve-type bioelectrochemical system. Bioresource Technology 143: 669–673.

Koupaie, E.H., M.R. Alavi Moghaddam and S.H. Hashemi. 2013. Evaluation of integrated anaerobic/aerobic fixed-bed sequencing batch biofilm reactor for decolorization and biodegradation of azo dye Acid Red 18: Comparison of using two types of packing media. Bioresource Technology 127: 415–421.

Kuai, L., I. De Vreese, P. Vandevivere and W. Verstraete. 1998. GAC-amended UASB reactor for the stable treatment of toxic textile wastewater. Environmental Technology 19(11): 1111–1117.

Kumar, T.A. and S. Saravanan. 2009. Treatability studies of textile wastewater on an aerobic fluidized bed biofilm reactor (FABR): A case study. Water Science and Technology 59(9): 1817–1821.

Li, Y., H.Y. Yang, J.Y. Shen, Y. Mu and H.Q. Yu. 2016. Enhancement of azo dye decolourization in a MFC-MEC coupled system. Bioresource Technology 202: 93–100.

Liu, L., F.B. Li, C.H. Feng and X.Z. Li. 2009. Microbial fuel cell with an azo-dye-feeding cathode. Applied Microbiology and Biotechnology 85(1): 175–183.

Lotito, A.M., U. Fratino, A. Mancini, G. Bergna and C. Di Iaconi. 2012. Effective aerobic granular sludge treatment of a real dyeing textile wastewater. International Biodeterioration and Biodegradation 69: 62–68.

Lotito, A.M., M. De Sanctis, C. Di Iaconi and G. Bergna. 2014. Textile wastewater treatment: Aerobic granular sludge vs activated sludge systems. Water Research 54: 337–346.

Mata, A.M.T., H.M. Pinheiro and N.D. Lourenço. 2015. Effect of sequencing batch cycle strategy on the treatment of a simulated textile wastewater with aerobic granular sludge. Biochemical Engineering Journal 104: 106–114.

Mathur, N., P. Bhatnagar and P. Sharma. 2012. Review of the mutagenicity of textile dye products. Univers. Journal of Environmental Research and Technology 2: 1–18.

Miran, W., K. Rasool, M. Nawaz, A. Kadam, S. Shin, J. Heo, J. Jang and D. Sung Lee. 2016. Simultaneous electricity production and Direct Red 80 degradation using a dual chamber microbial fuel cell. Desalination and Water Treatment 57(19): 9051–9059.

Mohanty, S., N. Dafale and N. Rao. 2006. Microbial decolourization of reactive black 5 in a two-stage anaerobic–aerobic reactor using acclimatized activated textile sludge. Biodegradation 17: 403–413.

Mu, Y., K. Rabaey, R.A. Rozendal, Z. Yuan and J. Keller. 2009. Decolorization of azo dyes in bioelectrochemical systems. Environmental Science and Technology 43(13): 5137–5143.

Muda, K., A. Aris, M.R. Salim, Z. Ibrahim, A. Yahya, M.C.M. van Loosdrecht, A. Ahmad and M.Z. Nawahwi. 2010. Development of granular sludge for textile wastewater treatment. Water Research 44: 4341–4350.

Murali, V., S.A. Ong, L.N. Ho and Y.S. Wong. 2013. Evaluation of integrated anaerobic–aerobic biofilm reactor for degradation of azo dye methyl orange. Bioresource Technology 143: 104–111.

Nancharaiah, Y.V., N. Schwarzenbeck, T.V.K. Mohan, S.V. Narasimhan, P.A. Wilderer and V.P. Venugopalan. 2006. Biodegradation of nitrilotriacetic acid (NTA) and ferric-NTA complex by aerobic microbial granules. Water Research 40(8): 1539–1546.

Nancharaiah, Y.V., G. Kiran Kumar Reddy, T.V. Krishna Mohan and V.P. Venugopalan. 2015a. Biodegradation of tributyl phosphate, an organophosphate ester, by aerobic granular biofilms. Journal of Hazardous Materials 283: 705–711.

Nancharaiah, Y.V., S. Venkata Mohan and P.N.L. Lens. 2015b. Metals removal and recovery in bioelectrochemical systems: A review. Bioresource Technology 195: 102–114.

Nancharaiah, Y.V. and G. Kiran Kumar Reddy. 2018. Aerobic granular sludge technology: mechanisms of granulation and biotechnological applications. Bioresource Technology 247: 1128–1143.

Naresh Kumar, A., C. Nagendranatha Reddy, R. Hari Prasad and S. Venkata Mohan. 2014. Azo dye load-shock on relative behavior of biofilm and suspended growth configured periodic discontinuous batch mode operations: Critical evaluation with enzymatic and bio-electrocatalytic analysis. Water Research 60: 182–196.

Oh, Y.-K., Y.-J. Kim, Y. Ahn, S.-K. Song and S. Park. 2004. Color removal of real textile wastewater by sequential anaerobic and aerobic reactors. Biotechnology Bioprocess Engineering 9(5).

Oon, Y.S., S.A. Ong, L.N. Ho, Y.S. Wong, Y.L. Oon, H.K. Lehl, W.E. Thung and N. Nordin. 2017. Microbial fuel cell operation using monoazo and diazo dyes as terminal electron acceptor for simultaneous decolourisation and bioelectricity generation. Journal of Hazardous Materials 325: 170–177.

Paul, S.A., S.K. Chavan and S.D. Khambe. 2012. Studies on characterization of textile industrial waste water in Solapur city. International Journal of Chemical Sciences 10(2): 635–642.

Pereira, L. and M. Alves. 2012. Dyes—Environmental Impact and Remediation BT—Environmental Protection Strategies for Sustainable Development. A. Malik and E. Grohmann (eds.). Dordrecht: Springer Netherlands. 111–162.

Rajaguru, P., K. Kalaiselvi, M. Palanivel and V. Subburam. 2000. Biodegradation of azo dyes in a sequential anaerobic–aerobic system. Applied Microbiology and Biotechnology 54: 268–273.

Razo-Flores, E., B. Donlon, J. Field and G. Lettinga. 1996. Biodegradability of N-substituted aromatics and alkylphenols under methanogenic conditions using granular sludge. Water Science and Technology 33: 47–57.

Rozendal, R.A., H.V.M. Hamelers, K. Rabaey, J. Keller and C.J.N. Buisman. 2008. Towards practical implementation of bioelectrochemical wastewater treatment. Trends in Biotechnology 26(8): 450–459.

Saratale, R.G., G.D. Saratale, J.S. Chang and S.P. Govindwar. 2011. Bacterial decolorization and degradation of azo dyes: A review. Journal of the Taiwan Institute of Chemical Engineers 42: 138–157.

Sarvajith, M., G. Kiran Kumar Reddy and Y.V. Nancharaiah. 2018. Textile dye biodecolourization and ammonium removal over nitrite in aerobic granular sludge sequencing batch reactors. Journal of Hazardous Materials 342: 536–543.

Sen, S. and G.N. Demirer. 2003a. Anaerobic treatment of synthetic textile wastewater containing reactive azo dye. Journal of Environmental Engineering ASCE 129(7): 595–601.

Sen, S. and G.N. Demirer. 2003b. Anaerobic treatment of real textile wastewater with a fluidized bed reactor. Water Research 37(8): 1868–1878.

Senthilkumar, M., G. Gnanapragasam, V. Arutchelvan and S. Nagarajan. 2011. Treatment of textile dyeing wastewater using two-phase pilot plant UASB reactor with sago wastewater as co-substrate. Chemical Engineering Journal 166(1): 10–14.

Solanki, K., S. Subramanian and S. Basu. 2013. Microbial fuel cells for azo dye treatment with electricity generation: A review. Bioresource Technology 131: 564–571.

Somasiri, W., X.F. Li, W.Q. Ruan and C. Jian. 2008. Evaluation of the efficacy of upflow anaerobic sludge blanket reactor in removal of colour and reduction of COD in real textile wastewater. Bioresource Technology 99(9): 3692–3699.

Sponza, D.T. and M.I. Isik. 2002. Decolorization and azo dye degradation by anaerobic/aerobic sequential process. Enzyme and Microbial Technology 31: 102–110.

Sponza, D. and M. Isik. 2005a. Reactor performances and fate of aromatic amines through decolorization of Direct Black 38 dye under anaerobic/aerobic sequentials. Process Biochemistry 40: 35–44.

Sponza, D. and M. Isik. 2005b. Toxicity and intermediates of C.I. Direct Red 28 dye through sequential anaerobic/ aerobic treatment. Process Biochemistry 40: 2735–2744.

Suja, E., Y.V. Nancharaiah and V.P. Venugopalan. 2012. P-nitrophenol biodegradation by aerobic microbial granules. Applied Biochemistry and Biotechnology 167(6): 1569–1577.

Sun, J., Y. you Hu, Z. Bi and Y. qing Cao. 2009. Simultaneous decolorization of azo dye and bioelectricity generation using a microfiltration membrane air-cathode single-chamber microbial fuel cell. Bioresource Technology 100(13): 3185–3192.

Sun, J., Z. Bi, B. Hou, Y. qing Cao and Y. you Hu. 2011. Further treatment of decolorization liquid of azo dye coupled with increased power production using microbial fuel cell equipped with an aerobic biocathode. Water Research 45(1): 283–291.

Talarposhti, A.M., T. Donnelly and G.K. Anderson. 2001. Colour removal from a simulated dye wastewater using a two-phase anaerobic packed bed reactor. Water Research 35(2): 425–432.

Vanderzee, F. and S. Villaverde. 2005. Combined anaerobic–aerobic treatment of azo dyes—a short review of bioreactor studies. Water Research 39: 1425–1440.

Viliesid, F. and M.D. Lilly. 1992. Influence of dissolved oxygen tension on the synthesis of catechol 1,2-dioxygenase by *Pseudomonas putida*. Enzyme and Microbial Technology 14: 561–565.

Wang, Y.Z., A. Jie Wang, A. Juan Zhou, W. Zong Liu, L. Ping Huang, M. Ying Xu and H. Chun Tao. 2014. Electrode as sole electrons donor for enhancing decolorization of azo dye by an isolated *Pseudomonas* sp. WYZ-2. Bioresource Technology 152: 530–533.

Wesenberg, D., I. Kyriakides and S.N. Agathos. 2003. White-rot fungi and their enzymes for the treatment of industrial dye effluents. Biotechnology Advances 22: 161–187.

Yang, H.Y., C.-S. He, L. Li, J. Zhang, J.-Y. Shen, Y. Mu and H.-Q. Yu. 2016. Process and kinetics of azo dye decolourization in bioelectrochemical systems: effect of several key factors. Scientific Reports 6: 27243.

Zimmermann, T., H. Kulla and T. Leisinger. 1982. Properties of purified Orange II azo reductase the enzyme initiating azo dye degradation by *Pseudomonas* KF46. European Journal of Biochemistry 129: 197–203.

CHAPTER **6**

Selenium Remediation Using Granular and Biofilm Systems

Lea C. Tan,[1,]* Joyabrata Mal[1,]* and Piet N.L. Lens[1,2]

1. Introduction

Selenium (Se) is part of the 16th group of the periodic table and is a non-metal or chalcogen with strong metal-like properties. Selenium shares many similar physiological properties and chemical behaviour with arsenic and sulfur, respectively. Selenium as a component has wide applications, from raw material to various product manufacturing such as shampoos, and semi-conductor for its intrinsic opto-electrical property to additive for agricultural fertilizer (Tan et al. 2016). It was reported by USGS in 2016 that the global consumptions and usages of selenium as raw material were in electrolytic manganese and metallurgy, glass manufacturing, agriculture, chemicals and pigments, and electronics, in decreasing order of consumption (Anderson 2016).

Selenium is a unique element that is both an essential nutrient and a toxic element to living organisms with a narrow range (Sakr et al. 2018). Selenium was accidentally discovered by Jakob Berzelius in 1817 in a sulfuric acid sludge due to reported poisoning of workers involved in the production line. In 1937, Moxon discovered a plant containing high concentration of selenium that was the cause of livestock poisoning. Later, during 1957, Klaus Schwartz was the first to report that selenium also has beneficial properties as an essential mineral nutrient, albeit in small dosages. Excess of selenium in humans have caused various ailments such as selenosis—associated with hair loss, nail brittleness and neurological abnormalities—while deficiency of selenium has resulted in Kashin-Beck disease—an endemic disease of the bones causing problems with the joints (Dinh et al. 2018). The toxicological mechanism of selenium poisoning is attributed to the substitution of selenium for sulfur, since selenium has close similarity with sulfur. This substitution causes biological function disruption and leads to the generation of reactive oxygen species, i.e., hydroxyl radical and superoxide radical (Hoffman 2002). Interestingly enough, selenium can also prevent oxidative damage to DNA through scavenging free radicals and has been explored as a component in cancer development as a carrier of therapeutic agents (Sakr et al. 2018).

---

[1] National University of Ireland Galway (NUIG), Galway, Ireland, H91 TK33.
[2] UNESCO-IHE Delft Institute for Water Education, Delft, Netherlands, 2611 AX.
* Corresponding author: lea.tan@nuigalway.ie; joyabrata.mal@nuigalway.ie

Industrial selenium pollution can come from various anthropogenic activities and leads to the generation of various selenium-laden wastewaters such as acid mine drainage from mining, flue gas desulfurization (FGD) waste from coal combustion/petrochemical industries, agricultural drainage and leachate from seleniferous soils (Figure 1). Most wastewater associated with selenium has a complex matrix containing multitudes of other anions (i.e., sulfate, nitrate, carbonate), cations (i.e., calcium, magnesium) and metals (i.e., iron, copper) that are typically much higher in concentration than selenium. Due to this, selenium was not considered to be an issue in remediation systems. However, when selenium is not removed and released into various lakes/water reservoir, it bioaccumulates over time resulting into severe cases of environmental water pollution and human health issues. Many well-known environmental issues have been documented in North America (Lemly 2014a, b, Brandt et al. 2017) while various inverse effects to humans were sighted in numerous locations in China and India (Dinh et al. 2018, Ullah et al. 2018). This has resulted in setting a strict selenium discharge limit with many nations adopting a concentration of 5 µg L^{-1} as the effluent regulatory discharged limit (Tan et al. 2016).

The Kesterson National Wildlife Refuge ponds in California is one of the known areas of selenium poisoning due to agricultural drainage discharges from San Joaquin Valley farms in large quantities. Due to the small concentration of selenium in the drainage discharged, it was not thought to be an issue when it started during 1971 and continued for more than a decade. This led to the discovery of thousands of birds' and fishes' deformities and high mortality rate in 1983, which eventually led to the shutdown of the reservoir. The awareness of the selenium pollution impact started in 1983 and, even to this present day, is a water pollution issue, with the US government agency struggling to find a solution (Weiser 2018). If the selenium discharged concentration levels are not monitored and controlled properly, it could further affect other nearby areas such as the Grasslands Ecological Area, the largest collection of wetlands in Western US, and become a big threat to migratory waterfowl and other wildlife. More recent example has been documented by Lemly (2014a, b) where wildlife selenium poisoning was observed at Lake Sutton in North Carolina as the direct result of coal ash impoundment disposal by Duke Energy Progress from its electric generating facilities. The author has reported cases of increasing levels of selenium in fish tissues and have documented evidence of teratogenic effect on various fish species resulting to an estimated $8.6 million annual loss due to fishery decline. Similarly, Elk River watershed, a transboundary water shared between Canada and

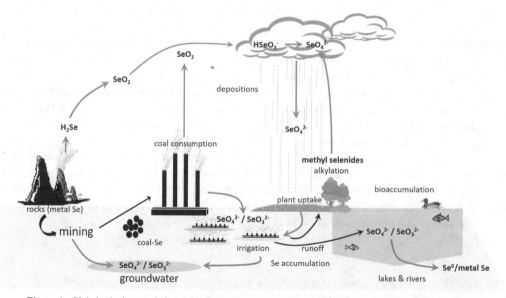

Figure 1. Global selenium cycle involving both natural and anthropogenic source of selenium to wastewater.

US, has recently been reported to have dangerously high levels of selenium caused from various coal-mining discharge operated by Teck resources. It has been reported that Teck Resource has been fined $1.4 million for selenium discharges and are currently undergoing further investigation (The Canadian Press 2018).

Despite the urgent need to properly handle selenium contaminated waters, removing selenium is a big challenge for the industry due to (i) generation of large wastewater volume with low selenium concentration, (ii) complex matrix with higher concentrated contaminants and (iii) low stringent discharged limits that are hard to achieve in most treatment methods. Conventional and advanced treatment methods such as chemical–adsorption and precipitation—and physical–membrane filtration and ion exchange resins—methods have been employed by industry to remove selenium. Technical reports by Golder Associates Inc. (2009) and CH2M HILL (2010, 2013) provide a good summary on physical methods while Santos et al. (2015) provide a detailed presentation on chemical adsorption and ion exchange strategies for selenium removal. Despite the success of some of these technologies, i.e., reverse osmosis and ferrihydrite adsorption, these methods are not sustainable, very costly in terms of operation and are very dependent on the concentration levels and selenium species present in the wastewater. Depending on the origin of the wastewater, selenium exists in several oxidation states (+6, +4, 0, −2) that dictates its water solubility, toxicity levels and bioavailability. The oxidation state can also affect the success of the treatment methods. For example, flue-gas desulfurized (FGD) wastewater contains about 90% selenate (Cordoba and Staicu 2018) which is the most oxidized state and very stable thus making it difficult to remove. On the other hand, selenite is more bioavailable and therefore more toxic but readily reactive and can be easily taken care of by chemical adsorption methods.

Increased interest in utilizing bioremediation has gained popularity with wastewater treatment. Bioremediation using various reactor configurations and microbial inoculum is a green sustainable alternative to physical and chemical methods and can be tailored to adapt to various wastewater characteristics and source. Bioremediation approach also has the advantage of coupling removal technology with resource recovery via reduction of soluble and toxic selenium oxyanions to insoluble and less toxic elemental selenium that can be recovered for further re-use. Compared to other methods, this approach shows promises in reducing selenium pollution burdens from environmental waters while providing an additional source of profit through valuable resource recovery that can offset the wastewater treatment cost.

To establish and select the appropriate bioremediation technique for wastewater impacted with selenium contamination, determination of appropriate reactor configuration must be understood and identified. This chapter focuses on the different reactor configurations used for selenium removal focusing on both granular (aggregated) and biofilm (attached) systems. Different selenium microbial pathway and reduction process are also presented. Special properties of selenium as an anti-biofilm agent are detailed, especially how it can affect biofilm reactors. Finally, selenium recovery in bioreactors and future directions for selenium bioremediation are explored and presented.

2. Selenium microbial reduction and pathways

Reduction of Se-oxyanions is widespread in natural environments and is more likely that biotic mechanisms, such as assimilatory and dissimilatory selenium reduction (Figure 2), are responsible for the presence of selenide in the environment rather than abiotic reduction (Mal et al. 2016b, Nancharaiah and Lens 2015). Assimilatory reduction of Se-oxyanion is the uptake and reduction of selenate for the synthesis of selenomethionine and selenocysteine to be used in selenium-containing enzymes and compounds as cofactors in several enzymes. Assimilatory selenate reduction is generally used by both aerobes and anaerobes. More than 50 distinct selenoprotein families are currently known including glutathione peroxidase (GPx), thioredoxin reductase, tetraiodothyronine deiodinase, selenophosphate synthetase, and selenoprotein P (Labunskyy et al. 2014). Although their distribution varies greatly

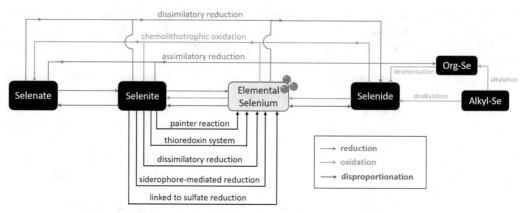

Figure 2. Selenium transformation involving different species and oxidation state (adopted and modified from Nancharaiah and Lens 2015). Transformation pathways are indicated through coloured arrows. Reduction pathway covers both assimilatory and dissimilatory reduction while oxidation is done through chemolithotroph pathway. Reduction pathways of selenite to elemental selenium pathway by microorganisms can occur through five different mechanisms.

Color version at the end of the book

among species, they are present in all three domains of life. Other organic forms of Se, e.g., dimethyl selenide (DMSe) and dimethyl diselenide (DMDSe), result from bacterial methylation processes.

Several microorganisms use selenate as a terminal electron acceptor in anaerobic respiration to conserve energy. Dissimilatory reduction of selenate through anaerobic respiration is a two-step process: selenate reduced sequentially to selenite and then to insoluble elemental selenium. Reduction of selenate to selenite is catalyzed by a trimeric molybdoenzyme, SerABC selenate reductase, located in the periplasmic space. Selenite formed in the periplasmic place is presumed to be transported to the cytoplasm via a sulfate transporter where selenite is reduced to elemental selenium (Nancharaiah and Lens 2015). Reduction of selenite in the cytoplasm was widely recognized to occur by thiol-mediated reduction as part of a microbial detoxification strategy (Kessi and Hanselmann 2004, Turner et al. 1998). Apart from that, enzymes such as nitrite reductase, sulfite reductase, fumarate reductase, and hydrogenase I can support the selenite reduction to elemental selenium in some microorganisms, e.g., *Thauera selenatis* AX, *Rhizobium sullae* strain HCNT1, and *Clostridium pasteurianum* (Basaglia et al. 2007, DeMoll-Decker and Macy 1993, Harrison et al. 1984). Fe (III) reducers such as *Shewanella oneidensis* and *Geobacter sulfurreducens* (Klonowska et al. 2005) reduce selenite to elemental selenium, while *Veillonella atypica* produces selenium nanospheres from selenite via a hydrogenase coupled reduction, mediated by ferredoxin (Pearce et al. 2009). Several other mechanisms have also been suggested for selenite reduction to elemental selenium in microorganisms (Figure 2), including (i) Painter-type reactions, (ii) the thioredoxin reductase system, (iii) siderophore-mediated reduction, (iv) sulfide-mediated reduction, and (v) dissimilatory reduction (Zannoni et al. 2008). Finally, respiration of selenate/selenite is often associated with the formation of brick red-coloured elemental selenium as a stable end product. Elemental selenium nanoparticles formed (Figure 3) during selenium reduction can be located both in the cytoplasm and/or outside the cells (Nancharaiah and Lens 2015). Selenate respiration either in *Enterobacter cloacae* or *Bacillus selenatarsenatis* SF-1 reveals that reduction of selenate to elemental selenium occurs primarily either in the periplasmic space or outside the cell, and the selenium nanospheres are expelled into the extracellular environment (Losi and Frankenberger 1997, Nancharaiah and Lens 2015, Ridley et al. 2006). It is reported that the extracellular accumulation of selenium nanospheres is much more than the intracellular accumulation, particularly during the dissimilatory reduction process. It is possible that the formation of internal and external selenium nanospheres are governed by two different and independent mechanisms. While the extracellular accumulation of selenium nanospheres mostly occurred during the anaerobic

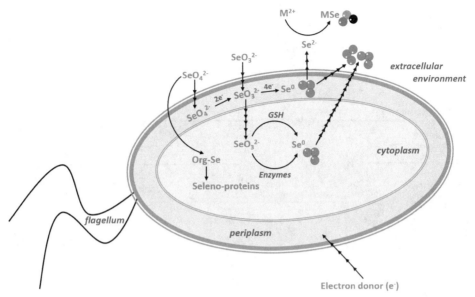

Figure 3. Microbial mechanism (adopted from Nancharaiah and Lens 2015) for selenium oxyanion reduction/detoxification/ assimilation and synthesis of biogenic selenium nanoparticles. The reduction mechanism can occur within the periplasm or cytoplasm while the biogenic elemental selenium can be localized inside or outside the cell.

respiration, intracellular accumulation might be the result of a detoxification of selenite (Nancharaiah and Lens 2015).

3. Selenium bioremediation in bioreactor system

In the last 30 years, biological treatment has emerged as the leading technology for selenium removal from wastewaters (Tan et al. 2016). Biological treatment of Se-laden wastewater has several advantages over chemical–physical remediation technologies, some of them being: the cost-effectiveness of microbial-based remediation approach, the avoidance in employing hazardous chemicals, and the possibility to recover elemental selenium in a recyclable form either as precipitates or as nanostructures, which are technologically and economically more valuable (Piacenza et al. 2017, Tan et al. 2016, van Hullebusch 2017). Indeed, several bacterial cultures capable of reducing selenate and/or selenite to Se(0) either through detoxification or anaerobic respiration mechanisms have been isolated, such as the sulfate transporter in *E. coli* (Springer and Huber 1973), the sulfate permease in *Salmonella typhimurium* (Brown and Shrift 1980), the sulfite uptake system in *Clostridium pasteurianum* (Bryant and Laishley 1989), and the polyol ABC transporter in *R. sphaeroides* (Bebien et al. 2001). However, using mixed microbial communities (i.e., biofilms or granular sludge) instead of pure culture as inoculum are more suitable for process development in bioreactors (Piacenza et al. 2017). Microbial aggregates (either in the form of biofilms, granules or flocs) offer great advantages such as the treating of larger volumes of wastewater, the ability of microbial communities to adapt to diverse conditions, the presence of synergic interactions among members within the consortium and the possibility to work in non-aseptic conditions (Piacenza et al. 2017). Though there are limited full-scale biological treatment plants for selenium treatment currently in place, multiple small-scale experiments (laboratory and pilot-scale) have been demonstrated (Figure 4) as a potential strategy for the decontamination of Se-polluted wastewater in reaching regulatory limits for selenium discharge into the aquatic environment (Piacenza et al. 2017, Tan et al. 2016).

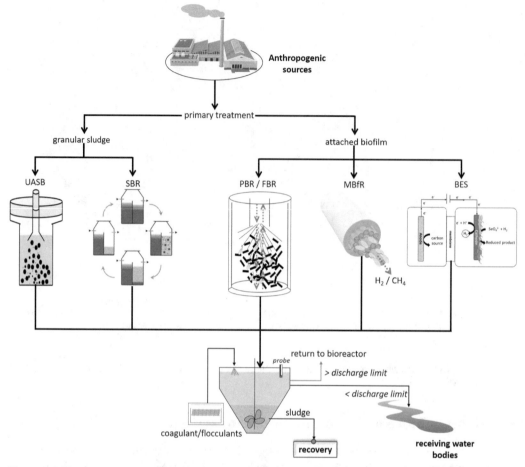

Figure 4. Different reactor configurations used for selenium-laden wastewater and recovery scheme for biological system.

Color version at the end of the book

3.1 Anaerobic/Aerobic granular system for selenium bioremediation

3.1.1 Upflow anaerobic sludge blanket (UASB) reactor

Upflow anaerobic sludge blanket (UASB) reactors are most commonly employed for anaerobic treatment of industrial effluents worldwide (Nnaji 2014, Seghezzo et al. 1998). Under certain conditions, natural aggregation of the bacteria forming flocs or granules leads to the formation of dense granules having good settling properties. Indeed, UASB reactors have also been employed for studying the treatment of Se-contaminated wastewaters which relies on the same concept. In an UASB reactor system, due to the natural aggregation of microbial biomass and suspended solids, a dense granular sludge bed at the bottom of the reactor is formed in which Se-oxyanions bioconversion occurs. Dense granulation also reduces the biomass wash-out from the system under full-scale reactor conditions which allows the reactor to have a high retention of active anaerobic sludge even at high organic loading rates (Bhunia and Ghangrekar 2007). The anaerobic microorganisms in granular sludge are also more protected towards toxic substances and have higher selenite-reducing activity than the suspended flocculent sludge (Soda et al. 2011). Additionally, it is easy to maintain good contact between the bacterial biomass and wastewater through the turbulence of the upflow influent flow and the biogas produced during the selenium reduction process by anaerobic microorganisms.

Moreover, majority of the selenium nanoparticles formed through reduction are retained in the granular sludge leading to high selenium removal efficiency. Association of biogenic elemental selenium nanoparticles with biomass/granular sludge also makes it easier to recover them for future use. Advantages of utilizing UASB reactors thus include reduced maintenance and operation costs (no oxygen supply and possibilities for energy recovery), and easy recovery of the biomass-associated selenium nanoparticles for future reuse (Piacenza et al. 2017, Tan et al. 2016).

Golder Associates Inc. (2009) reported that in the early 1900s, an UASB pilot test was performed for Se-removal from San Joaquin Valley agricultural drainage at the Adams Avenue Agricultural Drainage Research Center in San Joaquin Valley (California). The selenium removal efficiency varied from 58% to 90%, when the influent had a total Se content of 500 $\mu g\ L^{-1}$. The full-scale UASB system, however, faced a number of problems including short circuiting due to the accumulation of gas within the pilot reactor that became trapped in the sludge during its 2-years operation period. The development of stable biomass took approximately 6 months suggesting that long periods of time are required for start-up to acclimate the seed sludge. Additionally, the selenium removal efficiency was not consistent and largely depends on temperature. The removal efficiency reached 88% at 15°C but declined to only 35% when temperature dropped further at 7°C. A partial solution could be to insulate reactor tanks for the full-scale treatment system for maintaining the proper temperature for optimal reactor efficiency which may not be cost effective.

The use of UASB reactors to remove selenium oxyanions from contaminated waters was demonstrated by Lenz and co-workers in a series of studies by using microbial consortia under methanogenic, sulfate-reducing and denitrifying conditions (Lenz et al. 2008a, b, c). Using lactate as electron donor, selenate removal efficiencies in the sulfate-reducing and methanogenic reactor were 97 and 99%, respectively, at 30° and pH = 7.0 (Lenz et al. 2008a). While selenate was reduced by sulfate reducing bacteria in the sulfate-reducing reactor, a selenium-respiring microbial consortium possibly developed in the methanogenic reactor for selenate removal. The selenium removal efficiency in the methanogenic reactor also improved after 58 days of operation (Lenz et al. 2008a), suggesting the growth of specialized selenium converting population in the reactor and need for acclimatization of the seeds similar to Golder Associates Inc. report (2009). However, selenate and dissolved selenium removal were not affected by the sulfate load in the methanogenic reactor. But it was inhibited by sulphide formation (101 mg L^{-1}) suggesting to maintain the selenate to sulfate ratio carefully in the both reactor configurations to achieve optimal selenium removal efficiency (Lenz et al. 2008a). The sulfate-reducing reactor showed a higher selenate removal efficiency, but the total selenium removal was less due to wash out of high dissolved selenium. The total selenium removal efficiency was higher in the methanogenic reactor as the accumulation of elemental selenium in the methanogenic granular sludge was higher compared to the sulfate-reducing sludge (Lenz et al. 2008a).

Similarly, Tan et al. (2018a) also demonstrated that the presence of sulfate and nitrate can significantly influence selenium removal efficiency in UASB reactor using methanogenic granular sludge. In fact, the molar ratio of nitrate:selenate and sulfate:selenate proved to be an important factor in the bioreduction and speciation of selenate. A drop in total Se removal efficiencies was observed in the UASB reactor when sulfate and nitrate were sequentially removed from the influent (Tan et al. 2018). Contrarily, Dessì et al. (2016) demonstrated that the presence of nitrate in the influent reduce the selenium removal efficiency of UASB reactor as it leads to an excess release of colloidal elemental selenium in the effluent.

It is also important to focus on the details of selenium speciation during biological selenium removal process as the formation of selenium nanospheres, aqueous selenide and alkylated selenium species may significantly contribute to selenium effluent concentrations in an anaerobic bioreactor system (Lenz et al. 2008c, 2011). A detailed study on selenium associated with bioreactor granular sludge during treatment of selenium containing wastewater by direct, non-destructive X-ray absorption near edge structure (XANES) reveals that most of the selenium (86%) was present in elemental selenium and selenide and was immobilized to insoluble mineral phases (Lenz et al. 2008d). Heavy metals like cadmium and zinc are common co-contaminants in different selenium-containing waste streams and may also influence the microbial reduction processes as well as the speciation of bioreduced selenium (Mal et al. 2016a). Recently, Mal et al. (2016a) reported that selenite reduction by using anaerobic granular sludge and the formation of biogenic elemental selenium or selenide was influenced by the presence of heavy metals (e.g., cadmium, zinc and lead) and their initial concentration. In a recent study, Zeng et al. (2019) showed that both selenite and total Se had a higher removal efficiency in UASB reactors in the presence of cadmium and zinc than solely feeding selenite, both at 17.5° and 30°. To further evaluate the selenium-heavy metals interactions, especially whether the metals are sorbed onto the elemental selenium nanoparticles or bound as metal selenides (e.g., cadmium selenide), further speciation studies using XANES are required.

The formation of alkylated selenium compounds (dimethylselenide and dimethyldiselenide) during biological treatment of Se-laden wastewater is also not advantageous, as high amounts of alkylated selenium or seleno-sulfur species formed remain in the effluent which can limit the bioremediation in selenium laden waters (Lenz et al. 2008c). Choosing of suitable electron donor is also an important parameter for better reactor performance as high alkylation was reported after selenite reduction involving H_2/CO_2 as electron donor compared to lactate and acetate (Lenz et al. 2011). However, formation of alkylated selenium species is reported as not related to the temperature within the range of 30–55°C (Dessì et al. 2016). Dessì et al. (2016) also showed that temperature can also strongly influence the removal of selenate in UASB reactors using methanogenic granular sludge. 10–15% higher total Se removal efficiency was achieved at thermophilic conditions (55°C) compared to the mesophilic condition (30°C). Interestingly, the shape of biogenic selenium nanoparticles observed in both UASB reactors was significantly influenced due to high temperature: nanospheres and nanorods, respectively, in the mesophilic and thermophilic UASB reactors (Dessì et al. 2016). Additionally, characterization of the microbial consortia through a denatured gradient gel electrophoresis (DGGE) analysis revealed the differences between the bands of the anaerobic granules at different operating temperature. *Sulfurospirillum barnesii*, which produces extracellular biogenic selenium nanoparticles, was detected in the mesophilic UASB reactor which could be one of the reasons for its lower efficiency.

Nevertheless, these studies demonstrate that anaerobic granular sludge is a promising inoculum to be applied in the treatment of selenium containing wastewaters, since it can efficiently convert and immobilize selenate/selenite into insoluble elemental selenium phases (Piacenza et al. 2017, Tan et al. 2016). UASB reactors have the major advantage of allowing for high wastewater loading rates and have an improved selenium removal efficiency on prolonged reactor operation with no major loss of biomass. High biodiversity of sludge granules also makes the UASB reactor less vulnerable to disturbances than systems that rely on a single or a few specialized microorganisms. Finally, the granular sludge can convert high loads of COD into methane and offering energy recovery via biogas production makes the UASB process economically more favourable.

3.1.2 Sequential batch reactor (SBR)

Most of the previous research have focused particularly on the application of anaerobic biofilm or granular sludge in various types of bioreactors for Se removal from Se-containing wastewater. Recently, activated sludge (Mal et al. 2017a, Zhang et al. 2018a) or aerobic granular sludge (Nancharaiah et al. 2018) under aerobic or alternate aerobic-anoxic conditions have been employed to treat Se-containing wastewater and reported to have better Se removal efficiencies compared to Se removal under anaerobic condition. This suggests that the aerobic biological process could be more suitable for the development of Se-containing wastewater treatment system than the well-documented anaerobic processes. In sequencing batch reactor (SBR), the treatment is carried out in consecutive stages in the same tank: filling, reaction, sedimentation, draw and idle in a cyclic operation with a complete aeration or alternating anaerobic–aerobic phase during the reaction time period (Chan et al. 2009). Acclimation of desired microbial populations or the seed sludge can be achieved during the alternating anaerobic–aerobic phases. Although the SBR systems have mostly been used in the treatment of textile wastewater due to their efficient colour and COD removal performance, successful application of SBR for the treatment of Se-laden wastewater have also been reported recently (Mal et al. 2017a, Nancharaiah et al. 2018, Zhang et al. 2018a).

Rege et al. (1999) demonstrated the possibility of using SBR for treating selenium-containing wastewater. Denitrifying bacterial consortium obtained from the Pullman (Washington, USA) wastewater treatment facility was used as seed sludge in a 1.5 L volume SBR using acetate as the electron donor. Selenate or selenite reduction was observed only after a 150 h lag phase suggesting the acclimation of selenium reducing microbes in the microbial consortium. Nitrate and nitrite were completely denitrified during the log phase and reduction of nitrate and nitrite was concomitant with selenium reduction (Rege et al. 1999). The rate of selenium reduction was proportional to the initial concentration of selenium and to the biomass concentration. Furthermore, selenite reduction rate was determined to be approximately four times faster than selenate reduction rate. Recently, Mal et al. (2017a) investigated the feasibility of simultaneous removal of selenium and ammonium in a SBR and showed for the first time that a selenate removal efficiency of > 95% can be achieved using activated sludge by alternating anaerobic and aerobic phases. Jain et al. (2016) previously reported the failure of a continuously operated reactor for studying selenite removal using the similar activated sludge under aerobic condition. Although selenite reduction was observed, the reactor failed due to selenite toxicity to the activated sludge microorganisms. But stable selenate removal during long term reactor operation as reported by Mal et al. (2017a) suggests that the alternating anaerobic–aerobic phases inherent to SBR operation possibly played a significant role in nullifying the selenium toxicity to the microorganisms present in the activated sludge system. Increase in selenium removal efficiency due to an increase in the length of the anoxic period further confirms it and indicates the importance of the optimization of both the anoxic or aerobic phase to improve the selenium removal efficiencies.

Biological treatment of wastewater, e.g., mining effluents and agricultural drainage containing both ammonium and selenium, necessitates both aerobic and anaerobic conditions. SBR configuration allows the integration of both aerobic and anaerobic conditions in the same reactor tank without the need for another tank. Recently, Nancharaiah et al. (2018) showed that simultaneous removal of selenite and ammonium in a SBR using aerobic granular sludge. The effluent Se was significantly lower as compared to previous studies on selenite removal using activated sludge or anaerobic granular sludge and ranged from 0.02 to 0.25 mg Se L^{-1}, while treating up to 12.7 mg L^{-1} selenite, as selenium nanoparticles were entrapped in the granular sludge after selenite bioreduction. This study shows that aerobic granular sludge reactors are not only capable of removing toxic selenite, but have a better potential for treating Se-rich wastewaters than the well-documented anaerobic processes (Nancharaiah et al. 2018).

Zhang et al. (2018a) demonstrated the removal of selenite from artificial saline wastewater by activated sludge in a SBR under aerobic condition. Se-containing wastewaters are often characterized as having high salinity. Dilution of wastewater from Se refinery plant up to 2% salinity is necessary before biological treatment of Se under anaerobic conditions (Soda et al. 2011). For example, wastewater generated from a Se refinery contains 13.2–74.0 mg L^{-1} of Se and 6–7% salinity (Soda et al. 2011), while chemical leaching from kiln powder contains 2–39 mg/L of Se with 4.4–13.2% salinity (Soda et al. 2015). The study by Zhang et al. (2018a) was the first to demonstrate the possibility of using activated sludge in aerobic SBR for treating selenite (100 mg L^{-1}) in artificial wastewater under high-salinity conditions of 70 g L^{-1}. It is worth mentioning that Se was removed in this study mainly through biovolatilization by the activated sludge contrary to many previous researches when removal of soluble Se was achieved mainly via bioprecipitation and entrapment of selenium nanoparticles by activated or aerobic granular sludge (Mal et al. 2017a, Nancharaiah et al. 2018). All these studies shows that SBRs with activated sludge or aerobic granular sludge provide operational flexibility for simultaneous removal of ammonium and Se-oxyanions from mining effluents and agricultural drainage contaminated with both selenium and ammonium by integrating anaerobic and aerobic phases. However, complex operation set-up, difficulty in maintaining the anaerobic–aerobic microbial consortia and higher cost (aeration cost) can limit its application. It needs to be optimized to further improve the nitrogen and selenium removal efficiencies. A better understanding on the effect of the oxygen concentration and duration of aerobic/anaerobic phase on the selenium oxyanion bioreduction and localization of the elemental selenium in the sludge can help in improving the selenium removal efficiency.

3.2 *Attached biofilm reactors for selenium bioremediation*

Attached biofilm reactors require the use of external solid phase media (i.e., surface or biofilm carrier/packing material) for the formation and retention of active biomass. Attached biofilm reactors can be distinguished based on the degree of bed (material + biomass) expansion being employed. Compared to the granular sludge system, attached biofilm reactor is largely dependent on the growth and characteristics of inoculum used for selenium reduction. Figure 5 indicates the theoretical diagram of selenate removal in a biofilm system under a closed system when exposed to other co-contaminants. Thermodynamically, selenate reduction occurs in between nitrate and sulfate, where a steady reduction flow curve is expected at a negative higher redox state compared to nitrate but faster than sulfate. Additionally, interaction between the reduced products of nitrate and sulfate can occur with selenate species at different stages within the biofilm matrix leading to the formation of either elemental selenium or other poly-selenium compounds.

Tan et al. (2018a) investigated how the biofilm is formed under different feeding condition— selenate only and selenate with co-contaminants. Figure 6 gives a summary (data compiled from the study of Tan et al. 2018a) of how the biofilm formation occurred and selenium removal under different feeding condition. Interestingly, with selenate only condition, the biofilm formed poorly with low selenium removal performance but showed to have the highest cell viability. Opposite to this, when sulfate is present, biofilms grew thicker with better selenium removal performance but had lower cell viability. Cell viability could have been linked with the redox condition where the thicker the biofilm forms; the higher the negative redox condition was, the less the viable cells were. Further research should be conducted to investigate the response of biofilm to complex system and tracking selenium reduction within the biofilm layer using biosensors. Such studies could lead to the optimal operation of attached biofilm-based reactors. Among the reactors that are based on attached biofilm growth system, the following are utilized for selenium treatment and are discussed in separate sections: packed-bed, fluidized bed, membrane bioreactor and bioelectrochemical system.

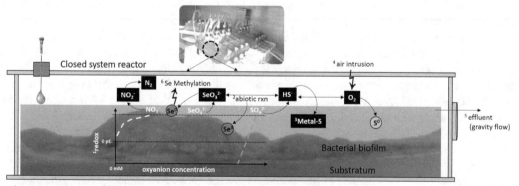

[1] Redox condition within the biofilm was not measured and was based on theoretical concept
[2] Formation of red color was observed when abiotic batch test was conducted between SeO_3^{2-} and HS^- (also observed by Hockin and Gadd 2003)
[3] Metal-sulfur precipitation can theoretical occur with iron (present in the growth medium at 7.5μM)
[4] Reactor was operated shut from open air but not in an anaerobic chamber (air intrusion is still possible)
[5] Reactor was at a 10° angle therefore no water accumulation occurred inside and effluent exits via gravity flow
[6] Volatilization of Se was considered negligible in this condition

Figure 5. Schematic representation of biofilm substrate consumption utilization exposed to nitrate, selenate and sulfate using DFR with possible interaction between oxyanions, by-products and biofilm detailed in the scheme.

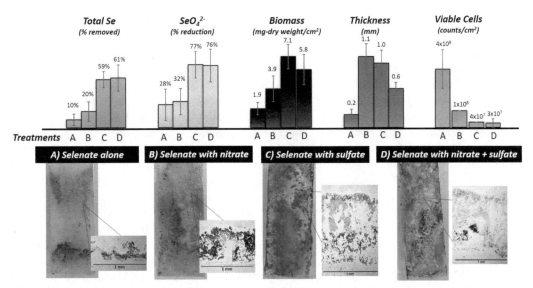

Figure 6. Impact of selenate reduction with co-contaminants on biofilm selenium removal performance, biofilm formation and growth. Data gathered and reconstructed from study conducted by Tan et al. (2018a) while biofilm images came from personal archive.

Color version at the end of the book

3.2.1 *Packed bed reactor (PBR)*

Packed bed reactor (PBR) is one of the most common biofilm reactors for selenium bioremediation with a commercial full-scale application already developed by Applied Biosciences and marketed by GE Water and Process Technology (CH2M HILL 2010). PBR utilizes any inert packing material, whether natural (i.e., rocks, sediments, sand) or synthetic (i.e., foam, silicon, glass/silica beads, plastic pall rings), as support for biofilm growth. PBR can be operated in both upflow and downflow mode with option of being continuous or intermittent operation. The packing material is fixed in the system and the fluid flows through the packing void spaces between the biofilm carrier with minimal recirculation flow. Low feed velocities were used to avoid biomass washout and packing material replacement was done by flushing and draining to avoid accumulation of solids/biomass overtime for reactor maintenance.

Application of a full-scale PBR has been employed for the site of San Joaquin Valley, California. The site has reported to produce selenium-laden wastewater through leaching of selenium from the irrigation of the seleniferous soils and is connected to Kesterson National Wildlife Reservoir via the San Luis Drain. The agricultural leachate contained both selenium oxyanions and nitrate at a concentration of < 1 mg L^{-1} and 45–92 mg L^{-1}, respectively. San Luis Drain has since been closed along with Kesterson Reservoir in 1985 due to the issue of selenium toxicity to wildlife with the leachate redirected for disposal to the evaporation ponds or in the San Joaquin River. To circumvent the problem of selenium discharge to the river, Cantafio et al. (1996) carried out the onsite feasibility study by using a selenate-respiring bacterium *Thauera selenatis* inoculated in a full-scale PBR for the removal of selenium oxyanions. *T. selenatis* is a Gram-positive, motile, rod, facultative organism that has a unique metabolism that is perfect for selenium bioremediation with co-contaminants. *T. selenatis* can use either selenate or nitrate as the terminal electron acceptor in different reductase located in the cell under different pH. Selenate can be reduced to selenite and nitrate to nitrite through the selenate reductase (located in the periplasm) and nitrate reductase (located in the cytoplasmic membrane), respectively. With this unique metabolism, *T. selenatis* can reduce both nitrate and selenate simultaneously without outcompeting each other.

Cantafio et al. (1996) constructed four medium-packed PBRs at a shed located in the Panoche Water District of San Joaquin Valley and operated the reactors for 186 days. The shed was maintained at a temperature of 22–28°C and the PBR volume for each reactor was approximately 800 L (3.8 m height by 50.7 cm diameter). PBR1 and PBR2 were filled with 2 inches plastic pall rings or Jaeger tri-packs, while PBR3 and PBR4 were filled with 136 kg of 60-mesh silica sand and at day 108, 90 kg of sand was replaced with Jaeger tri-packs to prevent clogging. The influent flowrate was at 7.6 L min^{-1} and residence time was approximately 8.3 h with recirculation flow of 300 L min^{-1} and 500 L min^{-1} for PBR1-2 and PBR3-4, respectively. As the drainage water lacks COD to support growth, an external acetic acid (varied from 1.0 to 17.0 M) was provided to the system as the sole electron donor with additional nutrient potassium phosphate dibasic (0.02 M) for enhancing the growth of *T. selenatis*. The pH of the reactor was maintained at 7.0 to facilitate optimal growth condition of *T. selenatis*. However, during day 1 to 136, it was observed that both nitrate and selenium oxyanion reduction was poor. Cantafio et al. (1996) then changed the pH of the feed water near the pK$_a$ of acetate, 6.5, and provided an acetate concentration of 5.0 M. After the decrease in pH (day 137 to 186), selenium oxyanion reduction efficiency increased to 98% with nitrate removal efficiency at 98%. The pH increased from 6.5 to 7.5 from PBR1 to PBR4 while the effluent temperature started from 23°C to 26°C. Biofilm growing on the Jaeger tri-pack from PBR2-3 were pinkish visual evidence of elemental selenium, while PBR1 and PB4 showed only white and red-brown to black colouration, respectively. Interestingly, upon checking the numbers of nitrate- and selenate-respiring bacteria per gram of biomass (PBR2 and PBR4 only), it was observed that nitrate-respiring bacteria were 200-fold higher compared to selenate-respiring bacteria. However, the selenate-respiring bacteria were mainly *T. selenatis*, which could be the reason for the effective selenium oxyanion reduction regardless of the lower numbers. Despite the effectiveness of using *T. selenatis* and the PBR4 in reducing selenium oxyanions, Cantafio et al. (1996) noted that addition of coagulant is needed to capture the elemental selenium released with the effluent. Additionally, external addition of acetate is required to facilitate the growth of *T. selenatis* and was estimated to cost $459 per day to treat 740,000 L per day. Furthermore, it was observed that once nitrate was consumed, selenium oxyanion reduction was lower with higher concentration being released into the effluent. This indicated that nitrate was needed to achieve higher efficiency for the biofilm to perform selenium removal.

Viamajala et al. (2006) used Bio-Sep beads as a packing material for a 60.0 cm by 2.5 cm glass PBR column. Bio-Sep beads are 2.5–5.0 mm, porous, adoptive, spherical beads made from 20–50% polyamide and 50–80% encapsulated activated carbon. The PBR was operated at a flowrate of 22.2 mL h^{-1} with a hydraulic retention time of 5.8 h. Viamajala et al. (2006) operated the PBR in 3 phases varying the feeding operation as well as concentration of nitrate (300 to 1,200 mg L^{-1}) and selenite (1 to 6 mg L^{-1}). Acetate was used as the electron donor at a range of 500 to 2,700 mg L^{-1}. The inoculum used for the bench-scale PBR was denitrifying bacteria enriched and taken from a shell oil refinery anaerobic waste treatment sludge in Martinez, California, USA. Results from the experiment conducted by Viamajala et al. (2006) showed that unlike what was observed by Cantafio et al. (1996), nitrate was a limiting factor in selenite reduction. When nitrate was fed continuously, < 30% selenite reduction occurred. When nitrate was fed in pulses, selenite concentration showed predictive reduction—spiked selenite concentration in the effluent which rapidly decreased when nitrate was completely consumed. It was also observed that there was nutrient depletion further along the column, therefore restricting the more metabolic active biofilms near the PBR entrance. The authors concluded that the generation and accumulation of nitrite inhibited selenite reduction suggesting that nitrite reductase by denitrifying biofilms could be responsible for selenite reduction.

Attached Growth Downflow Filter or ABMet® system has been developed and marketed commercially as a downflow PBR reactor using granular activated carbon as the packing material for biofilm growth (Somstegard et al. 2008, Sonstegard and Kennedy 2009, CH2M HILL 2010, Citulski et al. 2016). ABMet® process utilizes mesophilic heterotrophic microorganisms selected based on the site of operation, giving an advantage of also reducing nitrate along with selenium oxyanions. The carbon source for this process is molasses-based nutrient that is both a source of iron and phosphate for the biological system. ABMet® is a 2-stage bioreactor arranged in series and has an HRT of 4–8 hours. The 2-stage bioreactors are identical and allow for alternate operation during degassing and backwashing step. The process has been reported to work at a range of temperature from 7°C (winter time) to 30°C (summer). Excess sludge build-up is taken care of through degassing followed by dewatering and polishing step for the effluent using an aeration tank to treat any unconsumed carbon source before discharge. GE has marketed and demonstrated the full-scale application of ABMet® for selenium removal in mining/agriculture (South Dakota, USA and British Columbia, Canada), coal/power generation (North Carolina, USA) and oil/gas industry (only pilot-scale operated for 6 months in USA, location not mentioned).

Major application for ABMet® biological process was demonstrated in the coal combustion sector. Selenium removal is predominantly needed in the coal/power generation industry compared to other sectors due to the close association of selenium with coal mining operation. Burning of coal produces wastewater high in selenate concentration along with nitrate, sulfate and other heavy metals as the result of the flue-gas desulfurization (FGD) scrubber process. Two power plants in North Carolina, Duke Energy and Progress Energy have utilized ABMet® biological process for the removal of selenium and other contaminants from the FGD wastewater (Somstegard et al. 2008, Sonstegard and Kennedy 2009, CH2M HILL 2010). The ABMet® process was installed in 4 plants with 2-stage bioreactors of multiple parallel bioreactor cell trains with receiving/settling ponds for the effluent and a primary treatment before the bioreactors. ABMet® plant operation demonstrated successful long-term operation for low-level selenium FGD wastewater (maximum influent selenate concentration of 2 mg L^{-1}) with consistent removal performances. A reported total selenium and dissolved selenium removal of 95–99% and 98%, respectively, was recorded for the full-scale plant operated for one year with selenium concentration reaching 10 µg L^{-1} in the effluent. SEM imaging taken from the biofilm of ABMet® showed entrapment of biogenic selenium nanoparticles. The biological system also showed removal of heavy metals such as mercury, arsenic, copper and nickel with additional nitrate removal to < 32 mg L^{-1}. However, it was noted by GE that despite the effectiveness of the ABMet® process, the technology failed to meet the regulatory discharge limit of 5 µg L^{-1}.

An interesting application for PBR is through the inoculation of gel-immobilized bacteria. Tucker et al. (1998) immobilized *Desulfovibrio desulfuricans* to 4-mm gel cubes for the removal of selenium along with chromium, molybdenum and uranium. The immobilized cells were packed into a 100 ml anaerobic vials/PBR and operated at 37°C with 3–4 mL h^{-1} flowrate for 3 days using 350 mg L^{-1} lactate as the electron donor. The authors achieved 86–96% removal efficiency from an initial concentration of 1 mM or 14 mg L^{-1} selenate and other metals. Unfortunately, apart from molybdenum, selenate and the other metals showed a lower rate of reduction in immobilized system compared to suspension, possibly due to the transport limitation. However, diffusion of metal(loid) into the gel was confirmed when red colourization was observed within the immobilized hydro-environment.

3.2.2 *Fluidized bed reactor (FBR)*

Fluidized bed reactor (FBR) has close similarity with both UASB and PBR where both fluidization of the packing material and mixing of the liquid occur in the FBR. The biofilm growth mechanism is similar to FBR, but the physical design and hydro-dynamics/mass balance calculation is closer to a UASB or an expanded granular sludge blanket (EGSB) reactors. FBRs utilized packing materials for biofilm to attach and is expanded upwards or downwards by fluid water through the bed. FBR employs a high upflow velocity compared to UASB or EGSB at 20 m h^{-1} and has 100% bed expansion. The bed expansion and porosity of the packing material in the FBR can be controlled and varied through the flowrate of the fluid or recirculation. Due to the fluidization and expansion, FBR has more surface area to reactor volume available for biomass growth due to the use of finer or smaller biofilm carriers. FBR also has better mass transfer between the biofilm and fluid. FBR can take high organic loading rates and high biomass concentrations compared to PBR.

Lawson and Macy (1995) and Macy et al. (1993) utilized a bench- and pilot-scale, respectively, FBR coupled with SBR and inoculated with *T. selenatis* to remove selenium contamination in agricultural drainage wastewater. The two research studies were the foundation used by Cantafio et al. (1996) for up-scale (10,000 times larger) into a full-scale application with a different reactor configuration as discussed in section 3.2.1. The combination of SBR and FBR has been noted by the authors as the best combination for producing high removal efficiency. Macy et al. (1993) used 1 L volume capacity for both reactors and used sand as the biofilm carrier: 300 g for SBR and 400 g for FBR. Recirculation flowrate for the SBR was set at 260 ml min^{-1} while FBR was at 460 ml min^{-1}. Drainage water used had a concentration of selenium oxyanion of < 1 mg L^{-1} and nitrate of < 75 mg N L^{-1} with pH ranging from 5.9 to 7.9. External addition of 650 mg L^{-1} acetate (electron donor) and 155 mg L^{-1} ammonium chloride (growth nutrient for *T. selenatis*) was mixed with the feed solution prior to entering the FBR. The entire process was operated for 500 days. No information on the inoculum concentration was provided. Additionally, the authors noted that due to operational problems and breakdown (i.e., blocked feed line, broken FBR recirculation tube, etc.), the reactors were reinoculated 12 times over the course of the 500 days operation. The authors varied the flowrate from 6.5 and 13.0 mL min^{-1} but no significant changes in performances were observed for both SBR and FBR. It was noted that most of the reduction occurred in the SBR with 80–97% removal while FBR served as the catchment reduction tank for residual selenium oxyanions lowering the concentration to < 15 µg L^{-1} from 500 µg L^{-1}. As reconfirmed from the study by Cantafio et al. (1996), selenium oxyanion reduction was not hindered by the presence of nitrate due to the unique metabolism of *T. selenatis*. Additionally, compared to other reactor configuration, the residence time used by Macy et al. (1993) was relatively fast from 1.2 to 2.3 h and can be considered an advantage in coupling SBR with FBR. However, it was also noted that the effluent discharged from FBR still contains roughly 5–15 µg L^{-1} selenium concentration and would require a polishing step. No further discussion was mentioned regarding the advantage or novelty of using the SBR and FBR combination particularly for selenium bioremediation.

CH2M HILL (2013) reported several pilot studies using FBR technology for selenium removal from mining wastewaters and biologically treated refinery effluents, though it was not clearly mentioned which company was operating the studies. Most of the installed FBRs were in different parts of USA (i.e., West Virginia) and Canada (i.e., British Columbia) near the impacted lakes/rivers caused by release of selenium from the anthropogenic mining activities. The FBR pilot studies used either GAC or HDPE as carrier media. Interestingly, CH2M HILL (2013) reported that some of the pilot-scale FBR operated as low as 10°C while still showing about 90% removal of selenium as well as other contaminants, such as nitrate. Inoculum used for the FBR studies was heterotrophic organisms that can reduce both nitrate and selenium oxyanion. The organic carbon used was either molasses or MicroCg, an engineered organic carbon. Most of the FBR had a high residence time ranging from 1 to 4 h. Like other FBR studies, it was reported that selenium was effectively reduced to elemental selenium; however, a polishing step is required to capture the bioreduced product not attached to the biofilm and released into the effluent. Additionally, due to the fluctuating concentration entering the pilot-scale FBR, the effluent selenium concentration ranged from 2 to 20 µg L^{-1}.

Cheng et al. (2017) used an inverse FBR to treat synthetic mine wastewater containing high concentration of selenate at 1,400 g L^{-1} and ethanol as the electron donor. The inverse FBR had a 4 L volume capacity and used Floating Extendosphere particles (160 µm size and density of 0.7 g cm^{-3}) as the biofilm carrier. The inoculum used was mentioned to be selenate reducing microorganisms enriched from a freshwater sediment in Western Australia and anaerobic sludge from a municipal wastewater treatment plant. The inverse FBR was operated for 110 days in batch mode and continuous for 190 days with an HRT of 5 d. Cheng et al. (2017) reported a selenate removal efficiency of 94% with a removal rate of 251 mg L^{-1} d^{-1} during the steady-state period of the reactor. Microbial analysis revealed a diverse microbial community with the dominant organisms belonging to *Deltaproteobacteria* and *Gammaproteobacteria* that are known to utilize aliphatic compounds as electron donors. The results obtained were comparable to other studies in terms of removal efficiency; however, the inverse

FBR was able to take higher concentration of selenate concentration compared to the other studies. Unfortunately, the selenium effluent concentration was still high at approximately 76 mg L^{-1} and would require additional processing before being discharged.

3.2.3 Membrane biofilm reactor (MBfR)

Membrane biofilm reactor (MBfR) couples two different elements—biological and physical process—in one unit to accomplish wastewater treatment. Compared to other biofilm reactors, MBfR is considered more compact since it uses membrane separation technology within a bioreactor, thereby removing the need for tertiary process. The basis of MBfR is rooted in the different extraction process, membrane type and pore size used. Major drawbacks in using MBfR are (a) complex operational procedure and multiple system parameters required and (b) the formation of biofouling on the membrane leading to decreased removal performance. Nevertheless, using MBfR has become widespread and cheaper over the course of the years rising to its popularity and installation for wastewater bioremediation.

A series of bench-scale hydrogen-based MBfR (H_2-MBfR) have been performed in a research group in Arizona State University for selenate reduction along with other heavy metals and oxyanions. The H_2-MBfR consist of a glass tube fitted with 38 hollow-fibre membranes that are 25 cm long. Chung et al. (2006) showed that both nitrate- and sulfate-reducing organisms were used and the biofilms grown in the H_2-MBfR were capable of using hydrogen as electron donor for the reduction of selenate. Results showed that a removal efficiency of 94% from an influent concentration of 1 mg Se L^{-1} was obtained. Selenate reduction showed to be affected by the presence of nitrate due to competition in utilizing H_2; however, the process was unaffected by the range of pH from 7.0 to 9.0. Similar observation was noticed by Lai et al. (2014), where it was noted that the presence of nitrate shaped the selenate-reducing community grown in the biofilm of the H_2-MBfR. Opposite to this, presence of sulfate did not affect the selenate reduction process though sulfate reduction was affected by selenate concentration and H_2 pressure indicating interaction of selenate with sulfate reductase.

Chung et al. (2007) and van Ginkel et al. (2011) demonstrated the effectiveness of using H_2-MBfR in treating real agricultural drainage and FGD brine selenium-laden wastewater, respectively. Both wastewaters contain a range of contaminants apart from selenate such as perchlorate, nitrate and sulfate with high solid content. Both studies showed that the H_2-MBfR system was able to reduce multiple contaminants along with selenate simultaneously. An interesting aspect observed by Chung et al. (2007) was that biofilm grown and switched from chromate to selenate and vice versa did not hamper the H_2-MBfR removal performance despite not given an assimilation period which indicates the flexibility of the system. van Ginkel et al. (2011) observed that despite the high concentration of nitrate and sulfate in the wastewater, a maximum selenium removal flux of 362 mg Se m^{-2} d^{-1} was obtained which was higher compared to the previous experiment by Chung et al. (2006, 2007). However, van Ginkel et al. (2011) noted that the high salt and solid contents, such as calcium and magnesium, from the wastewater drastically decreased the membrane performance due to biofouling and is recommended to be precipitated before reduction of the oxyanions.

Apart from H_2, methane was also tested in MBfR as the source of electron donor for selenate reduction coupled with anaerobic oxidation of methane (AOM) organisms as shown by Luo et al. (2018). The MBfR was operated for 453 days with half of the phase fed with only nitrate to develop the biofilm before feeding the reactor with selenate at a concentration of 7 mg L^{-1}. The MBfR set-up used by Luo et al. (2018) consist of 8-bundles of hollow fibre membranes with an internal diameter of 200 μm containing 64 fibre with a length of 300 mm. The study showed that nitrate-reducing AOM organisms were successfully enriched in the biofilm of the MBfR and reduced selenate to elemental selenium with 90% removal rate. However, the selenate removal rate was very low compared to other biological system at the values of 0.4 to 1.8 mg L^{-1} d^{-1}. This is likely related to the slow growth rates of AOM microorganisms. Further research is needed to optimize the MBfR system in using methane as the electron donor which is a cheaper alternative to other sources such as lactate or acetate used for other biofilm reactors.

3.2.4 Bioelectrochemical system (BES)

Compared to the previously discussed biofilm reactors, bioelectrochemical system (BES) is a unique technology that uses microorganisms to reduce selenium oxyanion and other contaminants while simultaneously producing electric current. BES is an interesting platform combing wastewater treatment and resource recovery. Furthermore, BES system can be designed depending on the optimal oxidation and reduction reaction for wastewaters. Basic BES reactors typically consist of three parts—anode, cathode and separator. Wastewater high in COD or the carbon source would be placed in the anode chamber for oxidation process. The process is facilitated by the microorganisms resulting in the generation of electron that will flow to the cathode chamber for direct electricity generation or reduce contaminants such as selenium oxyanion. The BES system has the advantage of generating less sludge compared to other biofilm reactors and has been utilized for low strength wastewater. BES has only seen success in small-scale application. The BES system is not well employed in industry due to the difficulty in technical operation, issues with continual system, low power production and output instability (Koroglu et al. 2019). Currently, few studies have utilized BES system for selenium removal, but its application might rise in the future with the rising need for better reduction and recovery technology.

Catal et al. (2009) investigated the use of a 12 mL air single-chamber microbial fuel cell (MFC) for simultaneous electricity generation and selenite removal in synthetic water using acetate and glucose as the carbon source. The anode used 2 cm^2, type A, non-wet–proofed carbon cloth while the cathode was 7 cm^2 type 2, wet-proofed carbon cloth. The air-facing cathode was coated with carbon and PTFE layers while the liquid-facing cathode was coated with platinum at a concentration of 0.5 mg cm^{-2} cathode area with Nafion as the binding agent. An enriched inoculum taken from domestic wastewater was used. Catal et al. (2009) reported > 70% selenite removal when acetate was provided as electron donor at an influent selenite concentration of 100 mg Se L^{-1} with complete inhibition of electricity generation. In the case of glucose as the carbon source, power output was not affected up to 125 mg Se L^{-1} selenite concentration producing 2,900 mW m^{-2} power density. Interestingly, at 150 mg Se L^{-1} selenite concentration and fed with glucose, coulombic efficiency slightly increased from 25% to 38%. However, Catal et al. (2009) noted that further increase of selenite concentration (> 150 mg Se L^{-1}) caused a decrease in bacterial activities resulting in low power output and coulombic efficiency. The cathode showed bright red deposit indicative of the potential of electrogens to facilitate selenite reduction with power generation.

An interesting application of BES system is the discovery of unique organisms that can grow and reduce selenium oxyanion using conventional and unique substrates. Nguyen et al. (2016) investigated the use of an isolated bacterium THL1 strain that can perform both heterotrophic and electrotrophic selenite reduction. Electrophic organisms are those that can grow electrophically in a BES system using a solid-state electrode as a source of low-cost, unlimited electron donor for reduction processes. The use of such organisms would be useful for subsurface environment or wastewaters that have limited supply of electron donors and require external addition of carbon source, which is typical in most selenium-laden wastewater. The THL1 strain was able to grow under a wide range of carbon source and reduced selenite 80%. Furthermore, using a 300 mL H-type BES system and graphite felt (4 by 6 by 0.5 cm) as the electrode, Nguyen et al. (2016) was able to show that THL1 strain successfully utilized an electrode polarized at −0.3 V as the electron donor. Selenite was reduced 90% at a removal rate of 3.8 mg L^{-1} d^{-1} provided at an initial concentration of 79 mg Se L^{-1}. Though it took about 30 days for the full reduction of selenite, Nguyen et al. (2016) noted that a polarized electrode can supply virtually unlimited electron donor at the condition that the input voltage is maintained. Faster removal rates can be achieved by changing the applied current density and cathode potential.

Finally, Zhang et al. (2018b) reported for the first time the operation of continuous BES system for the reduction of selenate and production of extracellular biogenic elemental selenium. The authors operated the BES system at a constant flow of 0.2 L d^{-1} with a residence time of 1.45 d and a selenate surface loading of 330 mg Se m^{-2} d^{-1}. Acetate at a concentration of 10 mg C L^{-1} was provided and the inoculum used was the biofilm formed on the electrode enriched after 12 days incubation from activated sludge taken at a municipal wastewater treatment plant and landfill leachate. The pre-incubation for biofilm growth was conducted to ensure that only organisms capable of producing extracellular elemental selenium would be cultivated. Zhang et al. (2018b) concluded that successful production of only extracellular elemental selenium nanoparticles during the bioreduction allowing for quicker recovery by physical means. Additionally, the BES reactor can work at a high selenium loading rate while producing a high coulombic efficiency of 73%. Zhang et al. (2018b) noted that regular cleaning of the cation exchange membrane will further increase the efficiency.

3.2.5 *Impact of selenium anti-biofilm activity on biofilm reactor operation and performance*

The study by Tan et al. (2018b) observed a huge difference in reactor performance when utilizing different modes of microbial aggregates, i.e., granular sludge using UASB reactor and biofilm using biotrickling filter (BTF) reactor, and feeding condition, i.e., with selenate alone and selenate with sulfate. Results obtained by Tan et al. (2018b) showed that granular sludge did not show any changes in reduction profile when fed with either selenate alone or selenate with sulfate, showing a gradual increase in total selenium removal efficiencies from 20% (day 10) to 70%% (day 45). Opposite to this, BTF performance was poor under selenate only condition, achieving only 15% total selenium removal efficiency after 45 days of operation. On the other hand, when fed with both selenate and sulfate, BTF immediately showed 50% total selenium removal efficiencies after 10 days of operation and achieved 85% after 45 days. Furthermore, the amount of biofilm attached onto the carrier material in the BTF increased upon addition of selenate and sulfate by > 200%, when compared to the biofilm fed with selenate alone. Tan et al. (2018b) noted that the enhanced selenium removal performance of the BTF when exposed to selenate alone, compared to selenate with sulfate, can be strongly related to two major reasons: (i) alleviation of the anti-biofilm property of elemental selenium through formation of selenium-sulfide nanoparticle and (ii) resulting higher biofilm growth. Figure 7 illustrates the proposed difference in mechanisms of biofilm growth and resistance of biofilm compared to granular sludge.

Considering the anti-biofilm capability of elemental selenium and other organic selenium species, one major key factor for any biofilm reactor operation would be the initial condition for biofilm attachment. Careful process of the primary attachment phase of bacteria to the biofilm carrier

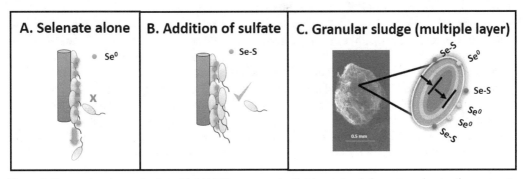

Figure 7. Proposed coping mechanism of (a-b) biofilm compared with (c) granular sludge system when treating wastewater containing either selenate or selenate with sulfate.

Color version at the end of the book

material must be observing to promote bacterial colonization and tenacious biofilm development. Based on the study by Tan et al. (2018b), selenate was fed to the BTF from the beginning of the continuous operation, after a 2-day attachment period conducted in batch operation. From day 1–15, the BTF showed ~ 16 mg Se $L^{-1} d^{-1}$ of total selenium entrapped within the biofilm, which decreased to ~ 3 mg Se $L^{-1} d^{-1}$ from day 16–45. Over time, selenate removal did not increase during selenate only condition and most of the reduced selenate was released in the effluent showing only < 10% total selenium removal efficiency. This could indicate that over time, the already entrapped biogenic selenium affected the biofilm formation and growth in the BTF. Lampis et al. (2017) determined the difference in selenite reduction rates by adding selenite to different growth phases of bacterial culture, i.e., lag (< 6 h), exponential (14 h < t < 20 h), or stationary phase (24 h < t < 48 h). The authors observed higher reduction rates when selenite was added during the stationary phase with corresponding higher elemental selenium formation. Selenate removal efficiencies of higher than 90% (with or without sulfate) have been achieved with a mature biofilm developed prior to selenate addition to a membrane biofilm reactor by Chung et al. (2006) and Ontiveros-Valencia et al. (2016). This indicates that pre-incubation of biofilm carrier and inoculum should be optimized to achieve a mature biofilm prior to exposure in selenium-laden wastewater.

Opposite to this, the removal performance of the UASB reactor was not affected by the feeding condition likely due to the inherent features of the granular sludge that allows it to be more resilient to selenium anti-biofilm activity, e.g., multi-layered structure. Granular sludge consists of quasi-spherical microbial aggregates being formed by an outer and inner zone with spatial organization of multiple microbial community layers (Sekiguchi et al. 1999, Gonzalez-Gil et al. 2016), and thus having a built-in protection layer that allows flexibility higher resistance to stress conditions or toxicant exposure.

Anti-biofilm activity of selenium, in the form of organo-selenium and selenium nanoparticles, has been reported in several literatures and has been applied as a coating agent in numerous fields such as for membrane fabrication, medical devices and other health-related products. Unlike silver nanoparticles, various studies have reported the effectiveness of selenium nanoparticles in preventing the growth of harmful biofilms/pathogens in medical device without causing adverse effect to the living organism (Sakr et al. 2018). Zonaro et al. (2015) showed successful biosynthesis of biogenic selenium nanoparticles *in vitro* and noted that complete inhibition and eradication of an established biofilm can occur at a low selenium concentration of 16–150 mg L^{-1}. Additionally, the authors observed that the smaller the nanoparticle size, the higher the anti-biofilm activity, suggesting that the main mechanism is linked to the high surface to volume ratio of nanoparticles. A photographic evidence for the anti-microbial activity mechanism of biogenic selenium nanoparticles was captured by Guisbiers et al. (2017) through a series of SEM images. The authors showed that biogenic selenium nanoparticles adhere onto the surface of the biofilm, penetrating the bacterial membrane, likely substituting selenium with sulfur, and consequently damaging the cell structure by inducing shrinking and folding of the outer cell membrane of a *Candida albicans* biofilm.

In industry, organo-selenium shows promising application to combat biofouling. Both Low (2010) and Vercellino (2013a, b) evaluated effectiveness of coating organo-selenium on the reverse osmosis (RO) surface membrane to reduce biofouling for better and maximized operation. In both researches, the authors modified the surface of the RO by attaching a selenium polymer as a passive biofilm control where it was theorized that superoxide production will be continuously generated on the surface preventing biofilm attachment and development. Low (2010) measured the efficacy of the selenium-modified RO membrane through two means: biologically through the growth/absence of Gram-positive bacteria *Staphylococcus aureus* and operationally through membrane permeate flow and salt rejection measurement. Low (2010) showed that attachment of monomer seleno-cystamine to the surface of the RO membrane successfully blocked the growth of *S. aureus* while conserving the RO permeate flux performance. Similarly, using model biofilm composed of either *Escherichia coli* or *S. aureus*, Vercellino (2013a, b) observed the effectiveness of coating both RO membrane and spacer with organo-selenium to prevent biofouling resulting in high removal rates and reduced operation costs. The authors observed that when the RO membranes and the feed spacer was coated by the organo-selenium complexes, a 5 log reduction in the biofilm formation and thickness was observed. Moreover, the flux loss decreased from 55 to 15% due to the biofouling attenuation.

Depending on the target application, the special anti-biofilm property of selenium can be either useful or detrimental for reactor operation. Using biofilm reactors that relies heavily on biofilm attachment for selenate/selenite reduction would require careful start-up operation to prevent any irreversible damage on the inoculum and subsequent biofilm growth. However, the anti-biofilm activity of selenium can also be very beneficial in membrane biofilm-based reactors. Despite the advances in biological system, it was still quite difficult to reach the low regulatory discharge selenium limit of 5 μg/L. Out of all the available technology, membrane system, particularly RO, has been reported and proven to be effective in removing selenium to < 5 μg L^{-1} at pilot- and full-scale operation (CH2M HILL 2010). However, RO has a disadvantage of high operational cost due to the frequent formation of biofouling and chemical cleaning. As such, the generation of selenium-based product to prevent growth of unwanted biofilm can be very useful in the membrane operation, particularly in dealing with membrane biofouling. Reduction of biofouling prolongs the usage of membrane filters, and lessens the operation cost and chemical usage. With further development, coupling reduction of selenate/selenite to generate biogenic elemental selenium or organo-selenium while simultaneously preventing membrane biofouling will improve the application of membrane-based technology for selenium bioremediation.

4. Selenium nanoparticles, recovery and future direction

The formation of amorphous red elemental nanoparticles has been found as the stable end-product following microbial reduction of Se oxyanion (Nancharaiah and Lens 2015). The association of bacterial proteins to selenium nanoparticles controls the size distribution and the morphology of the resultant selenium nanoparticles which is important in governing the fate of selenium nanoparticles in the environment (Buchs et al. 2013, Jain et al. 2015). Due to its colloidal properties, microbiologically produced selenium nanoparticles are always present in the effluent of bioreactors, which makes it difficult to achieve complete selenium removal, the ultimate objective of bioremediation or wastewater treatment. Hence, removal and recovery of colloidal selenium nanoparticles attached to the sludge and/or from the bioreactor effluents are necessary to meet the discharge limit and to reduce the environmental pollution and also for possible reusing of selenium nanoparticles (Figure 4).

Biogenic elemental selenium has a negative surface due to the presence of the biopolymer layer of extracellular polymeric substances (EPS) of granular sludge or biofilm that gives it a colloidal property (Buchs et al. 2013, Jain et al. 2015). Circumneutral pH, commonly applied in bioremediation, is not appropriate for gravitational separation due to its negative ζ-potential (in the range of –20 to –30 mV) as it prevents the agglomeration of the nanoparticles. Buchs et al. (2013) demonstrated that controlling the electrophoretic properties of biogenic elemental selenium suspensions by adjusting the pH and/or addition of counter cations (e.g., Ca^{2+} and Na$^+$) is one of the easiest way to enhance settleability. Slight change in chemical composition (i.e., pH and/or cation concentration) is the key for implementing Se bioremediation, both from the engineering viewpoint (i.e., achieving sufficient removal and recovery efficiency) and the economic viewpoint (i.e., achieving cost effective removal by small reactor footprint, gravitational settling) (Buchs et al. 2013). Physical methods including filtration or centrifugation with more advanced technique using electrocoagulation were employed to recover the biogenic elemental selenium from the effluent. Cantafio et al. (1996) simply added a polymer coagulant to the effluent after the 4-stage PBR biological treatment of agricultural drainage water containing < 1 mg L^{-1} selenium oxyanion and proved to be highly successful in recovery of the elemental selenium that is not attached to the biofilm. About 97.9% recovery of elemental selenium was achieved when the effluent water was treated with 100 mg of Nalmet 8072 per litre at a pH of 6.0. On the other hand, Staicu et al. (2015) utilized iron sacrificial electrodes in batch reactors under galvanostatic conditions. The use of iron electrode at a current density of 300 mA allowed for binding over 96% of the biogenic elemental selenium nanoparticle release from effluent to form stable iron-selenium complex sediments with a reticular structure.

Cordoba and Staicu (2018) noted that current price of pure selenium powder sells at approximately 81 euros for 5 g as marketed by Sigma-Aldrich. In wastewater such as FGD, selenium concentration reaches about 1 mg Se L^{-1}. Assuming complete reduction of selenium oxyanion to biogenic selenium nanoparticles, Cordoba and Staicu (2018) calculated that 5 m^3 of selenium-laden wastewater would generate 5 g of Se. Of course, in a non-ideal condition, more wastewater would be required to generate the same amount of selenium concentration. Additionally, this is only assuming that complete recovery can occur in the biological system. However, there is a lack of bioreactor system that incorporates recovery within its technology. Current practice in place for selenium bioremediation sees the disposal of selenium filled sludge instead of recovery the nanoparticles. As such, research practice should be focused on integrating the biogenic selenium nanoparticles embedded in the sludge slurry for resource recovery.

A recent publication by Zhang et al. (2018c) investigated a novel bioreactor configuration for the simultaneous reduction of selenate and recovery of biogenic elemental selenium nanoparticles. Zhang et al. (2018c) constructed a novel set-up consisting of three treatment units: (a) PBR for converting selenate to elemental selenium, (b) novel bacterium-nanoparticle separator using a titled polyethylene sheet and (c) UF module using tangential flow or TFU. Influent initially enters the PBR in a downward flow into the bacterium-nanoparticle separator where biomass and selenium nanoparticles are separated due to the higher surface free energy of the nanoparticles allowing it to attach to the polyethylene sheet. The separated biomass will then roll to the bottom of the separator tank. Water containing colloidal particles within the separator tank was pumped into the TFU unit. The TFU has been utilized for separating silver and nickel nanoparticles but have not been utilized in selenium recovery. The TFU unit allows for the separation of nanoparticles from bacterial through size difference of < 500 nm compared to > 500 nm, respectively. Apart from separating the nanoparticle from the biomass, the TFU unit serves to concentrate the selenate-reducing organisms leading to higher removal rates through recirculation of the rejected water back into the PBR. Zhang et al. (2018c) reported a good reduction of selenate with the permeate from the TFU attaining < 20 μg Se L^{-1} from a maximum initial concentration of 80 mg Se L^{-1}. The authors also reported that no nanoparticles and bacterium were detected in the effluent indicating the successful separation. Zhang et al. (2018c) also demonstrated that the use of polyethylene sheet with the appropriate tilting angle was able to successfully separate selenium nanoparticles from bacterial aggregates allowing for the retrieval of the nanoparticles without using chemical extraction. Readers are referred to the paper for full explanation on the separation mechanism proposed by Zhang et al. (2018c). Overall, the novel bioremediation set-up proposed from this study showed promising application for closing the loop from selenate reduction to recovery of biogenic selenium nanoparticles without compromising the biological reactor and usage of chemical methods.

The main challenges in microbial production and recovery of selenium nanoparticles from waste effluent are: (1) heterogeneity (particularly, size heterogeneity) of biogenic nanoparticles and (2) relatively large duration of synthesis (up to several days) (Tugarova and Kamnev 2017). But various advantages of biogenic selenium nanoparticles synthesis revealed recently also include: unique spectroscopic (optical) characteristics, their functionalisation and biological compatibility (Oremland et al. 2004, Tugarova and Kamnev 2017), which are substantially different from those of their chemically synthesized counterparts. Recently, Mal et al. (2017b) reported that biogenic elemental selenium formed by the anaerobic granular sludge in the bioreactors is comparatively much less toxic to aquatic organisms than selenite and chemically synthesized selenium nanoparticles. It appears that protein plays a significant role during the assembly of nanospheres mediated by microorganisms and influence the growth, crystallization and stabilization of selenium nanoparticles (Jain et al. 2015, Tugarova and Kamnev 2017). In case of bioremediation of Se-laden industrial effluents, the main waste is biogenic selenium nanoparticles that are currently being disposed of after exiting the bioreactor system. Hence, extracting a valuable resource like selenium nanoparticles from a wastewater could be considered a bonus that can help offset the treatment costs in addition to providing a service to the environment (Cordoba and Staicu 2018). However, despite the advantages of both the environmental which receives a cleaner effluent, and also the economical, by coupling the water treatment with resource recovery of biogenic selenium nanoparticles, there is a need to understand the mechanisms governing the formation and growth of nanoparticles during microbial respiration of selenium oxyanion (Nancharaiah and Lens 2015, Tugarova and Kamnev 2017). Hopefully in time, more research on molecular mechanisms together with genetic and proteomic approaches involved in the mechanisms governing the formation and growth of selenium nanoparticles during microbial respiration will find its place in the field of "green chemistry" (Cordoba and Staicu 2018, Nancharaiah and Lens 2015).

5. Conclusions

Recent reports have cited the toxic and harmful effect of selenium contamination in many water reservoirs, lakes and rivers as the result of mining and power generating industry leading to the demands for proper removal of selenium before being discharged. Due to the bioaccumulation property of selenium, a strict and low discharge limit has been set at 5 μg Se L^{-1}. Among the technologies available, biological approach gives an advantage compared to physical and chemical treatment due to the simultaneous reduction of selenium oxyanions and generation of biogenic elemental selenium. Use of bioreactors inoculated with biofilms and granular sludge allows the biogenic elemental selenium to be embedded intracellularly or extracellularly in a sludge matrix for easy recovery and further reuse. As such, the embedded biogenic elemental selenium can be harvested from the effluent and the sludge for resource recovery, potentially offsetting costly operation required in a biological system.

Among the bioreactors available, UASB and PBR have shown to be the best configuration for selenium remediation. Various full-scale using UASB and PBR have been demonstrated to be effective in treating selenium-laden wastewater taken from sites related to mining and agricultural activities. PBR-based technology, ABMet®, have been marketed and sold commercially by GE and deployed in various contaminated sites in USA and Canada with indication of successfully selenium oxyanion reduction. It should be noted that, due to the anti-biofilm property of selenium, attached biofilm-based reactors should have mature and stable biofilm matrix before feeding selenium-laden wastewater to avoid reactor failure. To date, there are no reported commercially sold UASB based technology being employed in the market for selenium bioremediation. Other reactor configurations such as SBR, MBfR and BES have only been reported in bench-scale studies. These types of configurations have shown to be useful in utilizing cheaper carbon source alternative compared to UASB and PBR and also under extreme conditions like high salinity. However, further research is required to properly develop all these reactors technology in large scale continuous operation.

Unfortunately, despite the advances seen in biological treatment, there is still a need for further optimization of bioreactor operations, since most studies reported were unable to reach the regulatory discharge limit. As such, most studies have indicated that a polishing step is required to further bring down the selenium concentration to 5 μg Se L^{-1} after the biological process. This could be achieved through coagulation process or membrane filtration. An interesting aspect to develop for the polishing step would be to utilize the anti-biofilm activity of selenium species by constructing a polymerized selenium modified membrane. Use of the modified membrane has the potential to have a higher resistance to biofouling allowing for better effluent quality and longer performance operation. Finally, future research should focus on incorporating a recovery process with the biological system to harvest the generated biogenic elemental selenium without disrupting the process or removing the enriched biomass from the reactor and biological treatment of Se-laden wastewater could meet the expectations of sustainability, operating on a more efficient and environment-friendly basis.

References

Anderson, C.S. 2016. Selenium and tellurium. USGS 2016 Mineral Yearbook. Access on January 21, 2018. URL: https://minerals.usgs.gov/minerals/pubs/commodity/selenium/myb1-2016-selen.pdf.

Basaglia, M., A. Toffanin, E. Baldan, M. Bottegal, J.P. Shapleigh and S. Casella. 2007. Selenite-reducing capacity of the copper-containing nitrite reductase of Rhizobium sullae. FEMS Microbiology Letter 269: 124–130.

Bebien, M., J.P. Chauvin, J.M. Adriano, S. Grosse and A. Vermeglio. 2001. Effect of selenite on growth and protein synthesis in the phototrophic bacterium Rhodobacter sphaeroides. Applied and Environmental Microbiology 67: 4440–4447.

Bhunia, P. and M.M. Ghangrekar. 2007. Required minimum granule size in UASB reactor and characteristics variation with size. Bioresource Technology 98: 994–999.

Brandt, J.E., E.S. Bernhardt, G.S. Dwyer and R.T. Di Guilio. 2017 Selenium ecotoxicology in freshwater lakes receiving coal combustion residual effluents: A North Carolina example. Environmental and Science Technology 5: 2418–2426.

Brown, T.A. and A. Shrift. 1980. Assimilation of selenate and selenite by Salmonella thyphimurium. Canadian Journal of Microbiology 26: 671–675.

Bryant, R.D. and E.J. Laishley. 1989. Evidence for proton motive force dependent transport of selenite by Clostridium paesteurianum. Canadian Journal of Microbiology 35: 481–486.

Buchs, B., M.W. Evangelou, L.H. Winkel and M. Lenz. 2013. Colloidal properties of nanoparticular biogenic selenium govern environmental fate and bioremediation effectiveness. Environmental Science and Technology 47(5): 2401–2407.

Cantafio, A.W., K.D. Hagen, G.E. Lewis, T.L. Bledsoe, K.M. Nunan and J.M. Macy. 1996. Pilot-scale selenium bioremediation of San Joaquin drainage water with *Thauera selenatis*. Applied and Environmental Microbiology 62(9): 3298–3303.

Catal, T., H. Bermek and H. Liu. 2009. Removal of selenite from wastewater using microbial cells. Biotechnology Letters 31: 1211–1216.

Chan, Y.J., M.F. Chong, C.L. Law and D.G. Hassell. 2009. A review on anaerobic–aerobic treatment of industrial and municipal wastewater. Chemical Engineering Journal 155: 1–18.

CH2M HILL. 2010. Review of Available Technologies for the Removal of Selenium from Water—Final Report Prepared for North American Metal Council.

CH2M HILL. 2013. Technical addendum: Review of available technologies for the removal of selenium from water. NAMC White Paper Report Addendum.

Chung, J., R. Nerenberg and B.E. Rittmann. 2006. Bioreduction of selenate using a hydrogen-based membrane biofilm reactor. Environmental Science and Technology 40: 1664–1671.

Chung, J., B.E. Rittmann, W.F. Wright and R.H. Bowman. 2007. Simultaneously bio-reduction of nitrate, perchlorate, selenate, chromate, arsenate, and dibromochloropropane using a hydrogen-based membrane biofilm reactor. Biodegradation 18: 199–209.

Chung, K.Y., M.P. Ginige and A.h. Koksonen. 2017. Microbially catalysed selenate removal in an inverse fluidised bed reactor. Solid State Phenomena 262: 677–681.

Citulski, J., G. Rajeev and S. Snowling. 2016. Optimization of ABMet biological selenium removal through advanced process modelling. Proceedings of the Water Environment Federation, WEFTEC 2016: Session 300 through session 309, USA 13: 949–961.

Cordoba, P. and L.C. Staicu. 2018. Flue gas desulfurization: An unexploited selenium resource. Fuel 223: 268–276.

DeMoll-Decker, H. and J.M. Macy. 1993. The periplasmic nitrite reductase of *Thauera selenatis* may catalyze the reduction of selenite to elemental selenium. Archive of Microbiology 160: 241–247.

Dessì, P., R. Jain, S. Singh, M. Seder-Colomina, E.D. van Hullebusch, E.R. Rene, S.Z. Ahammad, A. Carucci and P.N.L. Lens. 2016. Effect of temperature on selenium removal from wastewater by UASB reactors. Water Research 94: 146–154.

Dinh, Q.T., Z. Cui, J. Huang, T.A.T. Tran, W. Yang, F. Zhou, M. Wang, D. Yu and D. Liang. 2018. Selenium distribution in the Chinese environment and its relationship with human health: A review. Environment International 112: 294–309.

Golder Associates Inc. 2009. Literature review of treatment technologies to remove selenium from mining influenced water. Report to Teck Coal Limited, Calgary (08-1421-0034 Rev.2, AB).

Gonzalez-Gil, G., P.N.L. Lens and P.E. Saikaly. 2016. Selenite reduction by anaerobic microbial aggregates: Microbial community structure, and proteins associated to the produced selenium spheres. Frontier Microbiology 7: 571.

Guisbiers, G., H.H. Lara, R. Mendoza-Cruz, G. Naranjo, B.A. Vincent, X.G. Peralta and K.L. Nash. 2017. Inhibition of *Candida albicans* biofilm by pure selenium nanoparticles synthesized by pulsed laser ablation in liquids. Nanomedicine: Nanotechnology, Biology, and Medicine 13: 1095–1103.

Harrison, G., C. Curie and E.J. Laishley. 1984. Purification and characterization of an inducible dissimilatory type sulphite reductase from Clostridium pasteurianum. Archive of Microbiology 138: 72–78.

Hockin, S.L. and G.M. Gadd. 2003. Linked redox precipitation of sulfur and selenium under anaerobic conditions by sulfate-reducing bacterial biofilms. Applied and Environmental Microbiology 69: 7063–7072.

Hoffman, D.J. 2002. Role of selenium toxicity and oxidative stress in aquatic birds. Aquatic Toxicology 57: 11–26.

Jain, R., N. Jordan, S. Weiss, H. Foerstendorf, K. Heim, R. Kacker, R. Hübner, H. Kramer, E.D. van Hullebusch and P.N.L. Lens. 2015. Extracellular polymeric substances govern the surface charge of biogenic elemental selenium nanoparticles. Environmental Science and Technology 49(3): 1713–1720.

Jain, R., S. Matassa, S. Singh, E.D. van Hullebusch, G. Esposito and P.N.L. Lens. 2016. Reduction of selenite to elemental selenium nanoparticles by activated sludge. Environmental Science and Pollution Research International 23(2): 1193–1202.

Kessi, J. and K.W. Hanselmann. 2004. Similarities between the abiotic reduction of selenite with glutathione and the dissimilatory reaction mediated by Rhodospirillum rubrum and *Escherichia coli*. Journal of Biological Chemistry 279(49): 50662–50669.

Klonowska, A., T. Heulin and A. Vermeglio. 2005. Selenite and tellurite reduction by Shewanella oneidensis. Applied and Environmental Microbiology 71: 5607–5609.

Koroglu, E.O., H.C. Yoruklu, A. Demir and B. Ozkaya. 2019. Scale-up and commercialization issues of the MCFs: Challenges and implications. pp. 565–583. *In*: S.M. Mohan, A. Pandey and S. Varjani (eds.). Microbial Electrochemical Technology Sustainable Platform for Fuels, Chemicals and Remediation Biomass, Biofuels and Biochemicals. Elsevier.

Labunskyy, V.M., D.L. Hatfield and V.N. Gladyshev. 2014. Selenoproteins: molecular pathways and physiological roles. Physiological Reviews 94: 739–777.

Lai, C.-Y., X. Yang, Y. Tang, B.E. Rittmann and H.-P. Zhao. 2014. Nitrate shaped the selenate-reducing microbial community in a hydrogen-based biofilm reactor. Environmental Science and Technology 48: 3395–3402.

Lampis, S., E. Zonaro, C. Bertolini, D. Cecconi, F. Monti, M. Micaroni, R.J. Turner, C.S. Butler and G. Vallini. 2017. Selenite biotransformation and detoxification by *Stenotrophomonas maltophilia* SeITE02: Novel clues on the route to bacterial biogenesis of selenium nanoparticles. Journal of Hazardous Materials 324: 3–14.

Lawson, S. and J.M. Macy. 1995. Bioremediation of selenite in oil refinery wastewater. Applied Microbiology and Biotechnology 43: 762–765.

Lemly, A.D. 2014a. An urgent need for an EPA standard for disposal of coal ash. Environmental Pollution 191: 253–255.

Lemly, A.D. 2014b. Teratogenic effects and monetary cost of selenium poisoning of fish in Lake Sutton, North Carolina. Ecotoxicology and Environmental Safety 104: 160–167.

Lenz, M., E.D. van Hullebusch, G. Hommes, P.F. Corvine and P.N.L. Lens. 2008a. Selenate removal in methanogeic and sulfate-reducing upflow anaerobic sludge bed reactors. Water Research 42: 2184–2194.

Lenz, M., N. Janzen and P.N.L. Lens. 2008b. Selenium oxyanion inhibition of hydrogenotrophic and acetoclastic methanogenesis. Chemosphere 73: 383–388.

Lenz, M., M. Smit, P. Binder, A.C. van Aelst and P.N.L. Lens. 2008c. Biological alkylation and colloid formation of selenium in methanogenic UASB reactor. Journal of Environmental Quality 37: 1691–1700.

Lenz, M., E.D. van Hullebusch, F. Farges, S. Nikitenko, C. Borca, B. Grolimund and P.N.L. Lens. 2008d. Selenium speciation assessed by X-Ray absorption spectroscopy of sequentially extracted anaerobic biofilms. Environmental Science and Technology 42(20): 7587–7593.

Lenz, M., E.D. van Hullebusch, F. Farges, S. Nikitenko, P.F. Corvini and P.N.L. Lens. 2011. Combined speciation analysis by X-ray absorption near-edge structure spectroscopy, ion chromatography, and solid-phase microextraction gas chromatography-mass spectrometry to evaluate biotreatment of concentrated selenium wastewaters. Environmental Science and Technology 45(3): 1067–1073.

Losi, M.E. and W.T. Frankenberger. 1997. Reduction of selenium oxyanions by Enterobacter cloacae SLD1a-1: isolation and growth of the bacterium and its expulsion of selenium nanoparticles. Applied and Environmental Microbiology 63: 3079–3084.

Low, D.D. 2010. Evaluation and inhibition of biofilm attachment to membrane surfaces in water and wastewater treatment systems. Ph.D. Thesis, Texas Tech University, Texas, USA.

Luo, J.-H., H. Chen, S. Hu, C. Cai, Z. Yuan and J. Guo. 2018. Microbial selenate reduction driven by a denitrifying anaerobic methane oxidation biofilm. Environmental Science and Technology 52: 4006–4012.

Macy, J.M., S. Lawson and H. DeMoll-Decker. 1993. Bioremediation of selenium oxyanions in Sa Joaquin drainage water using *Thauera selenatis* in a biological reactor system. Applied Microbiology and Biotechnology 40: 588–594.

Mal, J., Y.V. Nancharaiah, E.D. van Hullebusch and P.N.L. Lens. 2016a. Effect of heavy metal co-contaminants on selenite bioreduction by anaerobic granular sludge. Bioresource Technology 206: 1–8.

Mal, J., Y.V. Nancharaiah, E.D. van Hullebusch and P.N.L. Lens. 2016b. Metal Chalcogenide quantum dots: biotechnological synthesis and applications. RSC Adv. 6: 41477–41495.

Mal, J., Y.V. Nancharaiah, E.D. van Hullebusch and P.N.L. Lens. 2017a. Biological removal of selenate and ammonium by activated sludge in a sequencing batch reactor. Bioresource Technology 229: 11–19.

Mal, J., W. Veneman, Y.V. Nancharaiah, E.D. van Hullebusch, W. Peijnenburg, M.G. Vijver and P.N.L. Lens. 2017b. A comparison of fate and toxicity of selenite, biogenically and chemically synthesized selenium nanoparticles to the Zebrafish (*Danio rerio*) embryogenesis. Nanotoxicology 11(1): 87–97.

Mirjafari, P. 2014. Complex biochemical reactors for selenium and sulphate reduction: Organic material biodegradation and microbial community shifts. Ph.D. Thesis, The University of British Columbia, British Columbia, Canada.

Nancharaiah, Y.V. and P.N.L. Lens. 2015. Ecology and biotechnology of selenium-respiring bacteria. Microbiology and Molecular Biology Reviews 79: 61–80.

Nancharaiah, Y.V., M. Sarvajith and P.N.L. Lens. 2018. Selenite reduction and ammoniacal nitrogen removal in an aerobic granular sludge sequencing batch reactor. Water Research 131: 131–141.

Nguyen, V.K., Y. Park, J. Yu and T. Lee. 2016. Microbial selenite reduction with organic carbon and electrode as sole electron donor by a bacterium isolated from domestic wastewater. Bioresource Technology 212: 182–189.

Nnaji, C.C. 2014. A review of the upflow anaerobic sludge blanket reactor. Desalination and Water Treatment 52(22-24): 4122–4143.

Ontiveros-Valencia, A., C.R. Penton, R. Krajmalnik-Brown and B.E. Rittmann. 2016. Hydrogen-fed biofilm reactors reducing selenate and sulfate: Community structure and capture of elemental selenium within the biofilm. Biotechnology Bioengineering 113: 1736–1744.

Oremland, R.S., M.J. Herbel, J.S. Blum, S. Langley, T.J. Beveridge, P.M. Ajayan, T. Sutto, A.V. Ellis and S. Curran. 2004. Structural and spectral features of selenium nanospheres produced by Se-respiring bacteria. Applied and Environmental Microbiology 70: 52–60.

Pearce, C.I., R.A.D. Pattrick, N. Law, J.M. Charnock, V.S. Coker, J.W. Fellowes, R.S. Oremland and J.R. Lloyd. 2009. Investigating different mechanisms for biogenic selenite transformations: Geobacter sulfurreducens, Shewanella oneidensis and Veillonella atypica. Environmental Technology 30: 1313–1326.

Piacenza, E., A. Presentato, E. Zonaro, S. Lampis, G. Vallini and R.J. Turner. 2017. Microbial-Based Bioremediation of Selenium and Tellurium Compounds, J. Derco and B. Vrana (eds.). Biosorption. IntechOpen.

Rege, M.A., D.R. Yonge, D.P. Mendoza, J.N. Petersen, Y. Bereded-Samuel, D.L. Johnstone, W. Apel and J.M. Barnes. 1999. Selenium reduction by a denitrifying consortium. Biotechnology and Bioengineering 62(4): 479–484.

Ridley, H., C.A. Watts, D. Richardson and C.S. Butler. 2006. Resolution of distinct membrane-bound enzymes from Enterobacter cloacae SLD1a-1 that are responsible for selective reduction of nitrate and selenate oxyanions. Applied and Environmental Microbiology 72: 5173–5180.

Sakr, T.M., M. Korany and K.V. Katti. 2018. Selenium nanomaterials in biomedicine—An overview of new opportunities in nanomedicine of selenium. Journal of Drug Delivery Science and Technology 46: 223–233.

Santos, S., G. Ungureanu, R. Bpaventura and C. Botelho. 2015. Selenium contaminated waters: An overview of analytical methods, treatment options and recent advances in sorption methods. Science of the Total Environment 521-522: 246–260.

Seghezzo, L., G. Zeeman, J.B. van Lier, H.V.M. Hamelers and G. Lettinga. 1998. A review: the anaerobic treatment of sewage in UASB and EGSB reactors. Bioresource Technology 65(3): 175–190.

Sekiguchi, Y., Y. Kamagata, K. Nakamura, A. Ohashi and H. Harada. 1999. Fluorescence *in situ* hybridization using 16S rRNA-targeted oligonucleotides reveals localization of methanogens and selected uncultured bacteria in mesophilic and thermophilic sludge granules. Applied and Environmental Microbiology 65: 1280–1288.

Soda, S., M. Kashiwa, T. Kagami, M. Kuroda, M. Yamashita and M. Ike. 2011. Laboratory scale bioreactors for soluble selenium removal from selenium refinery wastewater using anaerobic sludge. Desalination 279(1-3): 433–438.

Soda, S., A. Hasegawa, M. Kuroda, A. Hanada, M. Yamashita and M. Ike. 2015. Selenium recovery from kiln powder of cement manufacturing by chemical leaching and bioreduction. Water Science & Technology 72: 1294–1300.

Sonstegard, J., T. Pickett, J. Harwood and D. Johnson. 2008. Full scale operation of biological technology for the removal of selenium from FGD wastewaters. Proceeding for the Annual International Water Conference. USA 69: IWC 08–31.

Sonstegard, J. and W. Kennedy. 2009. ABMet: Setting the standard for selenium removal. Proceedings of the Water Environment Federation, Microconstituents and Industrial Water Quality USA 1: 485–485 IWC 10–18.

Springer, S.E. and R.E. Huber. 1973. Sulfate and selenate uptake and transport in wild and in two selenate-tolerant strains of *Escherichia coli* K12. Archives of Biochemistry and Biophysics 156: 595–603.

Staicu, L.C., E.D. van Hullebusch, P.N.L. Lens, E.A.H. Pilon-Smits and Mehmet A. Oturan. 2015. Electrocoagulation of colloidal biogenic selenium. Environmental Science and Pollution Research 22: 3127–3137.

Tan, L.C., Y.V. Nancharaiah, E.D. van Hullebusch and P.N.L. Lens. 2016. Selenium: Environmental significance, pollution, and biological treatment technologies. Biotechnology Advances 34: 886–907.

Tan, L.C., E.J. Espinosa-Ortiz, Y.V. Nancharaiah, E.D. van Hullebusch, R. Gerlach and P.N.L. Lens. 2018a. In Selenate removal in biofilm systems: Effect of nitrate and sulfate on selenium removal efficiency, biofilm structure and microbial community. Journal of Chemical Technology and Biotechnology 93: 2380–2389.

Tan, L.C., S. Papirio, V. Luongo, Y.V. Nancharaiah, P. Cennamo, G. Esposito, E.D. van Hullebusch and P.N.L. Lens. 2018b. Comparative performance of anaerobic attached biofilm and granular sludge reactors for the treatment of model mine drainage wastewater containing selenate, sulfate and nickel. Chemical Engineering Journal 345: 545–555.

The Canadian Press. 2018. Americans blame Canadians for delaying damning report on B.C. toxins in transboundary waters. CBC. Access on January 21, 2019. URL: https://www.cbc.ca/news/canada/british-columbia/koocanusa-reservoir-american-canadian-dispute-1.4738423.

Tucker, M.D., LL. Barton and B.M. Thomson. 1998. Reduction of Cr, Mo, Se and U by *Desulfovibrio desulfuricans* immobilized in polyacrylamide gels. Journal of Industrial Microbiology and Biotechnology 20: 13–19.

Tugarova, A.V. and A.A. Kamnev. 2017. Proteins in microbial synthesis of selenium nanoparticles. Talanta 174: 539–547.

Turner, R.J., J.H. Weiner and D.E. Taylor. 1998. Selenium metabolism in *Escherichia coli*. Biometals 11: 223–237.

Ullah, H., G. Liu, B. Yousaf, M.U. Alim S. Irshad, Q. Abbas and R. Ahmad. 2018. A comprehensive review on environmental transformation of selenium: Recent advances and research perspective. Environmental Geochemistry and Health. doi: 10.1007/s10653-018-0195-8.

van Ginkel, S.W., Z. Yang, B. Kim, M. Sholin and B.E. Rittmann. 2011. Effect of pH on nitrate and selenate reduction in flue gas desulfurization brine using H_2-based membrane biofilm reactor (MBfR). Water Science and Technology 63(12): 2923–2928.

van Hullebusch, E.D. 2017. Bioremediation of Selenium Contaminated Wastewater. Springer International Publishing.

Vercellino, T., A. Morse, P. Tran, L. Song, A. Hamood, T. Reid and T. Moseley. 2013a. Attachment of organo-selenium to polyamide composite reverse osmosis membranes to inhibit biofilm formation of *S. aureus* and *E. coli*. Desalination 309: 291–295.

Vercellino, T., A. Morse, P. Tran, A. Hamood, T. Reid, L. Song and T. Moseley. 2013b. The use of covalently attached organo-selenium to inhibit *S. aureus* and *E. coli* biofilms on RO membranes and feed spacers. Desalination 317: 142–151.

Viamajala, S., Y. Bereded-Samuel, W.A. Apel and J.N. Petersen. 2006. Selenite reduction by a denitrifying culture: Batch- and packed-bed reactor studies. Applied Microbiology and Biotechnology 71: 953–962.

Weiser, M. 2018. Pressure mounts to solve California's toxic farmland drainage problem. NewsDeeply: Water Deeply. Access on January 08, 2019. URL: https://www.newsdeeply.com/water/community/2018/05/01/to-manage-californias-groundwater-think-more-about-surface-water.

Zannoni, D., F. Borsetti, J.J. Harrison and R.J. Turner. 2008. The bacterial response to the chalcogen metalloids Se and Te. Advances in Microbial Physiology 53: 1–72.

Zeng, T., E.R. Rene, Q. Hub and P.N.N. Lens. 2019. Continuous biological removal of selenate in the presence of cadmium and zinc in UASB reactors at psychrophilic and mesophilic conditions. Biochemical Engineering Journal 141: 102–111.

Zhang, Y., M. Kuroda, Y. Nakatani, S. Soda and M. Ike. 2018a. Removal of selenite from artificial wastewater with high salinity by activated sludge in aerobic sequencing batch reactors. Journal of Bioscience and Bioengineering 127(5):618–624.

Zhang, Z., G. Chen and Y. Tang. 2018b. Towards selenium recovery: Biocathode induced selenate reduction to extracellular elemental selenium nanoparticles. Chemical Engineering Journal 351: 1095–1103.

Zhang, Z., I. Adedejji, G. Chen and Y. Tang. 2018c. Chemical-free recovery of elemental selenium from selenate-contaminated water by a system combining a biological reactor, a bacterium-nanoparticle separator, and a tangential flow filter. Environmental Science and Technology 52: 13231–13238.

Zonaro, E., S. Lampis, R.J. Turner, S.J.S. Qazi and G. Vallini. 2015. Biogenic selenium and tellurium nanoparticles synthesized by environmental microbial isolates efficaciously inhibit bacterial planktonic cultures and biofilms. Frontiers in Microbiology 6: 584.

CHAPTER 7

Role of Extracellular Polymeric Substances (EPS) in Cell Surface Hydrophobicity

Feishu Cao,[1,2,3] *Gilles Guibaud,*[2] *Isabelle Bourven,*[2]
Yoan Pechaud,[1] *Piet N.L. Lens*[3] and *Eric D. van Hullebusch*[3,4,]*

1. Introduction

Industrial activities, such as the manufacturing of textile, paper and pulp, iron-steel, petroleum, pesticide and pharmaceutics, generate large amounts of wastewater containing various hazardous heavy metals or organic micro-pollutants. Due to the non-biodegradable characteristics of the heavy metals as well as the potential mutagenicity and carcinogenicity of some organic micro-pollutants, biofilms or sludge are usually used in the wastewater treatment system to sorb and remove those pollutants from the wastewater (Wicke et al. 2008, Aksu 2005, Wang et al. 2002).

Microbial aggregation is one fundamental feature involved in the biofilm formation, granulation and sludge flocculation (Dufrêne 2015, Adav and Lee 2008). It can be regarded as a process starting with the individual cells adhering to each other via cell-to-cell interactions. The process of microbial aggregation is mediated by a multitude of molecular interactions including specific (i.e., molecular recognition between receptors and ligands) or non-specific (i.e., hydrogen bonding, hydrophobic, van der Waals, electrostatic, and macromolecular forces) interactions (Busscher et al. 2008). Hydrophobic interactions take place when particles or molecules are incapable of interacting electrostatically or forming hydrogen bonds with the water molecules (Magnusson 1980). These hydrophobic interactions between cells are usually evaluated in terms of cell surface hydrophobicity (Guo et al. 2011, Zita and Hermansson 1997). In wastewater treatment processes, sludge settling, granulation, dewatering as well as membrane fouling are strongly related to the cell surface hydrophobicity (Lin et al. 2014, Ras et al. 2013).

[1] Université Paris-Est, Laboratoire Géomatériaux et Environnement (EA 4508), UPEM, 77454 Marne-la-Vallée, France.
[2] Université de Limoges, Groupement de Recherche Eau Sol Environnement (EA 4330), Faculté des Sciences et Techniques, 123 Avenue A. Thomas, 87060 Limoges Cedex, France.
[3] Delft IHE Institute for Water Education, Westvest 7, 2611 AX Delft, the Netherlands.
[4] Université de Paris, Institut de physique du globe de Paris, CNRS, F-75005 Paris, France.
* Corresponding author: vanhullebusch@ipgp.fr

Analytical techniques usually applied in the hydrophobicity measurement are contact angle measurements (CAM) (Pen et al. 2015), hydrophobic interaction chromatography (HIC) (Ras et al. 2013), salt aggregation tests (SAT) (Rozgonyi et al. 1985, Lindahl et al. 1981), and microbial adhesion to hydrocarbon (MATH) tests (Gao et al. 2008, Rosenberg et al. 1980). Besides those techniques, hydrophobic/hydrophilic fractionation by XAD/DAX resin is frequently used to fractionate natural organic matter (NOM) in natural water. This technique can isolate the hydrophobic/hydrophilic fractions of NOM according to their polarity and thus, facilitates hydrophobicity studies (Peuravuori et al. 2002, Thurman et al. 1978).

Extracellular polymeric substances (EPS) are located outside the cells and account for the major part of the total organic carbon (TOC) in biofilms or sludge (Flemming et al. 2016). They exert a profound impact on the cell surface hydrophobicity (Liu et al. 2004), and enhance the affinity of heavy metals or organic micro-pollutants to the microorganisms (Wei et al. 2017, Jia et al. 2011). Two operationally defined fractions, soluble and bound EPS, are found to constitute the EPS (Nielsen et al. 1997). Soluble EPS (colloids, slimes) reside in the surrounding of microbial cells, which protect the cells against severe conditions. However, it is difficult to distinguish between produced soluble microbial products (SMP) and soluble EPS (Laspidou and Rittmann 2002). Bound EPS closely associate with the cell walls and consist of two layers. The outer layer is a loose and dispersible slime layer without an obvious edge called loosely bound EPS (LB-EPS), while the inner layer has a certain shape and binds tightly and stably with the cell surface called tightly bound EPS (TB-EPS) (Comte et al. 2006b, Nielsen and Jahn 1999).

The main organic fraction determined in the EPS of multispecies microbial aggregates include proteins (PN), polysaccharides (PS), humic-like substances (HS-like), and small amounts of nucleic acids, lipids or uronic acids (Sheng et al. 2010, Comte et al. 2006a). Besides organic macromolecules, mineral particles composed of Ca and/or Fe (calcium or ferric phosphates, carbonates, oxides, etc.) are also embedded in the EPS matrix (D'Abzac et al. 2012). The EPS composition is controlled by several factors, such as microbial type, nutrients and local environmental conditions (i.e., pH, temperature and metal concentrations) (Sheng et al. 2010). There is no standard method to extract EPS from biofilms and microbial aggregates. The extraction yield, organic composition, as well as the molecular structure and metal binding ability of the EPS largely differ according to the extraction method used (Zuriaga-Agustí et al. 2013, Domínguez et al. 2013, Comte et al. 2006a).

The hydrophilic/hydrophobic properties of EPS are ascribed to the different functional groups (i.e., carboxyl, hydroxyl, phosphate, amine, and sulfate) and non-polar regions (i.e., hydrophobic regions in *O*-methyl/acetyl polysaccharides; aromatics, and aliphatic regions in proteins) (Moran 2009). Nevertheless, the role of EPS in the cell surface hydrophobicity is not well identified. Zhang et al. (2014) found that after LB-EPS and TB-EPS extraction, the hydrophobicity of the cell suspensions was higher than before the extraction steps. In contrast, Guo et al. (2011) showed that after EPS extraction, the cell surface hydrophobicity of the aerobic granules decreased.

Monitoring of the organic micro-pollutant removal in wastewater treatment processes as well as the microbial aggregation (i.e., biofilm formation, granulation, and activated sludge flocculation) benefit from a better characterization of EPS from both quantitative and qualitative aspects. The factors governing microbial aggregation have been widely studied. Therefore, this chapter focuses on the role of EPS in the cell surface hydrophobicity of pure cultures (bacteria, fungi and micro-algae) and mixed-culture aggregates such as biofilms, aerobic/anaerobic granules and activated sludge. The role of cell surface hydrophobicity in the microbial aggregation, the relation between EPS and cell surface hydrophobicity, as well as the techniques applied in the characterization of hydrophobicity are described.

2. Hydrophobicity in microbial aggregation

Hydrophobic interactions can be viewed as the spontaneous tendency of nonpolar groups to aggregate or cluster to minimize their contact with the water molecules (Figure 1). Despite nonpolar solutes

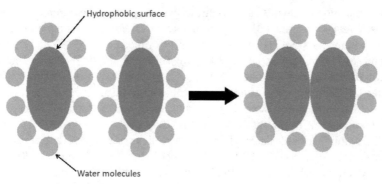

Figure 1. Schematic illustration of hydrophobic interactions in water.

having low solubility in water, the hydrophobic interactions are responsible for the adhesion between the hydrophobic surfaces. These interactions are also involved in the rapid coalescence or flocculation which is commonly observed in colloidal systems of hydrophobic liquid droplets or solid particles (Meyer et al. 2006). In a sense of process thermodynamics, microbial aggregation is driven by free energy decrease, thereby increasing the cell surface hydrophobicity which would cause a corresponding decrease in excess Gibbs energy of the surface. Therefore, cell surface hydrophobicity promotes the cell-to-cell interaction and further serves as a triggering force for the cells to aggregate in the hydrophilic liquid phase (Liu et al. 2004).

2.1 Microbial adhesion

Microbial aggregation, such as biofilm formation, activated sludge flocculation and granulation, is initiated by the microbial adhesion. The early stage of microbial aggregation begins with the attachment of planktonic microorganisms to a biotic or abiotic surface. These first colonists initially adhere to the surface through non-specific, long-range attractive Lifshitz–van der Waals forces as well as electrostatic, acid-base and hydrophobic forces (Chung et al. 2014, Israelachvili 2011).

Hydrophobic interactions affect the adhesion process by removing the vicinal water film between the cell surface and the substratum. In another words, hydrophobic groups use their dehydrating capacity to remove the water and aid the cell appendages to attach to the surface (Busscher et al. 2010, Busscher and Weerkamp 1987). In low ionic strength environment, cell surface hydrophobicity plays an important role in determining the adhesion of microorganisms to the surface (van Loosdrecht et al. 1987b), as the repulsion between the microorganisms and the surface increases with the decrease of the cell surface hydrophobicity (Muadcheingka and Tantivitayakul 2015, Pimentel-Filho et al. 2014, Stenström 1989). Uniform, flat, and thin biofilms can rapidly form on the hydrophobic surface of polyvinylidene fluoride (PVDF), whilst the hydrophilic surface of polyvinyl alcohol (PVA) prevents the adhesion of *Pseudomonas putida* and the biofilm formation (Saeki et al. 2016). However, Parkar et al. (2001), Flint et al. (1997) and Sorongon et al. (1991) demonstrated that there was little or no relationship between the cell surface hydrophobicity and the bacterial cell adhesion to the solid substratum.

The development of granular biofilms is a complex process that is initiated by the formation of smaller microbial aggregates (Adav and Lee 2008). Cell surface hydrophobicity induced by the culture conditions could serve as the triggering force for both anaerobic and aerobic granulation (Liu et al. 2004). With the increase in the cell surface hydrophobicity, the granulation process is accelerated (Liu et al. 2009). Xu and Tay (2002) used high surface hydrophobicity methanogen-enriched sludge to seed the upflow anaerobic sludge blanket (UASB) reactor. They found that the granulation process reached its post-maturation stage in about 15–20 days ahead of the control reactor in which normal seed sludge was used.

One of the models to elaborate anaerobic granulation is the local dehydration and hydrophobic interaction model proposed by Wilschut and Hoekstra (1984). This model assumes that irreversible adhesion occurs when the bacterial surface is strongly hydrophobic. Bacteria present in the different layers of anaerobic granules have different hydrophobicity: a surface layer with moderate hydrophobicity, a middle layer with extremely high hydrophobicity, and a core with high hydrophobicity (Daffonchio et al. 1995). Non-granular sludge washed out from anaerobic reactors was more hydrophilic than the sludge retained in the reactors (Mahoney et al. 1987). Likewise, aerobic granulation also follows a function of the cell surface hydrophobicity over the hydrophilicity, i.e., high cell surface hydrophobicity strongly favors microbial aggregation and results in a more compact granular structure (Liu et al. 2003). Therefore, a proper cell surface hydrophobicity is one of the important criteria to enhance sludge granulation (Yu et al. 1999).

2.2 Sludge flocculation, settling and dewatering

An effective wastewater treatment by using activated sludge is largely dependent on the sludge flocculation and settling. However, it is difficult to manipulate sludge flocculation and settling processes due to the complex nature and interactions among the microbial communities in the sludge (Wilén et al. 2003). Zita and Hermansson (1997) found that the cell surface hydrophobicity of *Escherichia coli* and the bacteria isolated from wastewater correlated well with the degree of the adhesion to the sludge flocs, and they postulated that a low level of cell surface hydrophobicity could be the reason why planktonic cells did not attach to the flocs. Liao et al. (2001) demonstrated that the sludge with higher cell surface hydrophobicity produced an effluent of lower turbidity, but it had no effect on the sludge volume index (SVI). Thus, they inferred that the cell surface hydrophobicity was crucial to the flocculation, but not the settling. Nevertheless, Urbain et al. (1993) reported that intrafloc hydrophobic interactions enhanced the sludge settling.

Sludge dewatering is the bottleneck in many wastewater treatment processes. The difficulty in monitoring sludge dewatering is also due to the physical, chemical and microbial complexity in the sludge (Liu and Fang 2003). Many efforts have been made to improve the sludge dewatering performance. For example, some hydrophobically modified polymers (e.g., cationic polyacrylamide) are often added as flocculants to improve the performance. One of the main mechanisms is that macromolecular flocculants can span and reduce the gap between the particles through a bridging effect. This effect takes place when the flocculant polymer chains interact with the hydrophobic groups on the activated sludge surface through the hydrogen bonding, van der Waals forces, and hydrophobic interaction (Sun et al. 2015). Huang et al. (2016) synthesized hydrophobic cationic chitosan (HTCC) flocculants by reacting chitosan with epoxy propyl trimethyl ammonium chloride (EPTAC) and (2, 3-epoxy propyl) dodecyl dimethyl ammonium chloride (EDC). They found that with the enhancement in the HTCC hydrophobicity, a better dewatering performance of the activated sludge was achieved. In addition, a good dewatering performance by using thermo-sensitive polymers (N, N-dimethylaminopropylacrylamide) in the activated sludge was observed by Sakohara et al. (2007). They ascribed the improvement to the hydrophobic interactions occurring between the thermo-sensitive polymer molecules and the activated sludge.

2.3 Membrane biofouling

Membrane biofouling is a result of biofilm formation on the solid substratum, in which membrane surface properties, i.e., hydrophobicity, play an important role (Bruggen et al. 2010). It is generally believed that a hydrophilic membrane leads to a lower membrane fouling potential rather than a hydrophobic one (Qu et al. 2014, Evans et al. 2008, Weis et al. 2005). However, since the cell surface hydrophobicity is not enough to evaluate the extent of fouling caused by the interaction between the cells and the membrane materials (Zhang et al. 2015a), some inconsistent results were also reported. More hydrophobic polyvinylidene fluoride (PVDF) membranes display lower fouling tendency compared to polysulfone (PSF) and cellulose acetate (CA) membranes (Choo and Lee 1996). Although the hydrophobicity of membranes PVDF, CA and polyether sulfones (PES) follows the order of PVDF > PES > CA, experimental fouling data from Chen et al. (2012) showed that the CA membrane was fouled more severely than PVDF and PES membranes.

3. Characterization of microbial hydrophobicity

Techniques that are usually applied in the hydrophobicity determination are CAM (Pen et al. 2015), HIC (Ras et al. 2013), SAT (Rozgonyi et al. 1985, Lindahl et al. 1981), MATH test (Gao et al. 2008, Rosenberg et al. 1980), and hydrophobic/hydrophilic fractionation by XAD/DAX resin (Jorand et al. 1998). It is noticed that the development of HIC and SAT methods are based on the hydrophobic properties of PN, while CAM and MATH methods give the general description of microbial hydrophobicity (Table 2). The comparison between the results obtained from SAT and HIC methods showed a good correlation ($p < 0.025$) for *Staphylococcus aureus* (Jonsson and Wadström 1984).

3.1 Contact Angle Measurement (CAM)

One of the advantages of CAM is that this technique can non-destructively characterize the cell surface hydrophobicity (Table 2). The measurement procedure includes: depositing a drop of liquid on a layer made from the studied cells, and then measuring the angle (θ) between the layer surface and the tangent (angle) to the drop at the solid-liquid-air meeting point (Figure 2) (Kwok and Neumann 1999). The liquid is often replaced by the deionized (DI) water droplets. CAM can also be applied to measure the contact angle of dried sludge cakes (Liao et al. 2001).

The preparation work follows the procedure described by Busscher et al. (1984). The washed cells are firstly filtered on a cellulose filter (normally 0.45 μm pore size) to form a layer of cells. Then, the filter with the layer of deposited cells is transferred into a solidified agar plate and dehydrated for some time to ensure the constant moisture content. Since the water evaporation is expected to influence the contact angle, the dehydration process should be carried out under controlled atmosphere (i.e., 20–22°C, 50–60% relative humidity) (Mozes and Rouxhet 1987). The following step is to place a drop of DI water on the film of cells, and a video image system is used to view the sessile drop from the top. The drop shape is captured as soon as no further shirking of the water drop is observed, and the image captured is then used to estimate the contact angle values. Considering the possible penetration of the water drop into the cell cake, the image recording is usually performed immediately (within 2–7 s) after depositing the water drop (Liao et al. 2001, Busscher et al. 1984).

According to the water contact angle, cell surface hydrophobicity is roughly classified into three categories: hydrophobic surface with a contact angle greater than 90°, medium hydrophobic surface with a contact angle range between 50° and 60°, and hydrophilic surface with a contact angle below 30° (Mozes and Rouxhet 1987). Most bacteria and activated sludge display a contact angle ranging from 20° to 45°, while some are higher than 50° or lower than 15° (Table 1). However, the layer of

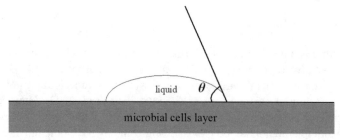

Figure 2. Schematic illustration of a sessile-drop contact angle system (adapted from Kwok and Neumann (1999)).

Table 1. Techniques applied for the determination of microbial hydrophobicity.

Type	Technique	Conditions	Relative hydrophobicity	Reference
Bacteria *Rhodococcus* RC291	CAM		20°–65°	Pen et al. 2015
Sludge from sequencing batch reactor (SBR)	CAM		10°–45°	Liao et al. 2001
Bacteria *Streptococcus salivarius* HB, *Streptococcus sanguis* CH3, *Streptococcus mitior* T6, and *Veillonella alcalescens* Vl	CAM	Static sessile method using de-ionized water	10°–65°	Busscher et al. 1984
Thermophilic spores from bacteria *Geobacillus* spp.	CAM		10°–30°	Seale et al. 2008
EPS extracted from a heterotrophic biofilm and a mixed autotrophic–heterotrophic biofilm	HIC	UV absorbance at 280 nm; elution gradient started off with 100% salt buffer (buffer B: 0.05 M KH_2PO_4/K_2HPO_4 with 3 M ammonium sulphate, pH 7) and ended with 100% elution buffer (buffer A: 0.05 M KH_2PO_4/K_2HPO_4, pH 7)	Hydrophobic EPS from both biofilms were mainly represented by two distinct fractions (dimensionless retention time (DRT) 0.3 and DRT 0.9)	Ras et al. 2013
EPS extracted from *Pseudomonas fluorescens* Biovar II and *Sagittula stelleta*	HIC	UV absorbance at 214 nm; eluted by 20 mM Tris buffer (pH 8.0) plus different ammonium sulphate concentrations (0.25, 0.5, 0.75 and 1.0 M)	–	Xu et al. 2011
EPS extracted from biofilms in the rotating biological contactor, and activated sludge	SAT	Absorbance at 280 nm; results expressed as absorbance reduction by adding ammonium sulphate	28–38% and around 15%, respectively	Martín-Cereceda et al. 2001
Mutants of *Staphylococcus aureus*	SAT	Results expressed as lowest molarity of ammonium sulphate in a mixture with bacterial cells which gives visual bacterial cell aggregation	1.4–1.6 M	Jonsson and Wadström 1984
Bacteria strains (including *Moniliella pollinis*, *Saccharomyces cerevisiae* and *Acetobacter acett*, etc.)	SAT		0.05 –>2 M	Mozes and Rouxhet 1987
Activated sludge and biofilms	MATH	Absorbance measured at 540 nm; mixed with *n*-hexadecane and agitated for 2 min	15–22%	Chao et al. 2014
Bacteria isolated from activated sludge	MATH	Absorbance measured at 600 nm; mixed with *n*-hexadecane and agitated for 2 min	3–77%	Xie et al. 2010
Aerobic granules	MATH	Mixed with *n*-hexadecane and agitated for 5 min; RH was expressed as a ratio of MLSS concentration in the aqueous phase	43–78%	Zhang et al. 2007

Table 1 contd. ...

...Table 1 contd.

Type	Technique	Conditions	Relative hydrophobicity	Reference
EPS extracted from activated sludge	XAD-8 and XAD-4 resins	Samples were acidified to pH 2 by concentrated HCl (37%) and filtered by 0.45 μm membranes	7% of the dissolved carbon and 12% of the proteins were considered as hydrophobic	Jorand et al. 1998
Organic matter from the liquid medium of an alga *Euglena gracilis* and a cyanobacteria *Microcystis aeruginosa*	DAX-8 and XAD-4 resin		Hydrophilic fraction was the major fraction (≥ 50% DOC content) of algal organic matter (AOM) whatever the growth phase was	Leloup et al. 2013
EPS extracted from anaerobic granular sludge	DAX-8 resin	Samples were acidified to pH 2 and pH 5 by 2M HCl and filtered by 0.45 μm membranes	Elution pH for hydrophobic fractionation was preferred at pH 5; the hydrophobic fraction of EPS retained by the resin was ascribed to a wide aMW range of > 440–0.3 kDa	Cao et al. 2017a

microbial cells on the filter consists not only of packed individual microorganisms but also of some unavoidably void spaces. The deposited drops on this thin porous cells layer can behave in different ways, depending on many characteristics of both the liquid and the microbial layer (Table 2), such as liquid viscosity, surface tension, porosity of the cells film, ratio of the porous film thickness to the drop radius and surface Gibbs energy of the individual particles (Marmur 1988). Therefore, the contact angle measured can only be viewed as a relative indicator of cell surface hydrophobicity (Gallardo-Moreno et al. 2011).

3.2 Hydrophobic Interaction Chromatography (HIC)

HIC differentiates the molecules based on their hydrophobic properties. It primarily adopts the reversible interactions that take place between the hydrophobic regions of protein-derived macromolecules and weakly hydrophobic ligands (i.e., butyl, octyl and phenyl) attached to the stationary phase (column packing matrix) under a high concentration of salt mobile phase. The working principle of HIC is the "salting-out chromatography" proposed by Shepard and Tiselius (1949). In general, the salts (e.g., 0.75 to 2.0 M of ammonium sulfate or 1.0 to 4.0 M of sodium chloride) in the mobile phase reduce the solvation of the sample solutes. As the solvation decreases, hydrophobic regions buried in the sample molecules become exposed and are adsorbed by the stationary phase. The more hydrophobic the molecules are, the longer the elution time is needed (Lienqueo et al. 2007).

Usually, a decreasing salt gradient is applied to elute samples from the column so that the hydrophobic interactions can be intensified. The salts frequently used are called anti-chaotropic salts (also kosmotropic or lyotropic salt), which includes sodium, potassium, ammonium sulfate, sodium nitrate, and sodium chloride in phosphate buffer at pH 7 (Baca et al. 2016). These salts have a higher polarity to bind water tightly, and their presence in the mobile phase benefits the stabilization of protein structure (Xia et al. 2004). By using these salts, the exclusion of water between the proteins and the ligand surface can be induced, and thus, promotes the hydrophobic interactions and protein

Table 2. Comparison of the techniques applied in the hydrophobicity determination.

Technique	Advantages	Limitations	References
CAM	Simple and rapid; non-destructive; provides information related to surface energetics; hydrated samples can be observed	Extreme experimental care is needed. Very minor vibrations can cause advancing contact angles to decrease, resulting in errors of several degrees: surface roughness; contacting time; liquid penetration; swelling of a solid by a liquid can change the chemistry of the solid	Pen et al. 2015, Marmur 1988, Busscher et al. 1984
HIC	Samples with high ionic strength can be used; the elution conditions promote the retention of the tertiary conformation and the biological activity of most proteins; good selectivity	Based on the difference in the hydrophobicity of amino acids; significant baseline changes during gradient elution and a requirement for non-volatile mobile phases; type and concentration of ligands, column packing materials, salts influence the performance, as well as pH, and temperature	Ras et al. 2013, Alpert 1988
SAT	A rapid and reproducible screening test	Hydrophobic bacterial cells will clump when ammonium sulfate is absent; poor visualization of the bacterial aggregates without a dark background; electrostatic interactions between the cells, temperature, pH, time and the bacterial cell concentration affect the results	Ljungh and Wadström 1982, Lindahl et al. 1981
MATH	Simple, no special equipment is needed; can be used for observing mixed cultures of the same species; bound bacteria to hexadecane can be desorbed by allowing the hexadecane to solidify at temperatures below 16°C	Under low ionic strength buffer, otherwise the hydrophobic interactions play a lesser role and electrostatic interactions increase; low discriminating power for hydrophilic strain; it measures the interplay of all the physicochemical forces between the cells and hydrocarbon	Rosenberg 2006
XAD/ DAX resin fractionation	Samples are fractionated according to their hydrophobicity; allows characterization of separated fractions by determination of their average structural and functional groups	This hydrophobic-hydrophilic distinction is artificial, since the pH of the tested samples is pre-adjusted; possible physical and chemical alteration during pH adjustment and fractionation process; irreversible adsorption of the samples onto the resins; contamination from the resin bleeding; size-exclusion effect induced by the resins	Malcolm 1991

precipitation (salting-out effect) (Kunz et al. 2004). However, studies have shown that the retention of hydrophobic molecules in HIC is affected by the interplay of different factors (Table 2), such as pH, salt type and concentration, ligand type and density in the column material, protein folding upon adsorption and kinetic of protein spreading (Nfor et al. 2011, Mahn et al. 2007, Haimer et al. 2007, Jungbauer et al. 2005, Xia et al. 2004, Perkins et al. 1997).

The hydrophobicity determined by HIC is usually evaluated by the dimensionless retention time (DRT) (Lienqueo et al. 1996). The calculation of the DRT is conducted according to Eq. (1):

$$DRT = \frac{t_R - t_0}{t_f - t_0}$$

Eq. (1)

where: t_R is the time corresponding to the peak on the chromatogram, t_0 is the time corresponding to the start of the salt gradient and t_f is the time corresponding to the end of the salt gradient.

When DRT equals to 1, this indicates the tested sample is extremely hydrophobic. Ras et al. (2013) utilized this method to characterize the hydrophobic EPS separated from biofilms. Their results showed that 97% of the hydrophobic EPS were represented by two distinct fractions: DRT 0.3 and DRT 0.9 (Table 1). They concluded that hydrophobic EPS were more abundant in the cohesive layers of the biofilms than in the top layer.

3.3 Salt Aggregation Test (SAT)

The SAT method was originally developed by Lindahl et al. (1981) and applied to measure the bacterial cell surface hydrophobicity (Polak-Berecka et al. 2014, Basson et al. 2007, Mozes and Rouxhet 1987). This method is also based on the salting-out effect: precipitation of cells by dosing salts (i.e., $(NH_4)_2SO_4$) at neutral pH, and highly hydrophobic cells are precipitated at low salt concentration. The grade of cell precipitation can be evaluated by comparing the absorbance of the supernatant of cell suspensions with/without ammonium sulfate at a certain wavelength (usually at 280 or 600 nm). The percentage of absorbance decrease can be used to reflect the cell surface hydrophobicity, and a greater absorbance decrease corresponds to a more hydrophobic cell surface (Martín-Cereceda et al. 2001, Urbain et al. 1993, Lindahl et al. 1981).

SAT can also be applied to investigate the cell surface hydrophobicity of sludge flocs. Martín-Cereceda et al. (2001) demonstrated that the percentage of absorbance decrease of biofilm EPS (28–38%) was twice higher than that of activated sludge EPS (around 15%), implying that the EPS extracted from the biofilm were two times more hydrophobic than those extracted from the activated sludge (Table 1). Jonsson and Wadström (1984) used the same technique but expressed the results in terms of the lowest molarity of ammonium sulfate (in the mixture with cells) at which it yields a visual cell aggregation. The same expression is also seen in Urbain et al. (1993), Mozes and Rouxhet (1987) and Rozgonyi et al. (1985) (Table 1). According to Balebona et al. (2001), if the lowest molarity of ammonium sulfate is below 1 M, cells are considered as highly hydrophobic; if the value is between 1 M and 2 M, cells are moderately hydrophobic; and if the value is above 2 M, cells are weakly hydrophobic or hydrophilic.

However, many hydrophobic bacterial cells will clump when ammonium sulfate is absent. Therefore, this technique only provides a qualitative estimation of the relative rank of microbial hydrophobicity (Liss et al. 2004). Besides, electrostatic interactions affect SAT results more than other hydrophobicity measurement techniques (Table 2). Temperature, pH, reaction time and the cell concentration also have impacts on the results (Rosenberg 2006).

3.4 Microbial Adhesion To Hydrocarbons (MATH) test

The MATH test is a simple and widely used method to determine the relative hydrophobicity (RH) of bacterial or fungal cells. When it is applied in mixed cultures like activated sludge or granules, a glass homogenizer or tissue grinder is often required to make homogeneous cell suspensions (Xie et al. 2010, Zhang et al. 2007). In the assay, the washed cells are suspended in PUM buffer at pH 7.1 (per liter: 22.2 g $K_2HPO_4·3H_2O$, 7.26 g KH_2PO_4, 1.8 g urea, 0.02 g $MgSO_4·7H_2O$), and then vortexed in the presence of liquid hydrocarbon (i.e., n-hexadecane, n-octane, p-xylene). During the vortex procedure, the liquid hydrocarbon is dispersed into droplets and can adhere to hydrophobic microbial cells. When the adhesion takes place, the cell-coated droplets will rise to form a stable foam (Figure 3) (Chao et al. 2014).

Phase separation is usually completed within one or two minutes. The extent of adhesion to the hydrocarbon droplets can be simply ascertained by the decrease in the cell-mediated turbidity of the lower aqueous phase. To measure the decrease, the lower aqueous phase is withdrawn by using a Pasteur pipette and transferred to a cuvette for absorbance (Abs) measurements at a wavelength (λ) of 400 or 600 nm (Rosenberg et al. 1980). Therefore, the difference in the Abs values of the aqueous

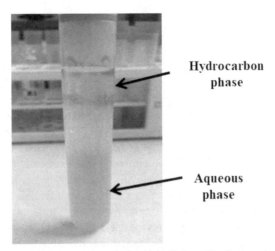

Figure 3. Phase separation of microbial cells by adding hydrocarbon.

Color version at the end of the book

phase before (Abs$_\lambda$(before)) and after (Abs$_\lambda$(after)) adding the hydrocarbon can be used to calculate RH (Eq. (2)). The proper value of Abs$_\lambda$(before) is usually controlled at 0.4–0.6.

$$\text{Relative hydrophobicity (\%)} = \frac{\text{Abs}_\lambda \text{(before)} - \text{Abs}_\lambda \text{(after)}}{\text{Abs}_\lambda \text{(before)}} \times 100\% \qquad \text{Eq. (2)}$$

It is generally considered that a cell surface corresponding to a RH (%) above 50% is highly hydrophobic; if the values ranges between 20% and 50%, the cell surface is moderately hydrophobic and if the value is below 20%, the cell surface is considered mostly hydrophilic (Balebona et al. 2001, Lee and Yii 1996). Nevertheless, the cells cannot only adhere to the hydrocarbons; they also adhere to each other. Therefore, the MATH test does not measure the absolute cell surface hydrophobicity, but rather the interplay of all the physicochemical forces including hydrophobic interactions, van der Waals forces and electrostatic interactions (Table 2). In addition, the interference of electrostatic interactions in the MATH test can be reduced by performing the test under ionic conditions in which either the cells or the hydrocarbon droplets (or both) are uncharged (Geertsema-Doornbusch et al. 1993). For this reason, several authors have carried out the test at a pH closer to the isoelectric points (IEP) of the solvents or the bacterial cultures to minimize the electrical interactions involved in the adhesion process, as well as to obtain a better evidence of the microbial surface hydrophobicity. Moreover, the kinetics of cell adhesion also depends on the initial cell concentration. In *Candida albicans* M7, the mixing of cells at low concentrations with *n*-hexadecane led to a coalesce within seconds after vortex, whereas higher cell concentrations yield stable cell-coated droplets and relatively high adhesion levels (Rosenberg et al. 1991).

3.5 Hydrophobic/hydrophilic fractionation by XAD/DAX resins

Besides the above-mentioned techniques, hydrophobic/hydrophilic fractionation by XAD/DAX resins is frequently used to fractionate natural organic matter (NOM). Classical XAD/DAX resin fractionation procedures are composed of: (i) acidification, in order to protonate the samples to an uncharged state and (ii) hydrophobic/hydrophilic fractionation by resin XAD-8/DAX-8 and XAD-4 in tandem use. As the production of Amberlite® XAD-8 resin (Rohm and Haas) has been ceased some years ago, Supelite™ DAX-8 resin (Sigma-Aldrich), also referred to as poly(methyl methacrylate) resin, has been substituted for XAD-8 resin and become popular in recent studies (Cao et al. 2017a, Leloup et al. 2013, Marhaba et al. 2003, Peuravuori et al. 2002, Bolto et al. 1999).

As water is a polar solvent, it forms strong hydrogen bond with polar organics and ionic groups which are, thus, referred to as hydrophilic. Nonpolar organics (termed hydrophobic) are unable to interact in this way and are partially separated from the aqueous phase. The resin fractionation exploits the differences in polarity to separate NOM into different fractions (Peuravuori et al. 2002, Thurman et al. 1978).

Disassociation tendencies of the functional groups (e.g., $-COOH$, $-NH_3^+$) in the NOM can be controlled by pH. Organic acids ($-COOH$) become protonated in acidic condition, and with the aid of certain hydrophobic carbon skeletons, these acidic organic solvents are sorbed by the resin via hydrophobic interactions (Figure 4) and thus, considered as hydrophobic (Nollet and Gelder 2013). Generally, samples are separated into three fractions by passing through the XAD/DAX resin series: hydrophobic fraction (HPO) is the part adsorbed on the XAD-8/DAX-8 resin at pH 2, the transphilic fraction (THP) is the elute of XAD-8/DAX-8 which then adsorbs to XAD-4 at pH 2, and the hydrophilic fraction (HPI) is the elute from both columns in series at pH 2 (Figure 5). The main advantage of this technique is that the fractionation facilitates the post-analyses of the hydrophobic/hydrophilic properties of the NOM (Aiken et al. 1992, Leenheer 1981, Thurman et al. 1978).

Aqueous NOM samples are commonly acidified to pH 2, as it maximizes the hydrophobicity of NOM without leading to the precipitation of humic acids (Malcolm 1991). Nevertheless, when this method is applied to EPS studies, the conformation of EPS molecules may be changed by adjusting the pH. The pH variation could result in major rearrangements of EPS structure, i.e., swelling of polymeric chains in the EPS (Dogsa et al. 2005).

Leloup et al. (2013) characterized the algal organic matter (AOM) produced by algae *Euglena gracilis* and cyanobacterium *Microcystis aeruginosa* by using DAX-8 and XAD-4 resins. The results showed that the HPI was the major fraction in the AOM whatever the growth phase was, and it was almost the only fraction produced during the lag and exponential phases. Also, Jorand et al. (1998) fractionated the EPS extracted from activated sludge by XAD-8 and XAD-4 resins. The results demonstrated that at least 7% of the dissolved carbon and 12% of the proteins were hydrophobic (Table 1).

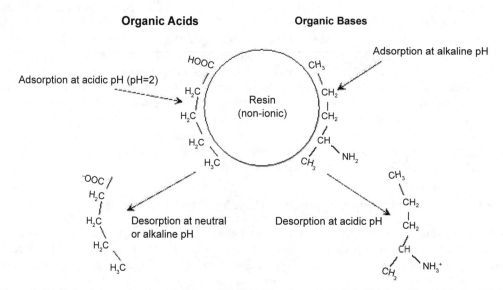

Figure 4. Interaction between organic acids/bases and XAD/DAX resins controlled by pH (adapted from Malcolm (1991)).

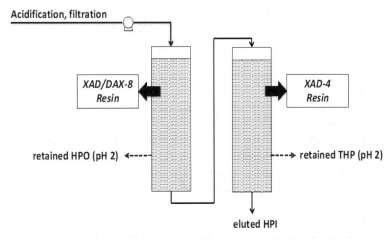

Figure 5. Classical hydrophobic/hydrophilic fractionation by XAD/DAX resins.

XAD/DAX fractionation, combined to different analytical techniques, i.e., mass spectroscopy and size partition, allows determining a specific molecular fingerprint for each organic matter extract. To obtain a comprehensive understanding of the fouling potential of SMPs in a membrane bioreactor (MBR) and their molecular size, Shen et al. (2010) fractionated MBR supernatant into different hydrophilic/hydrophobic fractions by using DAX-8 resin with joint size partition (ultrafiltration membrane). Results showed that hydrophilic fraction was the dominant foulant responsible for the flux deterioration in the MBR. The apparent molecular weight (aMW) of this hydrophobic fraction was higher than 100 kDa. Cao et al. (2017) investigated the hydrophobic properties of EPS extracted from anaerobic granular sludge by DAX-8 resin, and the EPS fingerprints recorded by size exclusion chromatography (SEC) demonstrated that the highly hydrophobic EPS molecules retained by the resin displayed an aMW of 31–175 kDa. After separation by XAD resins (XAD-8 and XAD-4), the algal (*Euglena gracilis*) organic fractions were studied by laser desorption ionization (LDI) and matrix-assisted laser desorption ionization time-of-flight mass spectrometry (MALDI–TOF). By using this developed protocol, a structural scheme and organic matter composition of the algae were proposed by Nicolau et al. (2015).

4. Factors influencing cell surface hydrophobicity

Microbial aggregation can also be viewed as a result of the cell response to a stressful environment (Liu et al. 2010). When the microbes enter an unfavorable environment, their cell surfaces will start a series of adaption changes. It is expected that the change of cell surface hydrophobicity can strengthen the cell-to-cell interactions resulting in a stronger surface structure to protect the cells from the unfavorable conditions. The hydrophobicity change is influenced by many factors such as substrates, microbial growth phase, external conditions (pH, temperature, metal cations and organic pollutants), as well as the EPS production (Liu et al. 2004) .

4.1 Substrates

Hazen et al. (1986) compared the cell surface hydrophobicity of both *Candida albicans* and *Candida glabrata* grown in different growth media, and a greater cell surface change was found in Auto-Pow minimum essential medium with Earle salts supplemented with biotin, glucose, glycine and HEPES (*N*-2-hydroxyethylpiperazine-*N'*-2-ethanesulfonic acid) (AP-MEM). A low percentage of cells had germ tubes when incubated 24 h in AP-MEM medium, while such cell form transition was not observed in the cells incubated in others media. Hazen et al. (1986) ascribed the variation of cell surface hydrophobicity in AP-MEM medium to the cell surface compositional modifications that occur during the form transition. Anaerobic granules grown in protein-rich media were reported to have a lower cell surface hydrophobicity and slower granulation process, when compared with those grown in carbohydrate-rich media (Thaveesri et al. 1994). Protein-rich substrates can lower the surface tension of the reactor liquid, which leads to the formation of anaerobic granules displaying low hydrophobicity (Thaveesri et al. 1995).

The cell surface hydrophobicity of *Corynebacterium glutamicum* became higher under the phosphate-saturated growth conditions than under the phosphate-depleted conditions (Büchs et al. 1988). Phosphate limitations induce the synthesis of teichuronic acids instead of teichoic acids in Gram-positive bacteria (Swoboda et al. 2010), and in turn, some additional changes could occur between these two hydrophilic polymers and the cell surface (Büchs et al. 1988).

When exposed to a toxic or inhibitory substrate, microorganisms are still able to regulate their cell surface properties. Many toxic hydrocarbons or heavy metals that are present as environmental pollutants are also potential substrates for the bacteria. To increase the accessibility of these low bioavailability compounds, one specific adaptive mechanism developed by the bacteria is the modification of cell surface hydrophobicity to permit direct hydrophobic interactions with those substrates (Heipieper et al. 2010). The Gram-positive bacterium *Bacillus licheniformis* reduced the cell surface hydrophobicity when cultured with the organic solvent isoamyl alcohol, and exhibited little affinity towards 3-methylbutan-1-ol (Torres et al. 2009). In contrast, *Mycobacterium frederiksbergense* increased its cell surface hydrophobicity in the presence of anthracene. High affinity of the cells to anthracene for utilizing it as unique carbon source requires an augmented cell surface hydrophobicity. This augmentation was induced by a degradative aromatic pathway which is developed by the bacterium to metabolize anthracene (Yamashita et al. 2006, Wick et al. 2002). The mutants of the Gram-negative bacterium *E. coli* K-12 presented a less hydrophobic cell surface when compared to the parental cells, and displayed a higher tolerance to organic solvents (Aono and Kobayashi 1997).

4.2 Microbial growth phase

In a study regarding the impact of brewing yeast cell age on fermentation performance, Powell et al. (2003) reported that the flocculation potential of cells and cell surface hydrophobicity increased in conjunction with the cell age. Malmqvist (1983) showed that the cell surface hydrophobicity of the bacterium *Staphylococcus aureus* increased 4–5 times during the exponential growth phase of the culture. Generally, cells in the stationary growth phase were more hydrophobic than those in the exponential growth phase (Allison et al. 1998, Hazen et al. 1986). However, van Loosdrecht et al. (1987a) observed that 23 bacterial strains became more hydrophobic during the exponential growth phase by studying their cell surface hydrophobicity. Similar results were also obtained by Jana et al. (2000) who concluded that the early- to mid-log exponential cells of 18 isolates from *Pseudomonas fluorescens* were more hydrophobic than those in the stationary phase. This contradictory observation could be ascribed to the cell surface hydrophobicity that is also influenced by other conditions such as temperatures, pH and the presence of the micro-pollutants (Muda et al. 2014, Correa et al. 1999, Mattarelli et al. 1999, Blanco et al. 1997).

Some studies showed that starvation conditions could enhance cell surface hydrophobicity and in turn, facilitate microbial adhesion and aggregation (Bossier and Verstraete 1996). It is shown by Chiesa et al. (1985) that the periodic starvation cycle increased the cell surface hydrophobicity of activated sludge. Nevertheless, some studies showed different effects that were caused by the starvation. Castellanos et al. (2000) demonstrated that upon transfer from a rich growth medium to famine condition, the cell surface hydrophobicity of *Azospirillum lipoferum* dropped sharply but recovered its initial value within 24–48 h. Sanin et al. (2003) reported that the cell surface hydrophobicity of xenobiotic degrading bacteria (two *Pseudomonas* strains and *Rhodococcus corallines*) stayed constant during carbon starvation, whereas a significant decrease in the hydrophobicity was observed when all these cultures were starved for nitrogen.

4.3 Other external conditions

Cell surface hydrophobicity is also affected by some other external conditions, such as pH, temperature, oxygen and the presence of multivalent cations or phosphate. Palmgren et al. (1998) tested the influence of oxygen on the cell surface hydrophobicity of 4 bacterial strains isolated from activated sludge, and found that oxygen limitation caused the decrease in the cell surface hydrophobicity of the studied strains. Blanco et al. (1997) found that the majority number of 42 strains of *Candida albicans* were hydrophobic at 22°C, but hydrophilic at 37°C. An opposite trend was reported for *Candida glabara*, which demonstrated higher cell surface hydrophobicity at 37°C than 25°C (Hazen et al. 1986). Sludge retention time (SRT) also influences sludge surface hydrophobicity. Liao et al. (2001) demonstrated that activated sludge at higher SRT (12, 16, 20 d) was more hydrophobic than that at lower SRT (4, 9 d).

The negatively charged functional groups such as hydroxyl and carboxyl groups embedded in the EPS have provided many sorption sites for metal cations (Flemming and Leis 2003). The interactions between metal cations and these functional groups alter the cell surface properties by decreasing the negative charges of the cell surface (Li 2005, Higgins and Novak 1997, Urbain et al. 1993). Cell surface with less negative charges was observed to possess higher hydrophobicity (Liao et al. 2001). Divalent cations, such as Ca^{2+} and Mg^{2+}, are widely considered to impact the bacterial self-immobilization and microbial aggregation (Ding et al. 2008, Yu et al. 2001, Schmidt and Ahring 1993), as well as the expression of cell surface hydrophobicity (Fattom and Shilo 1984). The removal of Ca^{2+} resulted in a disintegration of anaerobic granular sludge (Grotenhuis et al. 1991). Yu et al. (2001) found that an increase of Ca^{2+} concentrations from 150 to 300 mg/L enhanced the anaerobic granulation process, and Jiang et al. (2003) reduced the granulation time of seed activated sludge by dosing 100 mg/L of Ca^{2+}. Besides, the presence of divalent cations Ca^{2+} and Mg^{2+} notably improved the cell surface hydrophobicity of *Agrobacterium* and *Citrobacter* (Khemakhem et al. 2005). However, Singh and Vincent (1987) claimed that the expression of cell surface hydrophobicity of one clumping bacteria from sewage sludge, which was identified as *Pseudomonas* sp., was not influenced by the presence of metal cations.

5. Role of EPS in cell surface hydrophobicity

At the early stage of biofilm formation, once the microorganisms have firmly attached to the surface, they undergo a series of physiological changes to adjust their life on the surface. One common adaption is the expression of large quantities of EPS to form the three-dimensional architecture for hosting the cells. The subsequent biofilm development and the maintenance of structured multicellular microbial communities are also dependent on the production and quantity of EPS (Flemming and Wingender 2010, O'Toole et al. 2000). Since cells are embedded in the EPS matrix, their surface characteristics such as surface charge and hydrophobicity are driven by the EPS. Besides, the change of those properties mediate cellular recognition and promote initial cell adhesion onto the surface (Harimawan and Ting 2016).

EPS production and composition are also dependent on different parameters such as microorganisms, growth phase, substrates type and other external conditions (Sheng et al. 2010). EPS from the bacteria, fungi and micro-algae are mainly composed by PN and PS (Harimawan and Ting 2016, Turu et al. 2016, Ravella et al. 2010). In the mixed-culture aggregates, the composition of extracted EPS is more heterogeneous. LB-EPS from anaerobic granular sludge are found to be mainly composed of HS-like compounds (Yuan and Wang 2013), while PN are determined as the major constituent of TB-EPS from anammox sludge (Ni et al. 2015).

5.1 EPS in biological processes

5.1.1 Microbial adhesion

Tsuneda et al. (2003) investigated the impact of EPS on the bacterial adhesion onto glass beads by using 27 heterotrophic bacterial strains isolated from a wastewater treatment reactor. Their results showed that when the EPS content was relatively high, cell adhesion was enhanced by the hydrophobic interaction induced by the EPS. Meanwhile, four LB-EPS free bacteria (*Bacillus subtilis*, *Streptococcus suis*, *Escherichia coli* and *Pseudomonas putida*) exhibited lower cell surface hydrophobicity values (26.1–65.0%) when compared to the intact cells (47.4–69.3%). Moreover, Parker and Munn (1984) reported that *Aeromonas salmonicida* possessed an additional surface protein, i.e., protein-A, and the cell surface of this strain was more hydrophobic than that of the strains devoid of this protein. Harimawan and Ting (2016) found that the presence of PS promoted the adhesion strength of the EPS produced by *Pseudomonas aeruginosa* and *Bacillus subtilis*, while PN had a less adherence effect.

5.1.2 Granulation

During the aerobic granulation process, EPS production mainly occurs during the exponential phase, and they could serve as carbon and energy source during the starvation phase. Thus, the growth of bacteria in the interior and exterior of granules as well as the integrity of granules are regulated by the EPS (Wang et al. 2006). It was further found that the outer layer of aerobic granules exhibited a higher hydrophobicity than the core of the granules. The insoluble EPS present in the outer layer of the granule would play a protective role with respect to the structure stability as well as the integrity of aerobic granules (Wang et al. 2005). Generally, the production of EPS benefits the granulation process (Liu et al. 2004), and a higher EPS content may result in higher cell surface hydrophobicity (Wang et al. 2006). However, Yu et al. (2009) noticed that the aerobic granulation process was enhanced by EPS-free pellets; they assumed that EPS initially embedded in the seed sludge prior to granulation may sterically slow down subsequent cell-cell contact, thereby delaying aerobic granulation.

LB-EPS and TB-EPS have different contributions to the granulation. By analysing the interaction energy curves of aerobic and anaerobic granules before and after the EPS extraction using extended DLVO theory, Liu et al. (2010) concluded that the LB-EPS always display a positive effect on the granulation, while the role of TB-EPS in the granulation is dependent on the separation distance among the granules. In addition, with increased PN to PS ratio, the granule's surface became more hydrophobic after aerobic granulation than the flocculent seed sludge, which was observed by Zhang et al. (2007). The molecular support in the EPS that contributes to the granulation and structural stability has not been well elucidated, and the conclusions are contradictory to each other. McSwain et al. (2005) found that the stability of granules was dependent on a PN core, and Zhu et al. (2015) also claimed that PN significantly contributed to the formation of granular sludge via surface charge adjustment. In contrast, Seviour et al. (2012) and Wang et al. (2012) found that PS were responsible for the aggregation of flocs into granules.

5.1.3 Flocculation

The surface properties of activated sludge are affected by the EPS composition more than the number of filaments, and the EPS content was negatively correlated with the relative cell surface hydrophobicity and flocculation ability (Wilén et al. 2003). Li and Yang (2007) claimed that excessive EPS production in the form of LB-EPS could weaken cell attachment and the floc structure, which led to poor flocculation and retarded sludge–water separation. Li et al. (2016b) discovered that although the increase of EPS content was beneficial to the aggregation of larger flocs, the flocs with a high EPS content were susceptible to the large-scale fragmentation resulting in much smaller daughter-particles. However, Liao et al. (2001) found that the sludge surface hydrophobicity was influenced by individual EPS constituents instead of the total EPS content. Their results and other studies (Xie et al. 2010, Jorand et al. 1998) demonstrated that PN had a positive influence on the sludge surface hydrophobicity, but not PS. Overall, the cell surface hydrophobicity of the activated sludge increases with the PN concentration, and high cell surface hydrophobicity generally results in a better flocculation (Liu and Fang 2003).

5.1.4 Other biological processes

It is also found that EPS play a key role in binding a large volume of water (i.e., bound water) and influence sludge dewatering (Li and Yang 2007). The breakdown of EPS structure tends to reduce the relative hydrophobicity of the sludge, enabling the release of some water trapped within the floc to the bulk liquid phase, and thus, facilitate water removal (Raynaud et al. 2012). There exists a certain EPS content at which the sludge dewatering reached a maximum (Houghton et al. 2001). The findings of Murthy and Novak (1999) showed that PN are generally more important in sludge flocculation and dewatering than PS, whereby a high PN concentration is detrimental for the dewatering process. Moreover, Higgins and Novak (1997) demonstrated that sludge dewatering correlated well with the PN/PS ratio in sludge EPS.

Abundant negatively charged functional groups of the EPS provide various binding sites for the metal cations (Aquino and Stuckey 2004). This metal binding ability of EPS is also influenced by the hydrophobicity, as hydrophilic EPS fraction of anaerobic granular sludge adsorbed more Zn and Cu than the hydrophobic EPS fraction (Wei et al. 2017). Moreover, EPS are also considered as one of the membrane foulants (Lin et al. 2014), and the hydrophobic fraction in the EPS provides the sorption sites for the organic micro-pollutants such as polycyclic aromatic hydrocarbons (PAHs), or phenolic compounds (Nguyen et al. 2012, Yu et al. 2008, Wang et al. 2002, Liu et al. 2001). The interactions between the metal cations or organic micro-pollutants and EPS can alter cell surface hydrophobicity, and in turn, significantly influence the overall hydrophobicity of microbial aggregates in bioreactors (Liu and Fang 2003). Generally, a higher EPS content may result in a higher cell surface hydrophobicity (Wang et al. 2006), and statistical analyses showed a significantly high correlation between the ability to degrade organic pollutants and cell surface hydrophobicity (Obuekwe et al. 2009).

5.2 EPS hydrophobicity

The relative ratio between the hydrophilic and hydrophobic fractions of the EPS largely depends on the extraction method (Cao et al. 2017). The extraction is based on interrupting the interactions that are responsible for the stabilization of the three-dimensional structure of the EPS (such as hydrogen bonds, electrostatic, hydrophobic or van der Waals interactions) via physical forces or chemical reagents (Figure 6).

Physical extraction methods such as centrifugation, heating, and sonication break the low-energy bonds between the EPS and cell surface. Cationic exchange resin (CER) method, usually DOWEX® 50 × 8 (20–50 mesh in the sodium form), extracts EPS from the biomass via the sheer force provided by the shaking and the destabilization of the microbial aggregates structure by removing divalent cations such as Ca^{2+} (Frølund et al. 1996). Organic reagents like formaldehyde can bind with the functional groups (e.g., amine groups) on the cell walls and break the connection between EPS and these functional groups, which, as a result, has facilitated the extraction of EPS (D'Abzac et al. 2010). Hydrophobic extraction by using surfactants as sodium dodecyl sulfate (SDS) or Tween 20 are rarely reported in the literature, its extraction mechanism lies in the fact that these surfactants improve the solubility of EPS molecules by forming micelles via hydrophobic interactions to enhance the EPS extraction efficiency (Ras et al. 2008). Nevertheless, the breakage degree of those interactions

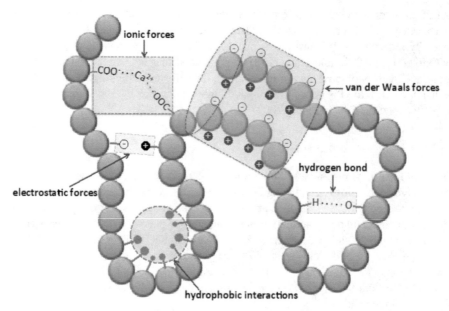

Figure 6. Schematic illustration of various interactions to maintain the EPS structure.

between EPS molecules cannot be estimated during the extraction, and results in unpredictable EPS extraction yield (Zuriaga-Agustí et al. 2013, Domínguez et al. 2010), and the left-over of organic reagents in the EPS solution can interfere with the characterization of the EPS hydrophobicity (Cao et al. 2017a).

Under various conditions (microbial type, pH, temperature and nutrient limitation), EPS exhibit different hydrophobic properties (Tourney and Ngwenya 2014, Ni and Yu 2012, Underwood and Paterson 2003). Some authors concluded that during aerobic granulation process, a more hydrophobic cell surface corresponded to a higher PN/PS ratio (Xie et al. 2010, Zhang et al. 2007). Limited information regarding the molecular support of EPS hydrophobicity can be found in the literature. Studies showed that mainly PN and HS-like contributed to the EPS hydrophobicity, whereas carbohydrates contributed more to the hydrophilic nature (Yu et al. 2008, Jorand et al. 1998). Li et al. (2016a) claimed that PN were responsible for the cell surface hydrophobicity of anaerobic granules, and nonpolar groups such as side chains of aromatic amino acids were pivotal factors deciding the EPS hydrophobicity. Several studies have demonstrated that acetylated PS form the hydrophobic moiety in the EPS structure and support the EPS hydrophobicity (Mayer et al. 1999, Neu et al. 1992). The presence of *N*-acetylamino sugars in the wastewater and treated water was confirmed by pyrolysis, and those molecules were originating from bacterial cell walls (Dignac et al. 2000). Besides, intra-molecular hydrogen bonding of PS can inhibit interactions between the hydroxyl groups and water, thus making the PS hydrophobic. Therefore, the gel-forming PS will influence the hydrophobicity of microbial aggregates as soon as they form a continuous network that potentially allows for hydrophobic components to be immobilized within the biofilm matrix as filler material (Seviour et al. 2012). In addition, HS-like compounds are also integral constituents of EPS extracted from the sludge used in wastewater treatment processes (Cao et al. 2017, Yuan and Wang 2013, D'Abzac et al. 2010, Riffaldi et al. 1982), but the hydrophobic features of these HS-like compounds in the EPS are usually neglected in EPS-related studies. Cao et al. (2017b) showed that the molecular support of EPS hydrophobicity in the anaerobic granular sludge was mostly ascribed to the HS-like compounds, and the aMW of these hydrophobic HS-like compounds mainly ranged from < 1 kDa to 8 kDa.

6. Future work

Attached growth processes (i.e., biofilm formation or granulation and activated sludge flocculation) start with microbial adhesion, where cell surface hydrophobicity acts as a triggering force when molecules are incapable of interacting electrostatically or establishing hydrogen bonds with water molecules (Muadcheingka and Tantivitayakul 2015, Liu et al. 2004). Since the secretion of EPS is a universal characteristic of microorganisms and abundant EPS molecules are found in-/outside granules or flocs, cell surface hydrophobicity of the microbial aggregates largely depends on the quantity and quality of EPS (Zhang et al. 2015b, More et al. 2014). Therefore, a thorough study of EPS hydrophobicity and the influencing factors in the EPS hydrophobicity becomes highly demanded.

Most of EPS studies are *ex situ* studies, in other words, investigating EPS characteristics after extracting them from the microorganisms. The EPS extraction methods are based on breaking the linkage between the cell surface and the EPS molecules, or within the EPS molecules (D'Abzac et al. 2010). This indicates that the extraction procedure may disrupt the original molecular structure of EPS and cell surface. Moreover, the EPS chemical extractants remaining in the extracted EPS solution or on the cell surface cannot be predicted or evaluated. Therefore, the results obtained from *ex situ* studies can only reflect the EPS or the cell surface characteristics under certain conditions. In a more complex microbial system such as biofilms, granules or activated sludge, the distribution of the microorganisms in the systems is more heterogeneous. As a result, the exposure of the nutrients or wastes to the microorganisms is unevenly assigned between the inner and outer layer of these aggregates. Therefore, the production and the distribution of EPS in these microbial aggregates is also heterogeneous (Guo et al. 2011). In turn, EPS extracted from these microbial aggregates can be considered as "average" EPS produced by all the microorganisms in the system. To make a more specific EPS characterization, different EPS extraction methods may be needed in the same study. The type of biomass and working conditions should be carefully controlled, as these two factors also influence the EPS production and their properties. Thus, by comparing the characteristics of EPS molecules extracted by different methods, more representative information about EPS extracted from certain type of biomass can be revealed.

A better characterization of EPS hydrophobicity will contribute to the understanding of microbial adhesion. Evidence showed that mainly PN and HS-like determined EPS hydrophobicity, and acetylated polysaccharides could form a hydrophobic "pocket" within the EPS structure and possibly contributed to the EPS hydrophobicity (Yu et al. 2008, Zhang et al. 2007, Jorand et al. 1998). As those organic compounds are macromolecules and usually link with each other (Bourven et al. 2015, Watanabe and Inoko 2013, Grotenhuis et al. 1991), the specific molecular structure in determining EPS hydrophobicity is still unknown, and thus, should be studied case by case. After determining the molecular support of the EPS hydrophobicity, the synthesizing pathway of these molecules as a function of the microbial species and the environmental conditions could be a subject of future studies.

Heavy metals such as Fe, Zn, Ni, Cu, and Co can act as micronutrients, and the microbial growth can be stimulated by adding low quantities of these metals (Thanh et al. 2016, Gikas 2008). Divalent cations like Ca^{2+} and Mg^{2+} are bridging ions to maintain the stability of the EPS structure (Frølund et al. 1996). Moreover, the release of some extracellular molecules is carried out as a function of the cell membrane enzymes. The normal enzymatic activity required for proper energy metabolism in living cells is also influenced by those metallic co-factors (Kosaric and Blaszczyk 1990). It is therefore essential to investigate whether the presence of those metal ions could change EPS hydrophobicity, aiming at the manipulation of engineered bioprocesses for enhancing metal bioavailability in case of trace metallic element dosing (Wei et al. 2017, Thanh et al. 2016). Likewise, many toxic hydrocarbons are present as environmental pollutants, but are also potential substrates for the bacteria (Xu et al. 2013, Jia et al. 2011, Heipieper et al. 2010, Torres et al. 2009). The removal of these organic compounds by the microorganisms is largely influenced by the cell surface hydrophobicity (Aksu 2005). Since the EPS content is positively correlated with the microbial cell surface hydrophobicity (Obuekwe et al. 2009), the interplay between EPS production and the removal ability of organic micro-pollutants by the EPS should also be investigated in the future.

Acknowledgements

The authors thank the EU for providing financial support through the Erasmus Mundus Joint Doctorate Programme ETeCoS³ (Environmental Technologies for Contaminated Solids, Soils and Sediments, grant agreement FPA no. 2010-0009). The authors also thank the Regional Council of Limousin for its financial support.

References

Adav, S.S. and D.J. Lee. 2008. Single-culture aerobic granules with *Acinetobacter calcoaceticus*. Applied Microbiology and Biotechnology 78: 551–557.

Aiken, G.R., D.M. McKnight, K.A. Thorn and E.M. Thurman. 1992. Isolation of hydrophilic organic acids from water using nonionic macroporous resins. Organic Geochemistry 18: 567–573.

Aksu, Z. 2005. Application of biosorption for the removal of organic pollutants: a review. Process Biochemistry 40: 997–1026.

Allison, D.G., B. Ruiz, C. SanJose, A. Jaspe and P. Gilbert. 1998. Extracellular products as mediators of the formation and detachment of *Pseudomonas fluorescens* biofilms. FEMS Microbiology Letters 167: 179–184.

Alpert, A.J. 1988. Hydrophobic interaction chromatography of peptides as an alternative to reversed-phase chromatography. Journal of Chromatogrphy A 444: 269–274.

Aono, R. and H. Kobayashi. 1997. Cell surface properties of organic solvent-tolerant mutants of *Escherichia coli* K-12. Applied and Environmental Microbiology 63: 3637–3642.

Aquino, S.F. and D.C. Stuckey. 2004. Soluble microbial products formation in anaerobic chemostats in the presence of toxic compounds. Water Research 38: 255–266.

Baca, M., J. De Vos, G. Bruylants, K. Bartik, X. Liu, K. Cook and S. Eeltink. 2016. A comprehensive study to protein retention in hydrophobic interaction chromatography. Journal of Chromatogrphy B 1032: 182–188.

Balebona, M., M. Moriñigo and J. Borrego. 2001. Hydrophobicity and adhesion to fish cells and mucus of *Vibrio* strains isolated from infected fish. International Microbiology 4: 21–26.

Basson, A., L.A. Flemming, and H.Y. Chenia. 2007. Evaluation of adherence, hydrophobicity, aggregation, and biofilm development of *Flavobacterium johnsoniae*-like isolates. Microbial Ecology 55: 1–14.

Blanco, M.T., J. Blanco, R. Sanchez-Benito, C. Pérez-Giraldo, F.J. Morán, C. Hurtado and A.C. Gómez-García. 1997. Incubation temperatures affect adherence to plastic of *Candida albicans* by changing the cellular surface hydrophobicity. Microbios 89: 23–28.

Bolto, B., G. Abbt-Braun, D. Dixon, R. Eldridge, F. Frimmel, S. Hesse, S. King and M. Toifl. 1999. Experimental evaluation of cationic polyelectrolytes for removing natural organic matter from water. Water Science and Technology 40: 71–79.

Bossier, P. and W. Verstraete. 1996. Triggers for microbial aggregation in activated sludge? Applied Microbiology and Biotechnology 45: 1–6.

Bourven, I., G. Bachellerie, G. Costa and G. Guibaud. 2015. Evidence of glycoproteins and sulphated proteoglycan-like presence in extracellular polymeric substance from anaerobic granular sludge. Environmental Technology 36: 2428–2435.

Büchs, J., N. Mozes, C. Wandrey and P.G. Rouxhet. 1988. Cell adsorption control by culture conditions. Applied Microbiology and Biotechnology 29: 119–128.

Busscher, H.J., A.H. Weerkamp, H.C. van der Mei, A.W. van Pelt, H.P. de Jong, and J. Arends. 1984. Measurement of the surface free energy of bacterial cell surfaces and its relevance for adhesion. Applied and Environmental Microbiology 48: 980–983.

Busscher, H.J. and A.H. Weerkamp. 1987. Specific and non-specific interactions in bacterial adhesion to solid substrata. FEMS Microbiology Letters 46: 165–173.

Busscher, H.J., W. Norde and H.C. van der Mei. 2008. Specific molecular recognition and nonspecific contributions to bacterial interaction forces. Applied and Environmental Microbiology 74: 2559–2564.

Busscher, H.J., W. Norde, P.K. Sharma and H.C. van der Mei. 2010. Interfacial re-arrangement in initial microbial adhesion to surfaces. Current Opinion in Colloid & Interface Science 15: 510–517.

Cao, F., I. Bourven, P.N.L. Lens, E.D. van Hullebusch, Y. Pechaud and G. Guibaud. 2017a. Hydrophobic features of EPS extracted from anaerobic granular sludge: an investigation based on DAX-8 resin fractionation and size exclusion chromatography. Applied Microbiology and Biotechnology 101: 3427–3438.

Cao, F., I. Bourven, P.N.L. Lens, E.D. van Hullebusch, Y. Pechaud and G. Guibaud. 2017b. Hydrophobic molecular features of EPS extracted from anaerobic granular sludge treating wastewater from a paper recycling plant. Process Biochemistry 58: 266–275. doi: 10.1016/j.procbio.2017.04.025.

Castellanos, T., F. Ascencio and Y. Bashan. 2000. Starvation-induced changes in the cell surface of *Azospirillum lipoferum*. FEMS Microbiol Ecology 33: 1–9.

Chao, Y., F. Guo, H.H.P. Fang and T. Zhang. 2014. Hydrophobicity of diverse bacterial populations in activated sludge and biofilm revealed by microbial adhesion to hydrocarbons assay and high-throughput sequencing. Colloids and Surfaces B: Biointerfaces 114: 379–385.

Chen, L., Y. Tian, C. Cao, J. Zhang and Z. Li. 2012. Interaction energy evaluation of soluble microbial products (SMP) on different membrane surfaces: Role of the reconstructed membrane topology. Water Research 46: 2693–2704.

Chiesa, S.C., R.L. Irvine and J.F. Manning. 1985. Feast/famine growth environments and activated sludge population selection. Biotechnology and Bioengineering 27: 562–568.

Choo, K.H. and C.H. Lee. 1996. Effect of anaerobic digestion broth composition on membrane permeability. Water Science and Technology 34: 173–179.

Chung, E., S. Yiacoumi and C. Tsouris. 2014. Interaction forces between spores and planar surfaces in aqueous solutions. Colloids and Surfaces A: Physicochemical and Engineering Aspects 443: 80–87.

Comte, S., G. Guibaud and M. Baudu. 2006a. Relations between extraction protocols for activated sludge extracellular polymeric substances (EPS) and EPS complexation properties: Part I. Comparison of the efficiency of eight EPS extraction methods. Enzyme and Microbial Technology 38: 237–245.

Comte, S., G. Guibaud and M. Baudu. 2006b. Relations between extraction protocols for activated sludge extracellular polymeric substances (EPS) and complexation properties of Pb and Cd with EPS: Part II. Consequences of EPS extraction methods on Pb^{2+} and Cd^{2+} complexation. Enzyme and Microbial Technology 38: 246–252.

Correa, O.S., E.A. Rivas and A.J. Barneix. 1999. Cellular envelopes and tolerance to acid pH in *Mesorhizobium loti*. Current Microbiology 38: 329–334.

D'Abzac, P., F. Bordas, E.D. van Hullebusch, P.N.L. Lens and G. Guibaud. 2010. Extraction of extracellular polymeric substances (EPS) from anaerobic granular sludges: comparison of chemical and physical extraction protocols. Applied Microbiology and Biotechnology 85: 1589–1599.

D'Abzac, P., F. Bordas, E. Joussein, E.D. van Hullebusch, P.N.L. Lens and G. Guibaud. 2012. Metal binding properties of extracellular polymeric substances extracted from anaerobic granular sludges. Environmental Science and Pollution Research 20: 4509–4519.

Daffonchio, D., J. Thaveesri and W. Verstraete. 1995. Contact angle measurement and cell hydrophobicity of granular sludge from upflow anaerobic sludge bed reactors. Applied and Environmental Microbiology 61: 3676–3680.

Ding, Y.X., W.C. Chin, A. Rodriguez, C.C. Hung, P.H. Santschi and P. Verdugo. 2008. Amphiphilic exopolymers from *Sagittula stellata* induce DOM self-assembly and formation of marine microgels. Marine Chemistry 112: 11–19.

Dogsa, I., M. Kriechbaum, D. Stopar and P. Laggner. 2005. Structure of bacterial extracellular polymeric substances at different pH values as determined by SAXS. Biophysics Journal 89: 2711–2720.

Domínguez, L., M. Rodríguez and D. Prats. 2010. Effect of different extraction methods on bound EPS from MBR sludges. Part I: Influence of extraction methods over three-dimensional EEM fluorescence spectroscopy fingerprint. Desalination 261: 19–26.

Domínguez, C.L., P.M. Rodríguez and D.P. Rico. 2013. Characterization of soluble and bound EPS obtained from 2 submerged membrane bioreactors by 3D-EEM and HPSEC. Talanta 115: 706–712.

Dufrêne, Y.F. 2015. Sticky microbes: forces in microbial cell adhesion. Trends in Microbiology 23: 376–382.

Evans, P.J., M.R. Bird, A. Pihlajamäki and M. Nyström. 2008. The influence of hydrophobicity, roughness and charge upon ultrafiltration membranes for black tea liquor clarification. Journal of Membrane Science 313: 250–262.

Fattom, A. and M. Shilo. 1984. Hydrophobicity as an adhesion mechanism of benthic cyanobacteria. Applied and Environmental Microbiology 47: 135–143.

Flemming, H.C. and A. Leis. 2003. Sorption properties of biofilms. pp. 2959–2966. *In*: Gabriel Bitton (ed.). Encyclopedia of Environmental Microbiology. John Wiley & Sons, Inc.. New York, USA.

Flemming, H.C. and J. Wingender. 2010. The biofilm matrix. Nature Reviews Microbiology 8: 623–633.

Flemming, H.C., J. Wingender, U. Szewzyk, P. Steinberg, S.A. Rice and S. Kjelleberg. 2016. Biofilms: an emergent form of bacterial life. Nature Reviews Microbiology 14: 563–575.

Flint, S.H., J.D. Brooks and P.J. Bremer. 1997. The influence of cell surface properties of thermophilic streptococci on attachment to stainless steel. Journal of Applied Microbiology 83: 508–517.

Frølund, B., R. Palmgren, K. Keiding and P.H. Nielsen. 1996. Extraction of extracellular polymers from activated sludge using a cation exchange resin. Water Research 30: 1749–1758.

Gallardo-Moreno, A.M., M.L. Navarro-Pérez, V. Vadillo-Rodríguez, J.M. Bruque and M.L. González-Martín. 2011. Insights into bacterial contact angles: Difficulties in defining hydrophobicity and surface Gibbs energy. Colloids and Surfaces B: Biointerfaces 88: 373–380.

Gao, B., X. Zhu, C. Xu, Q. Yue, W. Li and J. Wie. 2008. Influence of extracellular polymeric substances on microbial activity and cell hydrophobicity in biofilms. Journal of Chemical Technology and Biotechnology 83: 227–232.

Geertsema-Doornbusch, G.I., H.C. van der Mei and H.J. Busscher. 1993. Microbial cell surface hydrophobicity: The involvement of electrostatic interactions in microbial adhesion to hydrocarbons (MATH). Journal of Microbiological Methods 18: 61–68.

Grotenhuis, J.T.C., J.B. van Lier, C.M. Plugge, A.J.M. Stams and A.J.B. Zehnder. 1991. Effect of ethylene glycol-bis(β-aminoethyl ether)-N,N-tetraacetic acid (EGTA) on stability and activity of methanogenic granular sludge. Applied Microbiology and Biotechnology 36: 109–114.

Guo, F., S.H. Zhang, X. Yu and B. Wie. 2011. Variations of both bacterial community and extracellular polymers: The inducements of increase of cell hydrophobicity from biofloc to aerobic granule sludge. Bioresource Technology 102: 6421–6428.

Haimer, E., A. Tscheliessnig, R. Hahn and A. Jungbauer. 2007. Hydrophobic interaction chromatography of proteins IV: Kinetics of protein spreading. Journal of Chromatography A 1139: 84–94.

Harimawan, A. and Y.P. Ting. 2016. Investigation of extracellular polymeric substances (EPS) properties of *P. aeruginosa* and *B. subtilis* and their role in bacterial adhesion. Colloids and Surfaces B: Biointerfaces 146: 459–467.

Hazen, K.C., B.J. Plotkin and D.M. Klimas. 1986. Influence of growth conditions on cell surface hydrophobicity of *Candida albicans* and *Candida glabrata*. Infection and Immunity 54: 269–271.

Heipieper, H.J., S. Cornelissen and M. Pepi. 2010. Surface properties and cellular energetics of bacteria in response to the presence of hydrocarbons. pp. 1615–1624. *In*: K.N. Timmis (ed.). Handbook of Hydrocarbon and Lipid Microbiology. Springer Berlin Heidelberg.

Higgins, M.J. and J.T. Novak. 1997. Characterization of exocellular protein and its role in bioflocculation. Journal of Environmental Engineering ASCE 123: 479–485.

Houghton, J.J., J. Quarmby and T. Stephenson. 2001. Municipal wastewater sludge dewaterability and the presence of microbial extracellular polymer. Water Science and Technology 44: 373–379.

Huang, P., X. Zhao and L. Ye. 2016. Synthesis of hydrophobic cationic chitosan flocculant and its sludge dewatering property. Journal of Macromolecular Science Part B 55: 299–309.

Israelachvili, J.N. 2011. Interactions involving polar molecules. pp. 71–90. *In*: Intermolecular and Surface Forces (Third Edition). Academic Press, San Diego.

Jana, T.K., A.K. Srivastava, K. Csery and D.K. Arora. 2000. Influence of growth and environmental conditions on cell surface hydrophobicity of *Pseudomonas fluorescens* in non-specific adhesion. Canadian Journal of Microbiology 46: 28–37.

Jia, C., P. Li, X. Li, P. Tai, W. Liu and Z. Gong. 2011. Degradation of pyrene in soils by extracellular polymeric substances (EPS) extracted from liquid cultures. Process Biochemistry 46: 1627–1631.

Jiang, H.L., J.H. Tay, Y. Liu and S.T.L. Tay. 2003. Ca^{2+} augmentation for enhancement of aerobically grown microbial granules in sludge blanket reactors. Biotechnology Letters 25: 95–99.

Jonsson, P. and T. Wadström. 1984. Cell surface hydrophobicity of *Staphylococcus aureus* measured by the salt aggregation test (SAT). Current Microbiology 10: 203–209.

Jorand, F., F. Boué-Bigne, J.C. Block and V. Urbain. 1998. Hydrophobic/hydrophilic properties of activated sludge exopolymeric substances. Water Science and Technology 37: 307–315.

Jungbauer, A., C. Machold and R. Hahn. 2005. Hydrophobic interaction chromatography of proteins: III. Unfolding of proteins upon adsorption. Journal of Chromatography A 1079: 221–228.

Khemakhem, W., E. Ammar and A. Bakhrouf. 2005. Effect of environmental conditions on hydrophobicity of marine bacteria adapted to textile effluent treatment. World Journal of Microbiology and Biotechnology 21: 1623–1631.

Kosaric, N. and R. Blaszczyk. 1990. Microbial aggregates in anaerobic wastewater treatment. pp. 27–62. *In*: Bioprocesses and Applied Enzymology. Advances in Biochemical Engineering/Biotechnology, Vol 42. Springer Berlin Heidelberg.

Kunz, W., J. Henle and B.W. Ninham. 2004. Zur Lehre von der Wirkung der Salze (about the science of the effect of salts): Franz Hofmeister's historical papers. Current Opinion in Colloid and Interface Science 9: 19–37.

Kwok, D.Y. and A.W. Neumann. 1999. Contact angle measurement and contact angle interpretation. Advances in Colloid Interface Science 81: 167–249.

Laspidou, C.S. and B.E. Rittmann. 2002. A unified theory for extracellular polymeric substances, soluble microbial products, and active and inert biomass. Water Research 36: 2711–2720.

Lee, K.K. and K.C. Yii. 1996. A comparison of three methods for assaying hydrophobicity of pathogenic vibrios. Letters in Applied Microbiology 23: 343–346.

Leenheer, J.A. 1981. Comprehensive approach to preparative isolation and fractionation of dissolved organic carbon from natural waters and wastewaters. Environmental Science and Technology 15: 578–587.

Leloup, M., R. Nicolau, V. Pallier, C. Yéprémian and G. Feuillade-Cathalifaud. 2013. Organic matter produced by algae and cyanobacteria: Quantitative and qualitative characterization. Journal of Environmental Science 25: 1089–1097.

Li, J. 2005. Effects of Fe(III) on floc characteristics of activated sludge. Journal of Chemical Technology and Biotechnology 80: 313–319.

Li, X.Y. and S.F. Yang. 2007. Influence of loosely bound extracellular polymeric substances (EPS) on the flocculation, sedimentation and dewaterability of activated sludge. Water Research 41: 1022–1030.

Li, Y., P. Zheng, M. Zhang, Z. Zeng, Z. Wang, A. Ding and K. Ding. 2016a. Hydrophilicity/hydrophobicity of anaerobic granular sludge surface and their causes: An *in situ* research. Bioresource Technology 220: 117–123.

Li, Z., P. Lu, D. Zhang, G. Chen, S. Zeng and Q. He. 2016b. Population balance modeling of activated sludge flocculation: Investigating the influence of extracellular polymeric substances (EPS) content and zeta potential on flocculation dynamics. Separation and Purification Technology 162: 91–100.

Liao, B.Q., D.G. Allen, I.G. Droppo, G.G. Leppard and S.N. Liss. 2001. Surface properties of sludge and their role in bioflocculation and settleability. Water Research 35: 339–350.

Lienqueo, M.E., E.W. Leser and J.A. Asenjo. 1996. An expert system for the selection and synthesis of multistep protein separation processes. Computers and Chemical Engineering 20: S189–S194.

Lienqueo, M.E., A. Mahn, J.C. Salgado and J.A. Asenjo. 2007. Current insights on protein behavior in hydrophobic interaction chromatography. Journal of Chromatography B 849: 53–68.

Lin, H., M. Zhang, F. Wang, F. Meng, B.Q. Liao, H. Hong, J. Chen and W. Gao. 2014. A critical review of extracellular polymeric substances (EPSs) in membrane bioreactors: Characteristics, roles in membrane fouling and control strategies. Journal of Membrane Science 460: 110–125.

Lindahl, M., A. Faris, T. Wadström and S. Hjertén. 1981. A new test based on "salting out" to measure relative surface hydrophobicity of bacterial cells. Biochimica et Biophysica Acta 677: 471–476.

Liss, S.N., I.G. Droppo, G.G. Leppard and T.G. Milligan. 2004. Flocculation in Natural and Engineered Environmental Systems. CRC Press.

Liu, A., I.S. Ahn, C. Mansfield, L.W. Lion, M.L. Shuler and W.C. Ghiorse. 2001. Phenanthrene desorption from soil in the presence of bacterial extracellular polymer: Observations and model predictions of dynamic behavior. Water Research 35: 835–843.

Liu, X.W., G.P. Sheng and H.Q. Yu. 2009. Physicochemical characteristics of microbial granules. Biotechnology Advances 27: 1061–1070.

Liu, X.M., G.P. Sheng, H.W. Luo, F. Zhang, S.J. Yuan, J. Xu, R.J. Zeng, J.G. Wu and H.Q. Yu. 2010. Contribution of extracellular polymeric substances (EPS) to the sludge aggregation. Environ. Science and Technology 44: 4355–4360.

Liu, Y. and H.H.P. Fang. 2003. Influences of extracellular polymeric substances (EPS) on flocculation, settling, and dewatering of activated sludge. Critical Reviews in Environmental Science and Technology 33: 237–273.

Liu, Y., S.F. Yang, L. Qin and J.H. Tay. 2003. A thermodynamic interpretation of cell hydrophobicity in aerobic granulation. Applied Microbiology and Biotechnology 64: 410–415.

Liu, Y., S.F. Yang, J.H. Tay, Q.S. Liu, L. Qin and Y. Li. 2004. Cell hydrophobicity is a triggering force of biogranulation. Enzyme and Microbial Technology 34: 371–379.

Ljungh, A. and T. Wadström. 1982. Salt aggregation test for measuring cell surface hydrophobicity of urinary *Escherichia coli*. European Journal of Clinical Microbiology 1: 388–393.

Magnusson, K.E. 1980. The hydrophobic effect and how it can be measured with relevance for cell-cell interactions. Scandinavian Journal of Infectious Diseases Supplement 24: 131–134.

Mahn, A., M.E. Lienqueo and J.A. Asenjo. 2007. Optimal operation conditions for protein separation in hydrophobic interaction chromatography. Journal of Chromatography B 849: 236–242.

Mahoney, E.M., L.K. Varangu, W.L. Cairns, N. Kosaric and R.G.E. Murray. 1987. The effect of calcium on microbial aggregation during UASB reactor start-up. Water Science and Technology 19: 249–260.

Malcolm, R.L. 1991. Factors to be considered in the isolation and characterization of aquatic humic substances. pp. 7–36. *In*: P.B. Allard, D.H. Borén and P.A. Grimvall (eds.). Humic Substances in the Aquatic and Terrestrial Environment. Springer Berlin Heidelberg.

Malmqvist, T. 1983. Bacterial hydrophobicity measured as partition of palmitic acid between the two immiscible phases of cell surface and buffer. Acta Pathologica Microbiologica et Immunologica Scandinavica - Section B Microbiology 91B: 69–73.

Marhaba, T.F., Y. Pu and K. Bengraine. 2003. Modified dissolved organic matter fractionation technique for natural water. Journal of Hazardous Materials 101: 43–53.

Marmur, A. 1988. Penetration of a small drop into a capillary. Journal of Colloid and Interface Science 122: 209–219.

Martín-Cereceda, M., F. Jorand, A. Guinea and J.C. Block. 2001. Characterization of extracellular polymeric substances in rotating biological contactors and activated sludge flocs. Environmental Technology 22: 951–959.

Mattarelli, P., B. Biavati, M. Pesenti and F. Crociani. 1999. Effect of growth temperature on the biosynthesis of cell wall proteins from *Bifidobacterium globosum*. Research Microbiology 150: 117–127.

McSwain, B.S., R.L. Irvine, M. Hausner and P.A. Wilderer. 2005. Composition and distribution of extracellular polymeric substances in aerobic flocs and granular sludge. Applied and Environmental Microbiology 71: 1051–1057.

Meyer, E.E., K.J. Rosenberg and J. Israelachvili. 2006. Recent progress in understanding hydrophobic interactions. Proceedings of National Academy of Sciences of United States of America 103: 15739–15746.

Moran, A.P., O. Holst, P.J. Brennan and M.V. Itzstein. 2009. Microbial Glycobiology: Structures, Relevance and Applications. Academic Press, Amsterdam.

More, T.T., J.S.S. Yadav, S. Yan, R.D. Tyagi and R.Y. Surampalli. 2014. Extracellular polymeric substances of bacteria and their potential environmental applications. Journal of Environmental Management 144: 1–25.

Mozes, N. and P.G. Rouxhet. 1987. Methods for measuring hydrophobicity of microorganisms. Journal of Microbiological Methods 6: 99–112.

Muadcheingka, T. and P. Tantivitayakul. 2015. Distribution of *Candida albicans* and non-albicans *Candida* species in oral candidiasis patients: Correlation between cell surface hydrophobicity and biofilm forming activities. Archives of Oral Biology 60: 894–901.

Muda, K., A. Aris, M.R. Salim, Z. Ibrahim, M.C.M. van Loosdrecht, M.Z. Nawahwi and A.C. Affarm. 2014. Aggregation and surface hydrophobicity of selected microorganism due to the effect of substrate, pH and temperature. International Biodeterioration and Biodegradation 93: 202–209.

Murthy, S.N. and J.T. Novak. 1999. Factors affecting floc properties during aerobic digestion: Implications for dewatering. Water Environmental Research 71: 197–202.

Nfor, B.K., N.N. Hylkema, K.R. Wiedhaup, P.D.E.M. Verhaert, L.A.M. van der Wielen and M. Ottens. 2011. High-throughput protein precipitation and hydrophobic interaction chromatography: Salt effects and thermodynamic interrelation. Journal of Chromatography A 1218: 8958–8973.

Nguyen, T., F.A. Roddick and L. Fan. 2012. Biofouling of water treatment membranes: A review of the underlying causes, monitoring techniques and control measures. Membranes 2: 804–840.

Ni, B.J. and H.Q. Yu. 2012. Microbial products of activated sludge in biological wastewater treatment systems: A critical review. Critical Reviews in Environmental Science and Technology 42: 187–223.

Nicolau, R., M. Leloup, D. Lachassagne, E. Pinault and G. Feuillade-Cathalifaud. 2015. Matrix-assisted laser desorption/ionization time-of-flight mass spectrometry (MALDI–TOF–MS) coupled to XAD fractionation: Method to algal organic matter characterization. Talanta 136: 102–107.

Nielsen, P.H., A. Jahn and R. Palmgren. 1997. Conceptual model for production and composition of exopolymers in biofilms. Water Science and Technology 36: 11–19.

Nielsen, P.H. and A. Jahn. 1999. Extraction of EPS. pp. 49–72. *In*: D.J. Wingender, D.T.R. Neu and P.D.H.C. Flemming (eds.). Microbial Extracellular Polymeric Substances. Springer Berlin Heidelberg.

Nollet, L.M.L. and L.S.P.D. Gelder. 2013. Handbook of Water Analysis, Third Edition. CRC Press.

Obuekwe, C.O., Z.K. Al-Jadi and E.S. Al-Saleh. 2009. Hydrocarbon degradation in relation to cell-surface hydrophobicity among bacterial hydrocarbon degraders from petroleum-contaminated Kuwait desert environment. International Biodeterioration and Biodegradation 63: 273–279.

O'Toole, G., H.B. Kaplan and R. Kolter. 2000. Biofilm formation as microbial development. Annual Reviews in Microbiology 54: 49–79.

Palmgren, R., F. Jorand, P.H. Nielsen and J.C. Block. 1998. Influence of oxygen limitation on the cell surface properties of bacteria from activated sludge. Water Science and Technology 37: 349–352.

Parkar, S.G., S.H. Flint, J.S. Palmer and J.D. Brooks. 2001. Factors influencing attachment of thermophilic bacilli to stainless steel. Journal of Applied Microbiology 90: 901–908.

Parker, N.D. and C.B. Munn. 1984. Increased cell surface hydrophobicity associated with possession of an additional surface protein by *Aeromonas salmonicida*. FEMS Microbiology Letters 21: 233–237.

Pen, Y., Z.J. Zhang, A.L. Morales-García, M. Mears, D.S. Tarmey, R.G. Edyvean, S.A. Banwart and M. Geoghegan. 2015. Effect of extracellular polymeric substances on the mechanical properties of *Rhodococcus*. Biochimica et Biophysica Acta (BBA) - Biomembranes. 1848: 518–526.

Perkins, T.W., D.S. Mak, T.W. Root and E.N. Lightfoot. 1997. Protein retention in hydrophobic interaction chromatography: Modeling variation with buffer ionic strength and column hydrophobicity. Journal of Chromatography A 766: 1–14.

Peuravuori, J., T. Lehtonen and K. Pihlaja. 2002. Sorption of aquatic humic matter by DAX-8 and XAD-8 resins: Comparative study using pyrolysis gas chromatography. Analytica Chimica Acta 471: 219–226.

Pimentel-Filho, N. de J., M.C. de F. Martins, G.B. Nogueira, H.C. Mantovani and M.C.D. Vanetti. 2014. Bovicin HC5 and nisin reduce *Staphylococcus aureus* adhesion to polystyrene and change the hydrophobicity profile and Gibbs free energy of adhesion. International Journal of Food Microbiology 90: 1–8.

Polak-Berecka, M., A. Waśko, R. Paduch, T. Skrzypek and A. Sroka-Bartnicka. 2014. The effect of cell surface components on adhesion ability of *Lactobacillus rhamnosus*. Antonie Van Leeuwenhoek 106: 751–762.

Powell, C.D., D.E. Quain and K.A. Smart. 2003. The impact of brewing yeast cell age on fermentation performance, attenuation and flocculation. FEMS Yeast Research 3: 149–157.

Qu, F., H. Liang, J. Zhou, J. Nan, S. Shao, J. Zhang and G. Li. 2014. Ultrafiltration membrane fouling caused by extracellular organic matter (EOM) from *Microcystis aeruginosa*: Effects of membrane pore size and surface hydrophobicity. Journal of Membrane Science 449: 58–66.

Ras, M., E. Girbal-Neuhauser, E. Paul, D. Lefebvre. 2008. A high yield multi-method extraction protocol for protein quantification in activated sludge. Bioresource Technology 99: 7464–7471.

Ras, M., D. Lefebvre, N. Derlon, J. Hamelin, N. Bernet, E. Paul and E. Girbal-Neuhauser. 2013. Distribution and hydrophobic properties of extracellular polymeric substances in biofilms in relation towards cohesion. Journal of Biotechnology 165: 85–92.

Ravella, S.R., T.S. Quiñones, A. Retter, M. Heiermann, T. Amon and P.J. Hobbs. 2010. Extracellular polysaccharide (EPS) production by a novel strain of yeast-like fungus *Aureobasidium pullulans*. Carbohydrate Polymers 82: 728–732.

Raynaud, M., J. Vaxelaire, J. Olivier, E. Dieudé-Fauvel and J.C. Baudez. 2012. Compression dewatering of municipal activated sludge: Effects of salt and pH. Water Research 46: 4448–4456.

Riffaldi, R., F. Sartori and R. Levi-Minzi. 1982. Humic substances in sewage sludges. Environmental Pollution Series B, Chemical and Physical 3: 139–146.

Rosenberg, M., D. Gutnick and E. Rosenberg. 1980. Adherence of bacteria to hydrocarbons: A simple method for measuring cell-surface hydrophobicity. FEMS Microbiology Letters 9: 29–33.

Rosenberg, M., M. Barki, R. Bar-Ness, S. Goldberg and R.J. Doyle. 1991. Microbial adhesion to hydrocarbons (MATH). Biofouling 4: 121–128.

Rosenberg, M. 2006. Microbial adhesion to hydrocarbons: Twenty-five years of doing MATH. FEMS Microbiology Letters 262: 129–134.

Rozgonyi, F., K.R. Szitha, Å. Ljungh, S.B. Baloda, S. Hjertén and T. Wadström. 1985. Improvement of the salt aggregation test to study bacterial cell-surface hydrophobicity. FEMS Microbiology Letters 30: 131–138.

Saeki, D., Y. Nagashima, I. Sawada and H. Matsuyama. 2016. Effect of hydrophobicity of polymer materials used for water purification membranes on biofilm formation dynamics. Colloids and Surfaces A: Physicochemical and Engineering Aspects 506: 622–628.

Sakohara, S., E. Ochiai and T. Kusaka. 2007. Dewatering of activated sludge by thermosensitive polymers. Separation and Purification Technology 56: 296–302.

Sanin, S.L., F.D. Sanin and J.D. Bryers. 2003. Effect of starvation on the adhesive properties of xenobiotic degrading bacteria. Process Biochemistry 38: 909–914.

Schmidt, J.E. and B.K. Ahring. 1993. Effects of magnesium on thermophilic acetate-degrading granules in upflow anaerobic sludge blanket (UASB) reactors. Enzyme and Microbial Technology 15: 304–310.

Seale, R.B., S.H. Flint, A.J. McQuillan and P.J. Bremer. 2008. Recovery of spores from thermophilic dairy Bacilli and effects of their surface characteristics on attachment to different surfaces. Applied and Environmental Microbiology 74: 731–737.

Seviour, T., Z. Yuan, M.C.M. van Loosdrecht and Y. Lin. 2012. Aerobic sludge granulation: A tale of two polysaccharides? Water Research 46: 4803–4813.

Shen, Y., W. Zhao, K. Xiao and X. Huang. 2010. A systematic insight into fouling propensity of soluble microbial products in membrane bioreactors based on hydrophobic interaction and size exclusion. Journal of Membrane Science 346: 187–193.

Sheng, G.P., H.Q. Yu and X.Y. Li. 2010. Extracellular polymeric substances (EPS) of microbial aggregates in biological wastewater treatment systems: A review. Biotechnology Advances 28: 882–894.

Shepard, C.C. and A. Tiselius. 1949. The chromatography of proteins: The effect of salt concentration and pH on the adsorption of proteins to silica gel. Discussions of the Faraday Society 7: 275–285.

Singh, K.K. and W.S. Vincent. 1987. Clumping characteristics and hydrophobic behaviour of an isolated bacterial strain from sewage sludge. Applied Microbiology and Biotechnology 25: 396–398.

Sorongon, M.L., R.A. Bloodgood and R.P. Burchard. 1991. Hydrophobicity, adhesion, and surface-exposed proteins of gliding bacteria. Applied and Environmental Microbiology 57: 3193–3199.

Stenström, T.A. 1989. Bacterial hydrophobicity, an overall parameter for the measurement of adhesion potential to soil particles. Applied and Environmental Microbiology 55: 142–147.

Sun, Y., W. Fan, H. Zheng, Y. Zhang, F. Li and W. Chen. 2015. Evaluation of dewatering performance and fractal characteristics of alum sludge. PLoS ONE 10(6): e0130683.

Swoboda, J.G., J. Campbell, T.C. Meredith and S. Walker. 2010. Wall teichoic acid function, biosynthesis, and inhibition. ChemBioChem 11: 35–45.

Thanh, P.M., B. Ketheesan, Z. Yan and D. Stuckey. 2016. Trace metal speciation and bioavailability in anaerobic digestion: A review. Biotechnology Advances 34: 122–136.

Thaveesri, J., K. Gernaey, B. Kaonga, G. Boucneau and W. Verstraete. 1994. Organic and ammonium nitrogen and oxygen in relation to granular sludge growth in lab-scale UASB reactors. Water Science and Technology 30: 43–53.

Thaveesri, J., D. Daffonchio, B. Liessens, P. Vandermeren and W. Verstraete. 1995. Granulation and sludge bed stability in upflow anaerobic sludge bed reactors in relation to surface thermodynamics. Applied and Environmental Microbiology 61: 3681–3686.

Thurman, E.M., R.L. Malcolm and G.R. Aiken. 1978. Prediction of capacity factors for aqueous organic solutes adsorbed on a porous acrylic resin. Analytical Chemistry 50: 775–779.

Torres, S., M.D. Baigorí, S.L. Swathy, A. Pandey and G.R. Castro. 2009. Enzymatic synthesis of banana flavour (isoamyl acetate) by *Bacillus licheniformis* S-86 esterase. Food Research International 42: 454–460.

Tourney, J. and B.T. Ngwenya. 2014. The role of bacterial extracellular polymeric substances in geomicrobiology. Chemical Geology 386: 115–132.

Tsuneda, S., H. Aikawa, H. Hayashi, A. Yuasa and A. Hirata. 2003. Extracellular polymeric substances responsible for bacterial adhesion onto solid surface. FEMS Microbiology Letters 223: 287–292.

Turu, I.C., C. Turkcan-Kayhan, A. Kazan, E. Yildiz-Ozturk, S. Akgol and O. Yesil-Celiktas. 2016. Synthesis and characterization of cryogel structures for isolation of EPSs from *Botryococcus braunii*. Carbohydrate Polymers 150: 378–384.

Underwood, G.J.C. and D.M. Paterson. 2003. The importance of extracellular carbohydrate production by marine epipelic diatoms. Advances in Botanical Research 40: 183–240.

Urbain, V., J.C. Block and J. Manem. 1993. Bioflocculation in activated sludge: an analytic approach. Water Research 27: 829–838.

van der Bruggen, B., D. Segers, C. Vandecasteele, L. Braeken, A. Volodin and C. van Haesendonck. 2010. How a microfiltration pretreatment affects the performance in nanofiltration. Separation Science and Technology 39: 1443–1459.

van Loosdrecht, M.C., J. Lyklema, W. Norde, G. Schraa and A.J. Zehnder. 1987a. Electrophoretic mobility and hydrophobicity as a measured to predict the initial steps of bacterial adhesion. Applied and Environmental Microbiology 53: 1898–1901.

van Loosdrecht, M.C., J. Lyklema, W. Norde, G. Schraa and A.J. Zehnder. 1987b. The role of bacterial cell wall hydrophobicity in adhesion. Applied and Environmental Microbiology 53: 1893–1897.

Wang, B., S. Liu, H. Zhao, X. Zhang and D. Peng. 2012. Effects of extracellular polymeric substances on granulation of anoxic sludge in sequencing batch reactor. Water Science and Technology 66: 543–548.

Wang, W., W. Wang, X. Zhang and D. Wang. 2002. Adsorption of p-chlorophenol by biofilm components. Water Research 36: 551–560.

Wang, Z., L. Liu, J. Yao and W. Cai. 2006. Effects of extracellular polymeric substances on aerobic granulation in sequencing batch reactors. Chemosphere 63: 1728–1735.

Wang, Z.W., Y. Liu and J.H. Tay. 2005. Distribution of EPS and cell surface hydrophobicity in aerobic granules. Applied Microbiology and Biotechnology 69: 469.

Watanabe, Y. and Y. Inoko. 2013. Characterization of a large glycoprotein proteoglycan by size-exclusion chromatography combined with light and X-ray scattering methods. Journal of Chromatography A 1303: 100–104.

Wei, L., Y. Li, D.R. Noguera, N. Zhao, Y. Song, J. Ding, Q. Zhao and F. Cui. 2017. Adsorption of Cu^{2+} and Zn^{2+} by extracellular polymeric substances (EPS) in different sludges: Effect of EPS fractional polarity on binding mechanism. Journal of Hazardous Materials 321: 473–483.

Weis, A., M.R. Bird, M. Nyström and C. Wright. 2005. The influence of morphology, hydrophobicity and charge upon the long-term performance of ultrafiltration membranes fouled with spent sulphite liquor. Desalination 175: 73–85.

Wick, L., A. de Munain, D. Springael and H. Harms. 2002. Responses of *Mycobacterium* sp. LB501T to the low bioavailability of solid anthracene. Applied Microbiology and Biotechnology 58: 378–385.

Wicke, D., U. Böckelmann and T. Reemtsma. 2008. Environmental influences on the partitioning and diffusion of hydrophobic organic contaminants in microbial biofilms. Environ Science and Technology 42: 1990–1996.

Wilén, B.M., B. Jin and P. Lant. 2003. Relationship between flocculation of activated sludge and composition of extracellular polymeric substances. Water Science and Technology 47: 95–103.

Wilschut, J. and D. Hoekstra. 1984. Membrane fusion: from liposomes to biological membranes. Trends in Biochemical Science 9: 479–483.

Xia, F., D. Nagrath, S. Garde and S.M. Cramer. 2004. Evaluation of selectivity changes in HIC systems using a preferential interaction based analysis. Biotechnology and Bioengineering 87: 354–363.

Xie, B., J. Gu and J. Lu. 2010. Surface properties of bacteria from activated sludge in relation to bioflocculation. Journal of Environmental Science 22: 1840–1845.

Xu, C., S. Zhang, C. Chuang, E.J. Miller, K.A. Schwehr and P.H. Santschi. 2011. Chemical composition and relative hydrophobicity of microbial exopolymeric substances (EPS) isolated by anion exchange chromatography and their actinide-binding affinities. Marine Chemistry 126: 27–36.

Xu, H.L. and J.H. Tay. 2002. Anaerobic granulation with methanol-cultured seed sludge. Journal of Environmental Science Health Part A 37: 85–94.

Xu, J., G.P. Sheng, Y. Ma, L.F. Wang and H.Q. Yu. 2013. Roles of extracellular polymeric substances (EPS) in the migration and removal of sulfamethazine in activated sludge system. Water Research 47: 5298–5306.

Yamashita, S., M. Satoi, Y. Iwasa, K. Honda, Y. Sameshima, T. Omasa, J. Kato and H. Ohtake. 2006. Utilization of hydrophobic bacterium *Rhodococcus opacus* B-4 as whole-cell catalyst in anhydrous organic solvents. Applied Microbiology and Biotechnology 74: 761–767.

Yu, C.H., C.H. Wu, C.H. Lin, C.H. Hsiao and C.F. Lin. 2008. Hydrophobicity and molecular weight of humic substances on ultrafiltration fouling and resistance. Separation and Purification Technology 64: 206–212.

Yu, G.H., Y.C. Juang, D.J. Lee, P.J. He and L.M. Shao. 2009. Enhanced aerobic granulation with extracellular polymeric substances (EPS)-free pellets. Bioresource Technology 100: 4611–4615.

Yu, H.Q., J.H. Tay and H.H.P. Fang. 1999. Effects of added powdered and granular activated carbons on start-up performance of UASB reactors. Environmental Technology 20: 1095–1101.

Yu, H.Q., J.H. Tay and H.H.P. Fang. 2001. The roles of calcium in sludge granulation during UASB reactor start-up. Water Research 35: 1052–1060.

Yuan, D. and Y. Wang. 2013. Effects of solution conditions on the physicochemical properties of stratification components of extracellular polymeric substances in anaerobic digested sludge. Journal of Environmental Science 25: 155–162.

Zhang, L., X. Feng, N. Zhu and J. Chen. 2007. Role of extracellular protein in the formation and stability of aerobic granules. Enzyme and Microbial Technology 41: 551–557.

Zhang, M., B. Liao, X. Zhou, Y. He, H. Hong, H. Lin and J. Chen. 2015a. Effects of hydrophilicity/hydrophobicity of membrane on membrane fouling in a submerged membrane bioreactor. Bioresource Technology 175: 59–67.

Zhang, P., F. Fang, Y.P. Chen, Y. Shen, W. Zhang, J.X. Yang, C. Li, J.S. Guo, S.Y. Liu, Y. Huang, S. Li, X. Gao and P. Yan. 2014. Composition of EPS fractions from suspended sludge and biofilm and their roles in microbial cell aggregation. Chemosphere 117: 59–65.

Zhang, P., Y. Shen, J.S. Guo, C. Li, H. Wang, Y.P. Chen, P. Yan, J.X. Yang and F. Fang. 2015b. Extracellular protein analysis of activated sludge and their functions in wastewater treatment plant by shotgun proteomics. Scientific Reports 5: 12041.

Zhu, L., J. Zhou, M. Lv, H. Yu, H. Zhao and X. Xu. 2015. Specific component comparison of extracellular polymeric substances (EPS) in flocs and granular sludge using EEM and SDS-PAGE. Chemosphere 121: 26–32.

Zita, A. and M. Hermansson. 1997. Effects of bacterial cell surface structures and hydrophobicity on attachment to activated sludge flocs. Applied and Environmental Microbiology 63: 1168–1170.

Zuriaga-Agustí, E., A. Bes-Piá, J.A. Mendoza-Roca and J.L. Alonso-Molina. 2013. Influence of extraction methods on proteins and carbohydrates analysis from MBR activated sludge flocs in view of improving EPS determination. Separation and Purification Technology 112: 1–10.

CHAPTER 8

Removal of Pharmaceuticals from Wastewater by Membrane and Biofilm Reactors

Raju Sekar,[1,] Bharathi Ramalingam,[2] Yun Deng,[1,3] Qiaoli Feng[1] and Wenwei Li[2]*

1. Introduction

Pharmaceuticals are considered as emerging contaminants (ECs) along with other chemical compounds such as endocrine disturbing compounds (EDCs), personal care products (PCPs), plasticizers, surfactants, pesticides and flame retardants (Sauve and Desrosiers 2014, Ahmed et al. 2017) and their presence in natural environments is reported to cause various environmental issues and health risks (Ahmed et al. 2017, Esplugas et al. 2007, Rivera-Utrilla et al. 2013). Pharmaceuticals enter the environment through various ways and the major sources reported include treated effluent from industrial and municipal wastewater treatment plants (WWTPs), effluent discharged from pharmaceutical industries and hospitals, wastes generated through research activities and diagnostic medicine, expired medicines which are disposed through wastes, agriculture and animal wastes and landfills (Tiwari et al. 2017, Ahmed et al. 2017). The wastewater which enters the municipal WWTPs carry a wide variety of pharmaceutical products and other chemical residues, which negatively affect the biodegradation process (Pauwels and Verstraete 2006) and the potential of denitrifying microorganisms (Ozdemir et al. 2015) in the WWTPs. The treated effluent which enters the environment affects the life in aquatic systems (Santos et al. 2010, Arya et al. 2016, Tiwari et al. 2017).

[1] Department of Biological Sciences, Xi'an Jiaotong-Liverpool University, Suzhou 215123, China.
[2] Environmental Engineering Lab, Suzhou Institute for Advanced Study, University of Science and Technology of China, Suzhou 215123, China.
[3] Institute of Experimental Immunology, University of Zurich, Switzerland.
* Corresponding author: Sekar.Raju@xjtlu.edu.cn; rajusekar@yahoo.com

Pharmaceutical wastes include antibiotics, hormones, antiphlogistics, analgesics, lipid regulators, iodinated contrast media, antidepressants, non-steroid anti-inflammatory drugs (NSAIDS), psychiatric drugs, antiviral drugs, cosmetics, sun-screen agents and fragrances (Luis Martinez 2009, Tambosi et al. 2010, Mascolo et al. 2010, Dolar et al. 2012, Casas et al. 2015b, Yuan et al. 2015, Tiwari et al. 2017, Ahmed et al. 2017). These compounds are biologically active, persist in the environment with low biodegradation and cause harmful effects to the aquatic ecosystems and, therefore, are, considered as ECs. The increased accumulation of pharmaceutical compounds in the environment leads to the development of multi-drug resistant microorganisms, which is a serious threat to the health of humans and animals (Carvalho and Santos 2016, Gaze et al. 2008, Luis Martinez 2009).

Antibiotics, among all the pharmaceuticals, have received global attention in recent years due to their high consumption, existence in different environments and development of antibiotic resistance in clinically relevant bacteria (WHO 2016). It has been reported that from 2000 to 2010, the antibiotic usage increased by 36% in 72 countries, with US as the main consumer. However, a recent analysis on the global antibiotic consumption between 2000 and 2015 showed that India and China are the top two consumers (Klein et al. 2018). Moreover, the economic growth in BRICS (Brazil, Russia, India, China and South Africa) countries appears to have contributed to worldwide access to antibiotics (Talesnik 2016). There are several environmental pathways through which antibiotics enter into aquatic environments. In human and veterinary medicine, the antibiotics are widely used and the unabsorbed antibiotics are released into the environments, which contribute to the development of antibiotic resistance in bacteria. A recent review indicated that the antibiotics which are used in human or animal medicine enter into the environment through wastes excreted via urine and faeces (Carvalho and Santos 2016). Moreover, treated effluents from WWTPs have been reported to carry antibiotics and other pharmaceutical products, as conventional treatment process do not adequately remove these compounds from wastewater (Watkinson et al. 2007, Homem and Santos 2011, Santos et al. 2013, Xu et al. 2015, Verlicchi et al. 2015, Grandclement et al. 2017, Ahmed et al. 2017, Liu et al. 2017). Antibiotics such as tetracycline, sulfamethoxazole, roxithromycin and trimethoprim and analgesics such as acetaminophen are some of the most commonly found drugs in the wastewater (Kim et al. 2007, Tambosi et al. 2010, Abegglen et al. 2009, Sahar et al. 2011b, Gao et al. 2012, Xu et al. 2015, Zhang et al. 2009a). Analgesics and anti-inflammatory drugs, antibiotics, anti-epileptic drugs, anti-histamines, antiulcer agents, β-blockers, cholesterol-lowering statin drugs, lipid regulators, psychiatric drugs, antidiabetics and diuretic compounds were identified in the hospital, municipal, and industrial wastewaters (Radjenovic et al. 2007, Casas et al. 2015b, Yuan et al. 2015). Examples and chemical structure of some of the pharmaceuticals compounds commonly found in wastewater are given in Figure 1.

WWTPs treat wastewater by different processes: physical, chemical, biological and thermal methods. The widely used biological methods to treat wastewater include activated sludge process, aerated lagoons, wastewater stabilization lagoons, trickling filters, aerobic/anaerobic digestion and membrane bioreactor (MBR). MBR is one of the most efficient approaches used for removing organic and inorganic compounds from the wastewater (Judd 2008) by combining the filtration method with an activated sludge bioreactor system. MBR has been widely used in the treatment of municipal wastewater for many years (van der Roest et al. 2002, Du et al. 2011, Yoo et al. 2012). The advantages of this technology over the conventional approaches include small footprint, effluent with high quality, superior removal of toxic compounds and ease of operation (Judd 2011, Stephenson et al. 2007). The filter membrane, which is located inside the reactor, has approximately < 0.1 μm pore size and acts as microfiltration or ultrafiltration membrane in the process. This method removes larger particles and most microbes by physically excluding them outside of the membrane during the filtration process. Activated sludge contains enormous amount of aerobic microorganisms that require nutrients such as organic carbon, nitrogen and phosphate for their growth. By mixing activated sludge with wastewater, organic and biodegradable compounds will be absorbed by these microorganisms and precipitated along with other large molecules and finally removed from the wastewater (Sustarsic 2009). This chapter focuses on biological treatment methods, particularly in the context of removal of

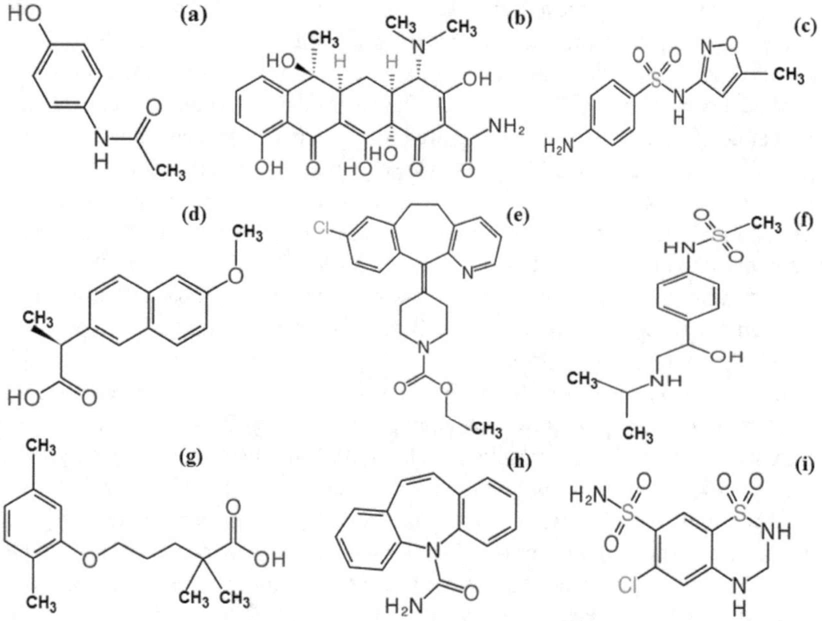

Figure 1. Chemical structure of some of the pharmaceutical compounds commonly found in wastewater; (a) acetaminophen (analgesics and anti-inflammatory drug); (b) tetracycline (antibiotic); (c) sulfamethoxazole (antibiotic); (d) naproxen (anti-inflammatory drug); (e) loratadine (antihistamine); (f) sotalol (β-blocker); (g) gemfibrozil (lipid regulator and cholesterol lowering drug); (h) carbamazepine (antiepileptic drug); (i) hydrochlorothiazide (diuretic drug). The chemical structures were drawn using ChemSpider (2001).

pharmaceuticals from the wastewater and the application of MBR for removal of antibiotic resistant bacteria (ARB) and antibiotic resistant genes (ARGs) from wastewater.

2. Membrane bioreactors for removing pharmaceuticals from wastewater

Membrane and biofilm based reactors used for removal of pharmaceuticals from wastewater in the past decade (representative publications) are summarized. Table 1 shows the removal methods used for treating different pharmaceuticals along with removal efficiencies of each method. Kim et al. (2007) used MBR to remove 14 pharmaceutical products, six hormones, three PCPs, two antibiotics and a flame retardant. This study reported that MBR was effective in eliminating hormones and some pharmaceutical products (acetaminophen and ibuprofen), but it had limited efficiency in removing other compounds. Similar removal efficiencies were observed by others (Abegglen et al. 2009); the removal of 25 pharmaceutical products including antibiotics, antiphlogistics, hormones, iodinated contrast media, lipid regulators were same as the removal at a centralized WWTP; they reported that the main elimination process was biological transformation. Dorival-Garcia et al. (2013) observed the influence of COD (chemical oxygen demand), MLSS (mixed liquor suspended solids), and temperature on the biodegradation process and removal of six quinolone antibiotics (ciprofloxacin, moxifloxacin, norfloxacin, ofloxacin, pipemidic acid and piromidic acid) using MBR. The study identified sorption as the main removal mechanism, followed by biodegradation, with a high degradation rate observed for easily biodegradable organic compounds. Another study reported that both adsorption and biodegradation were effective in removing trace organic contaminants (TrOCs) representing pharmaceuticals, hormones, UV filters, phytoestrogens and pesticides (Wijekoon et al. 2013). Hydrophilic TrOCs were mainly removed by biodegradation, while hydrophobic TrOCs were removed mainly by adsorption. Tambosi et al. (2010) demonstrated that higher removal efficiency was observed with MBR-30 (operated for 30 days) as compared to MBR-15 (15 days) for all the

Table 1. Removal of various pharmaceuticals from wastewater by membrane and biofilm reactors. (AnMBR—anaerobic membrane bioreactor; ARGs—antibiotic resistance genes; BERs—biofilm-electrode reactors; CAS—conventional activated sludge; DAF-MBR—dissolved air flotation-membrane bioreactor; EDCs—endocrine disrupting chemicals; ELAMBR—an external loop air lift membrane bioreactor; GAC—granular activated carbon; GMBR—granular sludge membrane bioreactor; HRT—hydraulic retention time; MBR—membrane bioreactor; MBBR—moving bed biofilm reactor; MBfR—Membrane biofilm reactor; MF—membrane filtration; NF—nanofiltration; MS2ALMBRs—multi-sparger multi-stage airlift loop membrane bioreactors; NSAIDS—non steroidal anti-inflammatory drugs; PCPs—personal care products; PMFs—polycyclic musk fragrances; SBBRs—sequencing batch biofilm reactors; SBC—suspended biofilm carriers; SRT—sludge retention time; SABF—submerged attached biofilter; TrOCs—Trace organic contaminants; UF/RO—ultrafiltration/reverse osmosis; WWTPs—wastewater treatment plants).

Sl. No.	Removal methods	Type of pharmaceuticals targeted	Removal efficiency and main findings	Reference(s)
1	MBR	14 pharmaceuticals, 6 hormones, 2 antibiotics, 3 PCPs, and 1 flame retardant	Effectively eliminated hormones and several pharmaceuticals (e.g., acetaminophen, ibuprofen); Limited efficiency in removal of other compounds	(Kim et al. 2007)
2	MBR	25 Pharmaceuticals (antiphlogistics, antibiotics, iodinated contrast media, lipid regulators and hormones)	Removal rate was same range as centralized WWTP; Biological transformation was found to be the main elimination process	(Abegglen et al. 2009)
3	MBRs with two different SRTs	3 NSAIDs and 3 antibiotics	For all compounds, MBR-30 showed higher removal efficiencies than MBR-15. Both MBRs showed the highest removal efficiencies for acetaminophen and ketoprofen, and lower efficiencies for sulfamethoxazole and roxithromycin	(Tambosi et al. 2010)
4	Anoxic/aerobic (A/O-MBR)	6 antibiotics	MBR accomplished high removal rate and antibiotics removal benefited from longer SRT	(Xia et al. 2012)
5	MBR	4 corticosteroids	MBR was effective at removing 4 corticosteroids with a removal efficiency of > 93%	(Lopez-Fernandez et al. 2012)
6	MBR with varying COD, MLSS, temperature	6 quinolones antibiotics (norfloxacin, ciprofloxacin, ofloxacin, pipemidic acid, moxifloxacin, and piromidic acid)	Highest removal efficiencies observed in aerobic MBR with high MLSS concentrations. High degradation rates observed by removal of easily biodegradable organics	(Dorival-Garcia et al. 2013)
7	MBR	29 TrOCs which included pharmaceuticals, UV-filters, steroid hormones, pesticides and phytoestrogens	Both biodegradation and adsorption were proved to be effective in removing TrOCs. The principal mechanism for hydrophilic TrOCs removal was biodegradation and adsorption for hydrophobic TrOCS	(Wijekoon et al. 2013)
8	MBRs with two different SRTs	3 NSAIDs (acetaminophen, ketoprofen and naproxen) and 3 antibiotics (roxithromycin, sulfamethoxazole and trimethoprim) in wastewater	Both MBRs showed higher removal effectiveness for NSAIDs than the antibiotics. MBR-30 was more effective in removal compared to MBR-15. The rates of removal varied from 100% (ketoprofen, acetaminophen) to 55% (sulfamethoxazole)	(Schroeder et al. 2012)

Table 1 contd. ...

...Table 1 contd.

Sl. No.	Removal methods	Type of pharmaceuticals targeted	Removal efficiency and main findings	Reference(s)
9	MBR with short HRT and different MLSS	11 pharmaceutical compounds	Most of the compounds were removed with high efficiency by MBR with a HRT of 3 h. Depending on the properties of the compounds, the removal by biodegradation and adsorption varied	(Prasertkulsak et al. 2016)
10	MBR and CAS	6 acidic pharmaceuticals	MBRs showed greater elimination rates. Higher elimination efficiency was found in MBR with a SRT of 65 days	(Kimura et al. 2007)
11	MBR and CAS	8 analgesics and anti-inflammatory drugs, 5 lipid regulators and cholesterol-lowering statin drugs, 5 antibiotics, 2 psychiatric drugs, an antiepileptic drug, 4 β-blockers, 2 anti-histaminics, 2 anti-ulcer agents, an anti-diabetic and a diuretic drug	MBR treatment efficiently removed most of the compounds; Removal efficiency of paroxetine and hydrochlorothiazide was appreciably higher in CAS treatment. Carbamazepine was not removed by either MBR or CAS treatment	(Radjenovic et al. 2007)
12	MBRs and CAS	31 pharmaceuticals	MBRs demonstrated considerably higher removal efficiencies for many compounds than CAS treatment. MBRs with longer SRTs showed poor removal rate for some compounds (e.g., ranitidine, β-blockers). Neither treatment could remove anti-epileptic drug diuretic hydrochlorothiazide and carbamazepine	(Radjenovic et al. 2009)
13	MBR and CAS	EDCs and antibiotics	Both treatments eliminated androgens and steroidal estrogens effectively. However, the target antibiotics were removed only partially. MBR showed higher removal efficiency than CAS system	(Le-Minh et al. 2010)
14	ELAMBR and CAS	Acetaminophen in synthetic pharmaceutical wastewater	MBR showed high efficiency in removing acetaminophen	(Shariati et al. 2010)
15	CAS and MBR with different MLSS, HRT and SRT	Commonly used 21 antimicrobials that belongs to 10 different classes	Higher removal efficiencies with MBR system were observed compared to CAS. Biological WWT processes removed beta-lactam, fluoroquinolone antibiotics and glycopeptide quiet easily. But the removal rates of lincosamides and trimethoprim were low	(Ngoc Han et al. 2016)
16	MBR and CAS	9 sulfonamides (SAs) and one of their acetylated metabolites	Removal rate was 100% for 3 SAs with MBR whereas the removal rate was below 55% and similar between CAS and MBR for rest of the SAs. Sulfamethizole showed negative removal efficiency in all the treatments	(Garcia Galan et al. 2012)
17	MBR/RO and CAS-UF/RO	6 antibiotics, 3 pharmaceuticals, 1 industrial product (BPA), and 1 hormone (cholesterol)	High elimination rates (93%–99%) were accomplished after RO although significant differences between selected OMPs were observed. The study concluded that RO can be used for efficient removal but cannot completely eliminate OMPs	(Sahar et al. 2011a)

No.	Treatment system	Target pharmaceuticals	Description	Reference
18	CAS+UF and MBR	Macrolide, sulfonamide, and trimethoprim antibiotics	Higher elimination efficiency was observed with MBR for all tested antibiotics as compared to CAS. The removal efficiency of antibiotics was highly improved by adding UF stage after CAS. Thus, removal efficiencies between the two systems were comparable	(Sahar et al. 2011c)
19	An integrated MBR-RO system	20 multiple-class of pharmaceuticals (including psychiatric drugs, antibiotics, anti-inflammatory drugs)	Integrated MBR-RO system accomplished higher removal rates for all the pharmaceuticals tested. Removal rates of MBR showed large variation (0–95%) based on the properties of compounds	(Dolar et al. 2012)
20	MBR, ozonation and integrated MBR-ozonation systems	Nalidixic acid; Synthetic wastewater	Ozonation removed Nalidixic acid completely, which went through MBR untouched. The combination of MBR-ozonation treatment achieved complete degradation in a better way	(Pollice et al. 2012)
21.	Suspended biofilm carriers and activated sludge	7 acidic pharmaceuticals (ibuprofen, diclofenac, naproxen, mefenamic acid, clofibric acid, ketoprofen and gemfibrozil)	Biofilm carriers showed higher removal rates for gemfibrozil, diclofenac, clofibric acid, ketoprofen and mefenamic acid compared to CAS. Naproxen and ibuprofen were eliminated with similar rates	(Falas et al. 2012)
22	MBR-only and integrated MBR-ozonation system	Acyclovir, diacetylacyclovir and other 28 organics	Integrated MBR-ozonation system showed results comparable to those achieved by two individual treatments placed one after the other (i.e., MBR after ozonation). Efficiency of MBR improved appreciably once the ozonation was incorporated to the recirculation stream of MBR. This integrated system also showed high effectiveness in elimination of by-product in the ozonation step	(Mascolo et al. 2010)
23	DAF-MBR-ozone oxidation.	Pharmaceuticals including ibuprofen, bezafibrate, naproxen	Removal rates by MBR varied from 50% to 99% based on the pharmaceuticals (> 95% for ibuprofen; 50–90% for bezafibrate). The majority of pharmaceuticals which passed through DAF-MBR un-degraded were either completely removed or reduced by ozone treatment. Bezafibrate and naproxen were not removed efficiently by ozone oxidation	(Choi et al. 2012)
24	MBR integrated with UV/H$_2$O$_2$ or ozonation	Nalidixic acid and other degradation products	Chemical oxidation step completely removed Nalidixic acid, which went through MBR undegraded. The integrated system accomplished higher removal efficiencies for 34 degradation products among 55	(Laera et al. 2012)
25	MBR integrated with solar Fenton oxidation	Sulfamethoxazole (SMX), clarithromycin (CLA) and erythromycin (ERY)	MBR removed ERY and SMX totally, but could not degrade CLA. ERY and SMX were completely removed from the spiked MBR wastewater effluent by Solar Fenton oxidation, while CLA went through undegraded. It showed that the removal of CLA from wastewater was quite challenging	(Karaolia et al. 2017)

Table 1 contd. ...

...Table 1 contd.

Sl. No.	Removal methods	Type of pharmaceuticals targeted	Removal efficiency and main findings	Reference(s)
26	Suspended activated sludge and MBBR	5 pharmaceuticals (ibuprofen, naproxen, ketoprofen, diclofenac and carbamazepine) and metabolite of the lipid regulating agent, clofibric acid	Coupling biological treatment (MBBR), HC/H_2O_2 process and UV treatment consecutively accomplished the highest removal efficiencies for all the compounds tested. Carbamazepine and diclofenac removal was > 98%, meanwhile the remaining ibuprofen, naproxen and ketoprofen were undetectable	(Zupanc et al. 2013)
27	MBR and biofilm layer	Ampicillin (AMP)	The retention of AMP was improved by 23% compared to suspended AS, which was mostly contributed by the compact structure	(Shen et al. 2014)
28	SBBRs	Tetracycline (TET)	The removal rate for TET was around 28% and biodegradation was found to be the main mechanism	(Matos et al. 2014)
29	Activated sludge reactor, two Hybas™ reactors and one MBBR	X-ray contrast media, analgesics, β-blockers, and antibiotics	Efficiently removed COD, nitrogen and also recalcitrant micropollutants	(Casas et al. 2015a)
30	3S-MBBR	6 compounds including β-blockers, X-ray contrast media, analgesics and antibiotics	Comparing to other technologies (i.e., MBR or activated sludge), MBBR reactors showed high removal efficiencies for pharmaceuticals, especially with certain compounds like X-ray contrast media	(Casas et al. 2015b)
31.	2S anaerobic fluidized MBR using GAC	26 pharmaceuticals	Various pharmaceuticals were removed with efficiencies of > 90%. Biological process, sorption on to GAC and MF were found to play a role in removing these pharmaceuticals	(Dutta et al. 2014)
32.	An aerobic GMBR	5 antibiotic and antiphlogistic PCPs in wastewater (prednisolone, naproxen, ibuprofen, amoxicillin and methyl alcohol)	GMBR showed higher efficiencies in removing prednisolone (98.46%), naproxen (84.02%), and ibuprofen (63.32%) and methanol. Degradation rate was insignificant on antibiotics such as amoxicillin	(Zhao et al. 2014)
33	MBR with a NF membrane	Spiramycin (SPM) and new spiramycin	The removal rates of both compounds were > 95%	(Wang et al. 2014)
34	MBR combined with a NF membrane	Spiramycin (SPM) and new spiramycin (NSPM)	A MBR-NF double membrane was found effective for the treatment of wastewater containing antibiotics. Removal of SPM and NSPM was higher than 95%, acute toxicity was considerably reduced too	(Wang et al. 2015)
35	An activated sludge system (AS-UF) or biofilm biological reactor (BBR-UF) with ultrafiltration	Treatment of hospital effluent	BBR-UF showed high removal of codeine, ketoprofen, diclofenac, naproxen, hydrochlorothiazide, roxithromycin, furosemide, metronidazole, pravastatin, iohexol and gemfibrozil, while AS-UF showed much lower removal for the same molecules	(Mousaab et al. 2015)

No.	System	Compounds	Findings	Reference
36.	MS2ALMBRs	Synthetic high-strength 7-ACA	All three Ms2ALMBRs were found effective on removal of COD and 7-ACA	(Chen et al. 2014)
37.	Aerated biofilters; MnOx ore and zeolite were used to pack them, named Mn-Biofilter and Zeo-Biofilter, respectively	Carbamazepine (CBZ), diclofenac (DFC), and sulfamethoxazole (SMX)	Mn- and Zeo-Biofilters were found effective on removal of DFC and SMX. Around one year was required by Zeo-Biofilter for the removal of DFC. MnOx oxidation was mainly responsible for the removal of DFC, meanwhile biological oxidation of manganese was mainly responsible for the removal of SMX	(Zhang et al. 2015)
38.	A combined anaerobic/anoxic/oxic (A/A/O) and MBBR and conventional C-Orbal oxidation ditch process	30 pharmaceuticals including sulfonamides, fuoroquinolones, tetracyclines, macrolides, β-blockers, antiepileptics, dihydrofolate reductase inhibitors, lipid regulators, and stimulants	Compared to C-orbal oxidation ditch treatment (84.5%), A/A/O-MBBR process was able to remove target pharmaceuticals (94.3%) with a considerably higher efficiency	(Yuan et al. 2015)
39.	A control AnMBR and a bioaugmented AnMBR	Wastewater	Bioaugmented AnMBR showed improved performance in high saline and complex pharmaceutical and more methane yield than the control AnMBR	(Ng et al. 2015)
40.	GMBR with SRT and HRT and different influent organic loading were used	Pharmaceuticals and PCPs	The removal rates were as follows for different compounds: prednisolone 98.56%; naproxen 84.02%; norfloxacin 87.85%; sulfamethoxazole 77.83% and ibuprofen 63.32%	(Xia et al. 2015)
41.	CAS, SABF and MBR Different OLR, SRT and HRT were investigated	Atenolol, gemfibrozil and ciprofloxacin	SABF showed the best performance. Biosorption was responsible for about 20% removal of ciprofloxacin in the reactors. Biological treatment with enriched culture was found to be effective in removal of pharmaceuticals	(Arya et al. 2016)
42.	MBfR. Different H_2 pressures and HRTs were investigated	Chlortetracycline .	96% chlortetracycline removal was achieved using **H2-MBfR** through biodegradation	(Aydin et al. 2016)
43.	Two-stage (2S)-MBR and 3S-MBR (using polyvinyl alcohol and sodium alginate beads of *Pseudomonas putida*) were used	10 different antibiotics	2S-MBR removal was dependent on the antibiotics and HRT. 3S-MBR showed higher efficiency of up to 90% CIP removal as compared to 58% removal using 2S-MBR. *P. putida* was found to be a potential candidate for CIP biodegradation via bioaugmentation of 3S-MBR process	(Hamjinda et al. 2017)

Table 1 contd. ...

...Table 1 contd.

Sl. No.	Removal methods	Type of pharmaceuticals targeted	Removal efficiency and main findings	Reference(s)
44	3D-BERs	sulfamethoxazole (SMX) and tetracycline (TC)	3D-BER showed removal efficiency of 88.9–93.5% for SMX, 95% for TC, and 89.3–95.6% for COD	(Zhang et al. 2016)
45	3D-BERs	sulfamethoxazole (SMX) and tetracycline (TET)	Removal rates of TC and SMX by the 3D-BERs were 82.6–97.3% and 72.2–93.2%, respectively	(Zhang et al. 2017)
46.	A H_2-MBfR with different H_2 pressures, HRTs, different influent TC concentrations	Tetracycline (TC)	TC removal of 80–95% was achieved at increased HRT and H_2 pressures	(Taskan et al. 2016)
47.	MBBR and modified anaerobic–anoxic–oxic technology	Vancomycin (VCM)	Total removal efficiencies were up to 99%; it showed biodegradation was the principal mechanism while sorption by sludge was negligible	(Qiu et al. 2016)
48.	GMBR	Prednisolone, norfloxacin, naproxen, sulfamethoxazole and ibuprofen	The higher removal rates were as follows: prednisolone 98.5%, norfloxacin 87.8% and naproxen 87.8%. The lower removal rates were: sulfamethoxazole 79.8% and ibuprofen 63.3%	(Wang et al. 2016)

tested compounds (3 NSAIDS and 3 antibiotics). In both MBRs, acetaminophen and ketoprofen were removed more effectively than roxithromycin and sulfamethoxazole. Similar removal efficiency with MBR-30 was observed by Shroeder et al. (2012), in which, NSAIDS were removed with higher efficiency as compared to MBR-15. The removal rates were 55% for sulfamethoxazole and 100% for acetoaminophen and ketoprofen. Interestingly, MBR with a longer sludge retention time (SRT) was found to be suitable for effectively removing 6 antibiotics (Xia et al. 2012).

2.1 Integrated MBR for removal of pharmaceuticals

2.1.1 MBR and conventional activated sludge process

MBR in combination with other methods were used for removal of pharmaceuticals from wastewater. MBR with conventional activated sludge (CAS) process was operated at longer SRT of 65 days, and showed greater elimination rates for six acidic pharmaceutical compounds (clofibric acid, diclofenac, ibuprofen, ketoprofen, mefenamic acid and naproxen) from municipal wastewater, as compared to the reactors operated at shorter SRTs (Kimura et al. 2007). Prasertkulsak et al. (2016) observed higher removal rate for majority of the pharmaceutical products from hospital wastewater studied by using MBR with shorter HRT of 3 hours. The rate of removal through adsorption and biodegradation was found to be different, depending on the properties of the pharmaceutical compounds. Many pharmaceutical products, including acetaminophen, were found to be removed completely by adsorption and degradation in MBR or CAS (Radjenovic et al. 2007, 2009). The removal of hydrochlorothiazide and paroxetine was found to be slightly better than with CAS treatment. However, carbamazepine was not removable either by MBR or CAS treatment. MBR operated at longer SRT under stable conditions was found to affect the elimination of some compounds such as β-blockers and erythromycin (Radjenovic et al. 2009).

External loop air lift membrane bioreactor (ELAMBR) and CAS were found to be efficient in removing acetaminophen from pharmaceutical wastewater (Shariati et al. 2010). Generally, antibiotics such as acetaminophen were completely removed from the wastewater by MBR treatment than CAS (Le-Minh et al. 2010). Garcia Galan et al. (2012) found that the removal rate was 100% for three sulfonamides and their metabolite with MBR, whereas the removal rate was only 55% with CAS. Similar to carbamazepine, the compounds such as trimethoprim and lincosamides showed poor removal efficiency with both the MBR and CAS even with different MLSS, HRT and SRT (Ngoc Han et al. 2016). All the above studies show that MBR system is more efficient in removing most of the antibiotics as compared to CAS, especially with micropollutants that are not readily biodegradable.

MBR was found to be a better removal method as compared to CAS due to its efficiency in removing most of the antibiotics and other pharmaceutical compounds. The ability of MBR to retain biomass, especially the slow-growing nitrifying microorganisms, hydrophobic compounds and production of high quality effluents, make MBR a better method as compared to CAS (Huang and Lee 2015). Although MBR treatment with membrane filtration (MF) is considered as an efficient method to treat various micropollutants, its large scale application is hindered by issues like membrane fouling (Deng et al. 2016, Guo et al. 2012). The wastewater from industries contains high concentrations of pollutants, particularly organic compounds; therefore, approaches such was ultrafiltration (UF) or reverse osmosis advanced oxidation have been suggested (Zhang et al. 2009b) along with MBR/CAS treatments.

2.1.2 MBR and CAS with UF/RO/NF

Sahar et al. (2011a) tested the removal of antibiotics from wastewater using CAS/UF and MBR and found that the removal efficiency was high as compared to CAS, while operated under stable conditions. However, the removal efficiency was significantly improved by incorporation of UF after CAS and the results were comparable to removal efficiency of MBR. These authors interpreted that the removal of antibiotics was enhanced by the biofilm formation on UF and MBR, which made its characteristics feature tighter to the organic compounds. MBR biomass removed 82% of sulfonamides and 92% for microlides at different MLSS concentrations through sorption. The study reported that the sorption by biomass attached to membrane and suspended biomass played a major role in the removal process. Later study (Sahar et al. 2011a) conducted by the same group reported that adding reverse osmosis (RO) after MBR or CAS/UF treatment could effectively remove several antibiotics, three pharmaceuticals and other industrial products. The removal rate for most of the organic micropollutants (OMPs) was nearly 99% and above 95% for diclofenac and 93% for sulfonamides. The authors concluded that adding RO to MBR and CAS/UF improved the removal efficiency, but not sufficient enough in some cases. Therefore, a process like adsorption to activated carbon or more advanced oxidation processes were suggested. Dolar et al. (2012) confirmed such a study with the addition of RO with MBR to remove antibiotics and other pharmaceutical products. Moscolo et al. (2010) used MBR and integrated ozonisation with the MBR to remove 8 organic antibiotics and they found that the removal efficiency of antibiotics was improved after integrating ozonisation with MBR. Pollice et al. (2012) found that nalidixic acid was completely removed in the membrane integrated with ozonisation process, whereas MBR integrated with ozonation or UV/H_2O_2 achieved complete antibiotic removal in the chemical oxidation step (Pollice et al. 2012). The removal efficiency of pharmaceuticals including ibuprofen, bezafibrate and naproxen by MBR varied from 50 to 99% (> 95% for ibuprofen, 50–90% for bezafibrate). The majority of the pharmaceuticals which escaped through Dissolved Air Filtration (DAF)-MBR were either reduced or completely removed by the ozone oxidation (Choi et al. 2012). However, bezafibrate and naproxen were not removed efficiently by the ozone oxidation step (Choi et al. 2012). The removal efficiency of four different pharmaceuticals, namely carbamazepine, diclofenac, ketoprofen and naproxen was tested by combing biological treatment approach (SAS-suspended activated sludge and MBB-moving bed biofilm) with chemical/physical treatment step (H_2O_2 and UV irradiation treatment). The removal efficiencies of these combined processes for carbamazepine and diclofenac were more than 98%. After the treatment, the levels of other pharmaceuticals (ketoprofen, ibuprofen and naproxen) were found to be lower than the detectable limits. Various pharmaceuticals (a total of 26) were removed with efficiencies of > 90% by the biological process (2 stage anaerobic fluidised MBR), sorption onto granular activated carbon (GAC) and membrane filtration (MF) (Dutta et al. 2014).

In recent years, nanofiltration (NF) in combination with MBR has been used for removing pharmaceuticals from wastewater as NF has high retention for salts, coloured compounds and organic matter (Choi et al. 2012, Wang et al. 2015). A double membrane MBR-NF was used for treatment of effluent of the anaerobic process of wastewater treatment plant of pharmaceutical company (Wang et al. 2015). The anaerobic treatment has removed spiromycin compounds from the raw wastewater produced during antibiotic production. However, the effluent from the anaerobic treatment still contained 1.99–2.79 mg/l of spiromycin. MBR-NF combination was studied for removing antibiotic compounds from the effluent of anaerobic process. This study (Wang et al. 2015) showed that the combined system was effective for treating wastewater containing antibiotics. The results were achieved by recycling the NF concentrate into MBR, which removed the antibiotics effectively and dramatically improved the quality of treated water. The major organics in the effluent were proteins, polysaccharides and humic acids, which were completely removed by the NF membrane and further biodegraded by MBR; however, these processes did not affect the microbial community composition.

Presence of 7-aminocephalosporanic acid (7-ACA) in wastewater was found to cause serious water pollution and potential health risks to humans (Chen et al. 2014). A treatment method called multi-sparger multi-stage airlift loop MBR (MS2ALMBR) was used for treating wastewater containing high strength 7-ACA under different HRTs, temperatures and pH conditions (Chen et al. 2014). The results showed that 7-ACA and COD removal rates were high, and the highest removal rate was obtained at 37°C temperature and pH range of 7–9 (Chen et al. 2014).

2.1.3 MBR and Biofilm

Recently, Zhang et al. (2015) proposed a cost effective method to treat pharmaceutical wastewater by using iron/manganese oxidising bacteria (I/MOB). I/MOB are widespread in nature and were detected in water distribution systems and drinking water (Cerrato et al. 2010). These bacteria were capable of degrading pharmaceutical compounds (Sabirova et al. 2008). Zhang et al. (2015) used aerated biofilters packed with MnOx and zeolite to remove antibiotics, carbamazepine (CBZ), sulfamethoxazole (SMX) and diclofenac (DFC). DFC was removed effectively (> 90%) by the biofilters, whereas the removal of CBZ and SMX was insignificant. The combination of anaerobic/ anoxic/oxic and moving bed biofilm reactor (A/A/O-MBBR) was used by Yuan et al. (2015) and compared with conventional C-orbital oxidation ditch process to remove 30 pharmaceuticals including sulfonamides, tetracycline and fluoroquinolones from municipal wastewater in China. The results showed that higher (93%) removal of pharmaceuticals can be achieved through A/A/O-MBBR, as compared to C-orbital oxidation ditch process (84%). Ng et al. (2015) compared two different treatment methods, i.e., control anaerobic MBR (AnMBR) and bioaugmented AnMBR seeded with coastal sediments, to treat pharmaceutical wastewater from a factory located in Singapore. The results showed that bioaugmented AnMBR exhibited improved performance in high saline condition and removal of complex pharmaceutical compounds was found to be high with more methane yield than that of control AnMBR (Ng et al. 2015).

Arya et al. (2016) studied the performance of activated sludge process (ASP) and attached growth bioreactors, i.e., submerged attached biofilters (SABF) and MBR for removing pharmaceutical compounds with cations and anions (atenolol, ciprofloxacin and gemfibrozil) in the wastewater and the results showed that the removal efficiencies of the reactors ranged between 75% and 95% for all three compounds tested. That study also showed that SABF was performing better for removing the tested pharmaceutical compounds as compared to ASP and MBR. The results indicated that biological treatment with enriched culture is effective for treating pharmaceutical wastes.

H_2-based membrane biofilm reactor (H_2-MBfR) is one of the new technologies used for removing multiple contaminants by bio-reduction using substrate H_2 gas as electron donor (Aydin et al. 2016, Taskan et al. 2016). Aydin et al. (2016) studied the removal of chlortetracycline using H_2-MBfR and they observed a complete denitrification and more than 96% removal of chlortetracycline using this method. It was concluded that biodegradation was the primary elimination process responsible for removing chlortetracycline. Two and three stages MBR configurations (2S-MBR and 3S-MBR) with *Pseudomonas putida* alginate beads were used for treating synthetic hospital wastewater, which contained high levels of recalcitrant antibiotic, ciprofloxacin. 3S-MBR was found to be more effective in removing ciprofloxacin (~ 90%) than 2S-MBR, in which the removal rate was only 58%. Bioaugmentation using *P. putida* beads was found to enhance CIP degradation in 3S-MBR (Hamjinda et al. 2017). Another method, 3D biofilm electrode reactor (BER), was used to remove the most challenging antibiotics such as sulfamethoxazole and tetracycline. In these systems, the removal efficiencies were higher and in the range of 80–90% (Zhang et al. 2016, 2017).

3. Removal of antibiotic resistant bacteria (ARB) and resistance genes (ARGs) using bioreactors

Wastewater is considered as a large reservoir for ARB and ARGs. The sewerage from industries, hospitals, human society, farming and agriculture (Deegan et al. 2011, Alonso et al. 2001, Zhu et al. 2013, Gao et al. 2012, Xu et al. 2015) contains high levels of antibiotics, as often these waters are discharged into the natural environment either with partial or without proper treatment. The bacteria which grow in such environments develop resistance to antibiotics and act as potential source for development of ARGs in clinically relevant bacteria (Dantas et al. 2008, D'Costa et al. 2006, Volkmann et al. 2004).

ARB and ARGs are now recognized as new contaminants and raised concerns by their potential health risks to the humans (Arias and Murray 2009). ARGs are extensively present in the wastewater (Gao et al. 2012, Xu et al. 2015, Li et al. 2010) and can be widely introduced into the natural environmental bacteria through horizontal gene transfer (Zhu et al. 2013). Specifically, the mobile genetic elements (MGEs) such as plasmids, transposons and integrons are capable of capturing the exogenous genes and inserting these genes into bacteria through recombination (Gaze et al. 2008). It is reported that the prevalence of MGEs is even higher in the bacteria exposed to high concentration of detergents or antibiotics (Gaze et al. 2011). However, it has not been fully proved that high concentration of antibiotics is the only reason for the bacteria to develop resistance genes (Subbiah et al. 2011). The correlation between the quantities of ARB and ARGs and respective antibiotic concentrations in sludge samples showed a weak correlation (Gao et al. 2012). In the same study, significant correlation was observed between the tetracycline resistant genes (tetO and tetW) and tetracycline concentration in the wastewater, but no significant correlation was observed between the concentration of sulfonamides and *sul1* gene (Gao et al. 2012).

Antibiotics such as tetracycline and sulfonamide are commonly used for various therapeutic treatments in hospitals (Olexy et al. 1979) and improvements for plants productivity (Schmitt et al. 2006, Brooks et al. 2007) for a long time. Both of these antibiotics are frequently detected in different stages of wastewater treatment and in the treated effluent and receiving waters (Gao et al. 2012, Xu et al. 2015). High abundance of tetracycline and sulfonamide resistant genes were also detected in the wastewater at various concentrations (Li et al. 2010, Gao et al. 2012, Xu et al. 2015, Auerbach et al. 2007).

Chlorine is a common disinfectant widely used for water and wastewater treatment for more than a century (Caravelli et al. 2003). It is recognized as an efficient chemical agent in removing the pathogenic/nonpathogenic microorganisms by killing them through the oxidation of cellular materials. An earlier study focused on studying the effect of chlorine on antibiotics and ARB (Wang et al. 2011) and it was reported that chlorine dioxide and free chlorine can breakdown tetracycline into smaller molecules. However, the presence of chlorine did not remove ARGs and ARB but increased their level in the wastewater. Another study reported that the presence of chlorine can induce antibiotic resistance in *Acinetobacter baumannii* (Karumathil et al. 2014) and this bacterium is considered as a multi-drug resistant and opportunistic human pathogen. The addition of advanced treatment such as ozonation and charcoal or sand filtration after the sewage treatment was reported to effectively reduce the total and antibiotic resistant bacteria from wastewater (Luddeke et al. 2015).

In this chapter, we report the efficiency of MBR treatment in removing ARB and ARGs by using culture based and culture independent molecular techniques, respectively.

3.1 Description of MBR and sample collection

A bench-scale MBR system comprising of a 20 L bioreactor tank and membrane filters with a pore size of 0.1 μm were used. The seeding sludge was collected from a WWTP in Suzhou, China and acclimatized with synthetic wastewater (Kimura et al. 2009) (Table 2) for more than one month. Water and sludge sampling was carried out after the system had reached a stable condition. Samples from the inlet (influent), bioreactor, membrane surface (foulant) and outlet (effluent) were collected in triplicate for further analyses (Figure 2).

3.2 Culture based methods

3.2.1 Viable bacterial count and antibiotic resistant bacterial count

The viable bacterial count was carried out using plate count agar medium (Lab M, UK) with or without the addition of antibiotics. Five different types of antibiotics were mixed separately with the media to get a final concentration of 50 μg/mL of ampicillin, 10 μg/mL of tetracycline, 5 μg/mL of

Table 2. Composition and concentration of the synthetic wastewater fed to the MBR (Kimura et al. 2009).

Chemicals	Concentration (g/m^3)
CH_3COONa	512
$(NH_4)_2SO_4$	189
KH_2PO_4	17.6
$FeCl_3 \cdot 6H_2O$	2.42
$CaCl_2 \cdot 2H2O$	0.37
$MgSO_4 \cdot 7H_2O$	5.08
$MnCl_2 \cdot 4H_2O$	0.28
$ZnSO_4 \cdot 7H_2O$	0.44
$CuSO_4 \cdot 5H_2O$	0.39
$CoCl_2 \cdot 6H_2O$	0.42
$Na_2MoO_4 \cdot 2H_2O$	1.26

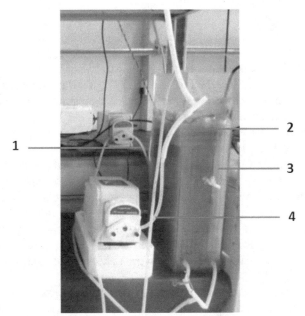

Figure 2. Experimental setup of the MBR system and the sample collection points: 1. Influent: synthetic wastewater; 2. Bioreactor: synthetic wastewater + activated sludge; 3. Foulant: biofilm sampled from the membrane; 4. Effluent (reclaimed water): achieved after MBR treatment.

Color version at the end of the book

ciprofloxacin, 128 μg/mL of sulfamethizole and 1 IU of penicillin. The antibiotic concentrations to determine the resistant bacterial count was selected based on the Clinical and Laboratory Standard Institute (CLSI 2010) manual for Antimicrobial Susceptibility Testing. Aqueous samples from influent, bioreactor tank and effluent were serially diluted 100, 100 and 10 times, respectively. A 0.1 g of the foulant was suspended in 9.9 mL of sterilized water and then diluted 100 times to make a final dilution of 10^4 times. Aliquots (100 μL) of each sample was aseptically spread on the plate and incubated at 29°C for 24 hours. Afterwards, the number of colonies was counted and converted into CFU per mL or CFU per gram. All measurements were done in triplicate.

3.3 Culture independent method

3.3.1 Sample collection and DNA extraction

200 mL of influent and effluent and 50 mL of bioreactor samples were filtered onto 0.22 µm pore size membrane filter (diameter 47 mm, Millipore Ltd., UK) to concentrate bacteria on membrane filter and stored the filters at –80°C until they were used for further molecular analyses. Genomic DNA was extracted using PowerSoil® DNA Isolation Kit (MO BIO Laboratories, CA) as per the manufacturer protocol. Half of the membrane filter and 50 mg of the foulant samples were used for DNA extraction. The DNA samples were quantified using a NanoDrop (ND-2000c, Thermo) and the quality was verified by 1% agarose gel electrophoresis.

3.3.2 Real-Time PCR assay

Quantification of 16S rRNA genes as well as tetracycline (*tetA*) and sulfonamide (*sul1*) resistant genes was performed by qPCR assays. The primers used in this study, annealing temperature for respective primers and amplicon size are given in Table 3. Reaction mixture was prepared to make a final volume of 10 µL which consisted of 5 µl of SYBR® Green, 2.8 µl of sterile water, 0.3 µM of each primer and 1 µl of DNA template. The PCR program was as follows: an initial denaturation for 10 min at 95°C, followed by 40 cycles of 15 s at 95°C, 30 s at respective annealing temperature and 30 s at 60°C for extension. Then, the melting curve was constructed by gradually increasing the temperature to 95°C. Each reaction was run in triplicate.

The standard curve for detection of each gene was generated by using plasmid DNA carrying the target gene. Explicitly, the purified PCR product with the target gene was first ligated into pMD19-T vector (TaKaRa), transformed into the competent *E. coli* cells and spread on the LB agar plates containing ampicillin (50 µg/mL) and X-gal. The white colonies that appeared on the plates were picked and screened by PCR to verify the presence of target gene and the positive clones were cultured overnight for plasmid DNA extraction. The copy numbers of the extracted DNA was calculated and the standard was generated by 10 times of serial dilution to make copy numbers ranging from 10^9 to 10^5 gene copies μL^{-1}. The R^2 values for 16S rRNA, *tetA* and *sul1* genes were 0.991, 0.996 and 0.994 and the average amplification efficiencies were 121.94, 117.40 and 114.45 and slope values

Table 3. Primers and amplification conditions used for qPCR assays.

Target group	Primers	Sequence (5' to 3')	Product size	Annealing temperature	Reference
16S rRNA gene	16S-U1048 16S-U1371R	5'-GTGITGCAIGGIIGTCGTCA-3' 5'-ACGTCITCCICICICCTTCCTC-3'	323 bp	60°C	Singh et al. (2014)
tetA gene	*tetA*-F *tetA*-R	5'-GCGCGATCTGGTTCACTCG-3' 5'-AGTCGACAGYRGCGCCGGC-3'	210 bp	60°C	Aminov et al. (2002)
Sul1 gene	*Sul1*-F *Sul1*-R	5'-CACCGGAAACATCGCTGCA-3' 5'-AAGTTCCGCCGCAAGGCT-3'	158 bp	60°C	Pei et al. (2006)

were –2.888, –2.965 and –3.018, respectively. The qPCR data were statistically analyzed by student's T-test using Excel and the results were considered statistically significant under the circumstances that p-values were less than 0.05. The data were log transformed prior to the analysis.

3.4 Antibiotic resistant bacterial count before and after MBR treatment

The resistant bacterial count in the presence of five antibiotics, namely ampicillin, ciprofloxacin, sulfamethizole, penicillin and tetracycline were recorded and compared with total viable count (TVC) obtained without the addition of antibiotics (Figure 3). As reflected in the results, ARB was observed in all the samples but their abundance varied between the samples tested. The ARB counts were significantly high in foulant sample (Figure 3b) and the count ranged from 8.33×10^5 cfu/mL (on tetracycline-amended plates) to 1.3×10^7 cfu/mL (on ciprofloxacin-amended plates) and followed by bioreactor and influent samples (Figure 3a). Statistical analysis indicated that the plates containing ampicillin, penicillin and tetracycline showed significant reduction ($p < 0.05$) in the effluent sample compared to the influent sample. However, ARB was still detected in all of the effluent samples. In terms of different antibiotics applied for testing, ARB was found to be the most abundant on plates containing ciprofloxacin followed by penicillin, ampicillin, and sulfamethizole, and the least abundance was noticed on tetracycline containing plates.

The prevalence of each type of ARB in all four samples was determined by comparing the number of resistant colonies vis-a-vis total bacterial count obtained without the addition of antibiotics (Table 4). In general, the highest percentage of ARB was observed in the bioreactor and foulant

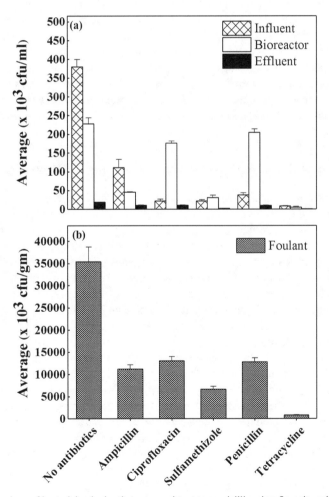

Figure 3. Average numbers of bacterial colonies that were resistant to ampicillin, ciprofloxacin, sulfamethizole, penicillin and tetracycline found in the samples. (A) Resistant count results for influent, bioreactor and effluent samples. (B) Resistant count result for foulant sample.

Table 4. Removal efficiency of the antibiotic resistant bacteria (ARB) in the effluent sample compared to the influent sample. Antibiotics tested include ampicillin, ciprofloxacin, sulfamethizole, penicillin and tetracycline.

Sample	Total viable count (cfu/ml)	% of total viable count resistant to				
		Ampicillin	**Ciprofloxacin**	**Sulfamethizole**	**Penicillin**	**Tetracycline**
Influent	3.80×10^5	29.09 ± 4.12	5.66 ± 1.68	5.59 ± 0.54	9.89 ± 0.92	2.28 ± 0.12
Bioreactor	2.28×10^5	20.35 ± 1.16	77.36 ± 2.82	13.50 ± 1.69	89.91 ± 3.41	2.31 ± 0.50
Foulant	3.54×10^7	31.67 ± 1.11	36.86 ± 0.77	18.84 ± 0.33	36.21 ± 0.84	2.36 ± 0.12
Effluent	1.90×10^4	61.29 ± 5.55	59.34 ± 1.68	10.41 ± 1.65	55.32 ± 2.15	2.94 ± 0.31

samples. Therefore, the activated sludge in the bioreactor can be considered as one of the reservoirs for the presence of antibiotics and development of ARB. The relative abundance of ARB in the effluent sample was higher than that in the influent sample; this was due to the addition of extra ARB in the sludge from the bioreactor.

The antibiotics tested in this study are commonly used in daily life for different purposes; for example, treating urinary tract infections (UTIs) by sulfamethizole (Kerrn et al. 2003). The un-metabolized antibiotics can flow into the natural environment, which lead to high probability to develop resistance to these antibiotics by natural bacteria. In this study, high resistance rate was observed for ciprofloxacin and penicillin, which could be due to high amount of these antibiotics released into the environment due to overuse or abuse of these antibiotics. Penicillin is one of the first discovered antibiotics and it has been used for nearly a century to cure diseases such as syphilis and to inhibit bacterial infections. Similarly, ciprofloxacin is also widely used for clinical purposes. It has been reported that both of these antibiotics were detected in great amount in the hospital wastewater (Tuc et al. 2017, Kimosop et al. 2016) and this is considered as one of the main sources for the presence of ARB and ARGs.

3.5 Detection of tetA and sul1 genes before and after MBR treatment

Figure 4 summarizes the detection and abundance of tetracycline (*tetA*) and sulfonamide (*sul1*) genes, as compared to the total abundance of 16S rRNA gene (total bacteria) in the samples collected from the bioreactor. The experimental results showed that gene copies of *tetA* and *sul1* found in the influent sample (*tetA*: $5.3 \pm 0.3 \times 10^3$ genes/mL; *sul1*: $7.5 \pm 0.1 \times 10^3$ genes/mL) and effluent sample (*tetA*: $1.6 \pm 0.5 \times 10^3$ genes/mL; *sul1*: $1.7 \pm 0.1 \times 10^3$ genes/mL) were comparable. In contrast, the gene copy numbers of *tetA* and *sul1* were significantly different ($p < 0.05$) in the bioreactor and foulant samples. This number was 1 or 2 order of magnitude higher in *sul1* gene than in *tetA* gene. Additionally, both *tetA* and *sul1* genes were highly abundant in bioreactor and foulant samples. The gene copy numbers for both of the samples are about 2 to 3 orders of magnitude higher than those in the effluent sample (Figure 4).

High amount of resistant gene copies in bioreactor and foulant samples were detected in this experiment. This implies that the wastewater sludge and the membrane biofoulant layer contained a large amount of ARGs, which is in agreement with previous report by Chen and Zhang (2013). Another study in China confirmed a significant correlation between *tet* gene copies and the 16S rRNA gene copies in the soils sampled near swine farms (Wu et al. 2010). Their results showed that overall amount of ARGs was removed along with the removal of the biomass and this is in agreement with the present study. Considering the wide usage of tetracycline and sulfamethizole, it was considered as the original source for antibiotics flow into the natural environment and thus, more resistant genes were developed and spread from one organism to another. For example, *tetA* genes have been widely found in both Gram-positive and Gram-negative bacteria as well as in *Streptomyces* (Arioli et al.

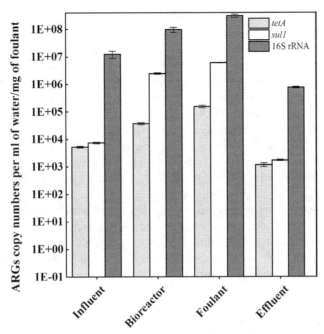

Figure 4. Abundance of ARGs (*tetA* and *sul1*) in influent, bioreactor, foulant and effluent samples in comparison to the total 16S rRNA gene.

2014). Based on our experimental results, the removal rate of this was significant ($p < 0.05$) in the effluent sample as compared to the bioreactor and foulant samples. *Sul1* gene was one of the earliest ARGs discovered, the drug resistance being transmitted by plasmids. The presence of this gene can make the organism resistant to sulfonamides (Akiba et al. 1960). As the result indicates in this study, a reduction of more than 3 orders of magnitude was obtained in the effluent sample by MBR treatment. By comparing the results between the influent and effluent samples, statistically significant reduction in *tetA* ($p < 0.001$) and *sul1* ($p < 0.05$) genes was observed. This result indicates that MBR treatment is efficient in removing the antibiotic resistance genes. However, these genes were still present in the effluent after the MBR treatment.

4. Conclusions

4.1 Removal of pharmaceuticals by bioreactors

CAS system was shown to remove surfactants and PCPs more effectively than other methods such as trickling filters, biofilm reactors, nitrification and denitrification processes. The **MBR and integrated approaches** were found to be efficient in removing pharmaceuticals from wastewater. In recent years, intensive research has been conducted on the biodegradation and removal of pharmaceutical compounds along with antibiotic resistant microbes with the aim of reducing ECs in the treated effluent. Based on the review, MBR method was found to be efficient in removing some of the pharmaceutical products such as antibiotics, antiphlogistics, steroids, iodinated contrast media, lipid regulators, and pesticides. The compounds acetoaminophen and ketoprofen were removed more efficiently than roxithromycin (ROX) and sulfamethoxazole (SMX). Optimizing the operating parameters in MBR such as COD, temperature, pH, MLSS, SRT and HRT were found to enhance the removal of pharmaceuticals from wastewater.

    Integrated MBR and CAS showed greater efficiency in removing most of the pharmaceutical products including analgesics, anti-inflammatory drugs, psychiatric drugs, beta-blockers and quinolone compounds. However, this combination was not particularly suitable to remove carbamazepine (anti-epileptic drug), hydrocholorothiazide (diuretic drug), trimethoprim (TMP) and lincosamides (antimicrobials).

MBR and CAS were reported to perform better in the treatment of pharmaceutical wastes; however, it was not efficient enough. Therefore, in some studies the removal efficiencies between the two methods, i.e., MBR and ultrafiltration (UF) membrane in CAS was found to be similar. However, the introduction of RO with CAS and UF with MBR, the removal of antibiotics such as erythromycin (ERY), roxithromycin (ROX), sulfamethoxazole (SMX), sulfamethazine (SMZ), trimetheoprim (TMP) and clarithromycin (CLA) was improved with an efficiency of > 95%. Notably, ROX, SMX and TMP were not removed efficiently by MBR or MBR-CAS. Spiramycin and new spiramycin were removed 95% using MBR combined with NF membrane. The acute toxicity of the effluent was considerably reduced after treating by MBR with NF membrane.

Nalidixic acid was completely removed by integrating **MBR with ozonisation** step. By combining MBR with DAF and ozonisation, most of the pharmaceuticals could be efficiently removed, including ibuprofen, bezafibrate and naproxen. MBR integrated with fenton oxidation completely removed SMX and ERY; however, CLA was quiet difficult to remove by this method. **MBR coupled with HC/ H_2O_2 and UV treatment** was highly efficient (> 98%) in removing carbamazepine and diclofenac including other antibiotics and lipid regulating agents. Carbamazepine, which was not removed by MBR/CAS, was removed by this method.

Using Granular sludge MBR, personal care products (PCPs) and pharmaceuticals such as naproxen, norfloxacin, ibuprofen, prednisolone, and SMX were removed at 80 to 98% efficiency by this method. Ampicillin (AMP) was removed 23% more effectively by MBR and biofilm layer (**Membrane biofilm bioreactor**), as compared to suspended activated sludge. Recalcitrant micropollutants, β-blockers, X-ray contrast media, analgesic and antibiotics were treated better with the 3S-MBBR along with activated sludge reactor and hybrid biofilm and activated sludge system (Hybas) reactors.

Hospital effluents were treated efficiently by removing most of the pharmaceutical compounds, including hydrochlorothiazide by using **biofilm biological reactor (BBR) with UF** (BBR-UF), which was not removed by CAS/MBR. Similarly, **membrane biofilm reactor (H_2-MBfR)** supplied with different pressure of H_2 gas at longer HRTs efficiently removed chlorotetracycline (96%) and tetracycline (80–95%).

Various pharmaceuticals were removed with efficiencies of > 90% due to sorption onto the GAC. A more advanced hybrid approach using **anaerobic/anoxic/oxic/MBBR** was very effective in removing most of the pharmaceutical products including sulfanamides, tetracycline and macrolides > 94.3%. Vancomycin removal efficiency was up to 99% when treated with modified anaerobic/anoxic/oxic MBR. Bioaugmented AnMBR showed improved performance in removing high saline and complex pharmaceutical wastes than AnMBR.

CBZ, diclofenac (DFC) and SMX were effectively removed by Mn and zeolite biofilters (**Aerobic MBR with biofilters**). However, removal of DFC required long time of ~ 1 year. SMX and TC were removed with 95% efficiency for TC and 93% efficiency for SMX using 3D BERs. However, more improved technologies and optimised MBR are required to increase the removal efficiency of recalcitrant pharmaceutical compounds.

4.2 *Removal of ARB and ARGs using bioreactors*

The viable bacterial count and total cell count results reflected a decrease in number of bacteria in the effluent sample compared to the influent sample. This result is further proved by the qPCR result through the detection of 16S rRNA genes in the water samples. High concentrations of antibiotic resistant genes and antibiotic resistant bacteria were found in the bioreactor and foulant samples. However, a significant reduction ($p < 0.05$) was shown by the results of resistant bacterial count and qPCR targeted to specific ARGs in effluent sample. The overall results indicate that MBR can efficiently remove most of the bacteria, ARB and ARGs from the wastewater. Future studies can focus on the application of MBR-integrated with other approaches to study removal efficiencies of more ARB and ARGs. Moreover, tracking of ARB and ARGs can be adopted for routine treatment process monitoring by WWTPs.

Acknowledgements

The authors would like to acknowledge the funding support provided by Xi'an Jiaotong-Liverpool University (XJTLU Research Development Fund, RDF 10-02-05) and Natural Science Foundation of the Jiangsu Higher Education Institutions of China (Jiangsu University Natural Science Programme, Grant No. 13KJB180022). Thanks to Williamson Gustave for his support with preparation of figures. The authors would also like to thank the Department of Biological Sciences, XJTLU and Suzhou Institute for Advanced Study, University of Science and Technology of China (USTC), for providing facilities to carry out experiments on the removal of ARB and ARGs by using MBR.

References

Abegglen, C., A. Joss, C.S. Mcardell, G. Fink, M.P. Schluesener, T.A. Ternes and H. Siegrist. 2009. The fate of selected micropollutants in a single-house MBR. Water Research 43: 2036–2046.

Ahmed, M.B., J.L. Zhou, N. Huu Hao, W. Guo, N.S. Thomaidis and J. Xu. 2017. Progress in the biological and chemical treatment technologies for emerging contaminant removal from wastewater: A critical review. Journal of Hazardous Materials 323: 274–298.

Akiba, T., K. Koyama, Y. Ishiki, S. Kimura and T. Fukushima. 1960. On the mechanism of the development of multiple-drug-resistant clones of Shigella. Japanese Journal of Microbiology 4: 219–27.

Alonso, A., P. Sanchez and J.L. Martinez. 2001. Environmental selection of antibiotic resistance genes. Environmental Microbiology 3: 1–9.

Arias, C.A. and B.E. Murray. 2009. Antibiotic-Resistant Bugs in the 21st Century—A Clinical Super-Challenge. New England Journal of Medicine 360: 439–443.

Arioli, S., S. Guglielmetti, S. Amalfitano, C. Viti, F. Marchi, F. Decorosi, L. Giovannetti and D. Mora. 2014. Characterization of tetA-like gene encoding for a major facilitator superfamily efflux pump in *Streptococcus thermophilus*. Fems Microbiology Letters 355: 61–70.

Arya, V., L. Philip and S.M. Bhallamudi. 2016. Performance of suspended and attached growth bioreactors for the removal of cationic and anionic pharmaceuticals. Chemical Engineering Journal 284: 1295–1307.

Auerbach, E.A., E.F. Seyfried and K. Mcmahon. 2007. Tetracycline resistance genes in activated sludge wastewater treatment plants. Water Research 41: 1143–1151.

Ayadin, E., M. Sahin, E. Taskan, H. Hasar and M. Erdem. 2016. Chlortetracycline removal by using hydrogen based membrane biofilm reactor. Journal of Hazardous Materials 320: 88–95.

Brooks, J.P., S.L. Maxwell, C. Rensing, C.P. Gerba and I.L. Pepper. 2007. Occurrence of antibiotic-resistant bacteria and endotoxin associated with the land application of biosolids. Canadian Journal of Microbiology 53: 616–622.

Caravelli, A., E.M. Contreras, L. Giannuzzi and N. Zaritzky. 2003. Modeling of chlorine effect on floc forming and filamentous micro-organisms of activated sludges. Water Research 37: 2097–2105.

Carvalho, I.T. and L. Santos. 2016. Antibiotics in the aquatic environments: A review of the European scenario. Environment International 94: 736–757.

Casas, M.E., R.K. Chhetri, G. Ooi, K. Hansen, K. Litty, M. Christensson, C. Kragelund, H.R. Andersen and K. Bester. 2015a. Biodegradation of pharmaceuticals in hospital wastewater by a hybrid biofilm and activated sludge system (Hybas). Science of the Total Environment 530: 383–392.

Casas, M.E., R.K. Chhetri, G. Ooi, K. Hansen, K. Litty, M. Christensson, C. Kragelund, H.R. Andersen and K. Bester. 2015b. Biodegradation of pharmaceuticals in hospital wastewater by staged Moving Bed Biofilm Reactors (MBBR). Water Research 83: 293–302.

Cerrato, J.M., J.O. Falkinham, A.M. Dietrich, W.R. Knocke, C.W. Mckinney and A. Pruden. 2010. Manganese-oxidizing and -reducing microorganisms isolated from biofilms in chlorinated drinking water systems. Water Research 44: 3935–3945.

Chemists, A. 2001. ChemSpider Synthetic Pages, http://cssp.chemspider.com/123.

Chen, H. and M. Zhang. 2013. Effects of advanced treatment systems on the removal of antibiotic resistance genes in wastewater treatment plants from Hangzhou, China. Environmental Science & Technology 47: 8157–8163.

Chen, Z.-B., Z.-W. He, C.-C. Tang, D.-X. Hu, Y.-B. Cui, A.-J. Wang, Y. Zhang, L.-L. Yan and N.-Q. Ren. 2014. Performance and model of a novel multi-sparger multi-stage airlift loop membrane bioreactor to treat high-strength 7-ACA pharmaceutical wastewater: Effect of hydraulic retention time, temperature and pH. Bioresource Technology 167: 241–250.

Choi, M., D.W. Choi, J.Y. Lee, Y.S. Kim, B.S. Kim and B.H. Lee. 2012. Removal of pharmaceutical residue in municipal wastewater by DAF (dissolved air flotation)-MBR (membrane bioreactor) and ozone oxidation. Water Science and Technology 66: 2546–2555.

CLSI. 2010. Performance standards for Antimicrobial Susceptibility Testing; Twetieth Informational Supplement; Clinical Laboratory and Standards Institute (CLSI) document M100-S20. Wayne, PA.

D'costa, V.M.., K.M. Mcgrann, D.W. Hughes and G.D. Wright. 2006. Sampling the antibiotic resistome. Science 311: 374–377.

Dantas, G., M.O.A. Sommer, R.D. Oluwasegun and G.M. Church. 2008. Bacteria subsisting on antibiotics. Science 320: 100–103.

Daze, W., C. O'Neill, E. Wellington and P. Hawkey. 2008. Antibiotic resistance in the environment, with particular reference to MRSA. *In*: A.L. Laskin and S. Sariaslani (eds.). Advances in Applied Microbiology, Vol 63.

Deegan, A.M., B. Shaik, K. Nolan, K. Urell, M. Oelgemoeller, J. Tobin and A. Morrissey. 2011. Treatment options for wastewater effluents from pharmaceutical companies. International Journal of Environmental Science and Technology 8: 649–666.

Deng, L., W. Guo, H.H. Ngo, H. Zhang, J. Wang, J. Li, S. Xia and Y. Wu. 2016. Biofouling and control approaches in membrane bioreactors. Bioresource Technology 221: 656–665.

Dolar, D., M. Gros, S. Rodriguez-Mozaz, J. Moreno, J. Comas, I. Rodriguez-Roda and D. Barcelo. 2012. Removal of emerging contaminants from municipal wastewater with an integrated membrane system, MBR-RO. Journal of Hazardous Materials 239: 64–69.

Dorival-Gracia, N., A. Zafra-Gomez, A. Navalon, J. Gonalez and J.L. Vilchez. 2013. Removal of quinolone antibiotics from wastewaters by sorption and biological degradation in laboratory-scale membrane bioreactors. Science of the Total Environment 442: 317–328.

Du, J.-R., K.-X. Li, J. Zhou, Y.-P. Gan and G.-Z. Huang. 2011. Sodium hypochlorite disinfection on effluent of MBR in municipal wastewater treatment process. Huan jing ke xue= Huanjing kexue 32: 2292–7.

Dutta, K., M.-Y. Lee, W.W.-P. Lai, C.H. Lee, A.-Y.-C. Lin, C.-F. Lin and J.-G. Lin. 2014. Removal of pharmaceuticals and organic matter from municipal wastewater using two-stage anaerobic fluidized membrane bioreactor. Bioresource Technology 165: 42–49.

Esplugas, S., D.M. Bila, L.G.T. Krause and M. Dezotti. 2007. Ozonation and advanced oxidation technologies to remove endocrine disrupting chemicals (EDCs) and pharmaceuticals and personal care products (PPCPs) in water effluents. Journal of Hazardous Materials 149: 631–642.

Falas, P., A. Baillon-Dhumez, H.R. Andersen, A. Ledin and J.L.C. Jansen. 2012. Suspended biofilm carrier and activated sludge removal of acidic pharmaceuticals. Water Research 46: 1167–1175.

Gao, P., M. Munir and I. Xagoraraki. 2012. Correlation of tetracycline and sulfonamide antibiotics with corresponding resistance genes and resistant bacteria in a conventional municipal wastewater treatment plant. Science of the Total Environment 421: 173–183.

Garcia Galan, M.J., M.S. Diaz-Cruz and D. Barcelo. 2012. Removal of sulfonamide antibiotics upon conventional activated sludge and advanced membrane bioreactor treatment. Analytical and Bioanalytical Chemistry 404: 1505–15.

Gaze, W.H., L. Zhang, N.A. Abdouslam, P.M. Hawkey, L. Calvo-Bado, J. Royle, H. Brown, S. Davis, P. Kay, A.B.A. Boxall and E.M.H. Wellington. 2011. Impacts of anthropogenic activity on the ecology of class 1 integrons and integron-associated genes in the environment. ISME Journal 5: 1253–1261.

Grandclement, C., I. Seyssiecq, A. Piram, P. Wong-Wah-Chung, G. Vanot, N. Tiliacos, N. Roche and P. Doumenq. 2017. From the conventional biological wastewater treatment to hybrid processes, the evaluation of organic micropollutant removal: A review. Water Research 111: 297–317.

Guo, W., H.-H. Ngo and J. Li. 2012. A mini-review on membrane fouling. Bioresource Technology 122: 27–34.

Hamjinda, N.S., W. Chiemchaisri and C. Chiemchaisri. 2017. Upgrading two-stage membrane bioreactor by bioaugmentation of Pseudomonas putida entrapment in PVA/SA gel beads in treatment of ciprofloxacin. International Biodeterioration & Biodegradation 119: 595–604.

Homem, V. and L. Santos. 2011. Degradation and removal methods of antibiotics from aqueous matrices—A review. Journal of Environmental Management 92: 2304–2347.

Huang, L.Y. and D.J. Lee. 2015. Membrane bioreactor: A mini review on recent R&D works. Bioresource Technology 194: 383–388.

Judd, S. 2008. The status of membrane bioreactor technology. Trends in Biotechnology 26: 109–116.

Judd, S. 2011. The MBR Book: Principles and Applications of Membrane Bioreactors for Water and Wastewater Treatment.

Karaolia, P., I. Michael-Kordatou, E. Hapeshi, J. Alexander, T. Schwartz and D. Fatta-Kassinos. 2017. Investigation of the potential of a Membrane BioReactor followed by solar Fenton oxidation to remove antibiotic-related microcontaminants. Chemical Engineering Journal 310: 491–502.

Karumathil, D.P., H.-B. Yin, A. Kollanoor-Johny and K. Venkitanarayanan. 2014. Effect of chlorine exposure on the survival and antibiotic gene expression of multidrug resistant Acinetobacter baumannii in water. International Journal of Environmental Research and Public Health 11: 1844–1854.

Kerrn, M.B., N. Frimodt-Moller and F. Espersen. 2003. Effects of sulfamethizole and Amdinocillin against *Escherichia coli* strains (with various susceptibilities) in an ascending urinary tract infection mouse model. Antimicrobial Agents and Chemotherapy 47: 1002–1009.

Kim, S.D., J. Cho, I.S. Kim, B.J. Vanderford and S.A. Snyder. 2007. Occurrence and removal of pharmaceuticals and endocrine disruptors in South Korean surface, drinking, and waste waters. Water Research 41: 1013–1021.

Kimosop, S.J., Z.M. Getenga, F. Orata, V.A. Okello and J.K. Cheruiyot. 2016. Residue levels and discharge loads of antibiotics in wastewater treatment plants (WWTPs), hospital lagoons, and rivers within Lake Victoria Basin, Kenya. Environmental Monitoring and Assessment 188: 9.

Kimura, K., H. Hara and Y. Watanabe. 2007. Elimination of selected acidic pharmaceuticals from municipal wastewater by an activated sludge system and membrane bioreactors. Environmental Science & Technology 41: 3708–3714.

Kimura, K., T. Naruse and Y. Watanabe. 2009. Changes in characteristics of soluble microbial products in membrane bioreactors associated with different solid retention times: Relation to membrane fouling. Water Research 43: 1033–1039.

Klein, E.Y., T.P. Van Boeckel, E.M. Martinez, S. Pant, S. Gandra, S.A. Levin, H. Goossens and R. Laximinarayan. 2018. Global increase and geographic convergence in antibiotic consumption between 2000 and 2015. Proceedings of the National Academy of Sciences of the United State of America 115(15): E3463–E3470.

Laera, G., D. Cassano, A. Lopez, A. Pinto, A. Pollice, G. Ricco and G. Mascolo. 2012. Removal of organics and degradation products from industrial wastewater by a membrane bioreactor integrated with ozone or UV/H2O2 treatment. Environmental Science & Technology 46: 1010–1018.

Le-Minh, N., H.M. Coleman, S.J. Khan, Y. Van Luer, T.T.T. Trang, G. Watkins and R.M. Struetz. 2010. The application of membrane bioreactors as decentralised systems for removal of endocrine disrupting chemicals and pharmaceuticals. Water Science and Technology 61: 1081–1088.

Li, D., T. Yu, Y. Zhang, M. Yang, Z. Li, M. Liu and R. Qi. 2010. Antibiotic resistance characteristics of environmental bacteria from an oxytetracycline production wastewater treatment plant and the receiving river. Applied and Environmental Microbiology 76: 3444–3451.

Liu, H.Q., J.C.W. Lam, W.W. Li, H.Q. Yu and P.K.S. Lam. 2017. Spatial distribution and removal performance of pharmaceuticals in municipal wastewater treatment plants in China. Science of The Total Environment 586: 1162–1169.

Lopez-Fernandez, R., L. Martinez and S. Villaverde. 2012. Membrane bioreactor for the treatment of pharmaceutical wastewater containing corticosteroids. Desalination 300: 19–23.

Luddeke, F., S. Hess, C. Gallert, J. Winter, H. Gude and H. Loffler. 2015. Removal of total and antibiotic resistant bacteria in advanced wastewater treatment by ozonation in combination with different filtering techniques. Water Research 69: 243–251.

Luis Martinez, J. 2009. Environmental pollution by antibiotics and by antibiotic resistance determinants. Environmental Pollution 157: 2893–2902.

Mascolo, G., G. Laera, A. Pollice, D. Cassano, A. Pinto, C. Salerno and A. Lopez. 2010. Effective organics degradation from pharmaceutical wastewater by an integrated process including membrane bioreactor and ozonation. Chemosphere 78: 1100–1109.

Matos, M., M.A. Pereira, P. Parpot, A. Brito and R. Nogueira. 2014. Influence of tetracycline on the microbial community composition and activity of nitrifying biofilms. Chemosphere 117: 295–302.

Mousaab, A., C. Claire, C. Magali and D. Christophe. 2015. Upgrading the performances of ultrafiltration membrane system coupled with activated sludge reactor by addition of biofilm supports for the treatment of hospital effluents. Chemical Engineering Journal 262: 456–463.

Ng, K.K., X. Shi and H.Y. Ng. 2015. Evaluation of system performance and microbial communities of a bioaugmented anaerobic membrane bioreactor treating pharmaceutical wastewater. Water Research 81: 311–324.

Ngoc Han, T., H. Chen, M. Reinhard, F. Mao and K.Y.-H. Gin. 2016. Occurrence and removal of multiple classes of antibiotics and antimicrobial agents in biological wastewater treatment processes. Water Research 104: 461–472.

Olexy, V.M., T.J. Bird, H.G. Grieble and S.K. Farrand.1979. Hosptial isolates of *Serratia marcescens* transferring ampicillin, carbenicillin and gentamicin resistance to resistance to other Gram-negative bacteria including *Pseudmonas aeruginosa*. Antimicrobial Agents and Chemotherapy 15: 93–100.

Ozdemir, G., E. Aydin, E. Topuz, C. Yangin-Gomec and D.O. Tas. 2015. Acute and chronic responses of denitrifying culture to diclofenac. Bioresource Technology 176: 112–120.

Pauwels, B. and W. Verstraete. 2006. The treatment of hospital wastewater: an appraisal. Journal of Water and Health 4: 405–16.

Pollice, A., G. Laera, D. Cassano, S. Diomede, A. Pinto, A. Lopez and G. Mascolo. 2012. Removal of nalidixic acid and its degradation products by an integrated MBR-ozonation system. Journal of Hazardous Materials 203: 46–52.

Prasertkulsak, S., C. Chiemchaisri, W. Chiemchaisri, T. Itonaga and K. Yamamoto. 2016. Removals of pharmaceutical compounds from hospital wastewater in membrane bioreactor operated under short hydraulic retention time. Chemosphere 150: 624–631.

Qiu, P., X. Guo, Y. Zhang, X. Chen and N. Wang. 2016. Occurrence, fate, and risk assessment of vancomycin in two typical pharmaceutical wastewater treatment plants in Eastern China. Environmental Science and Pollution Research 23: 16513–16523.

Radjenovic, J., M. Petrovic and D. Barcelo. 2007. Analysis of pharmaceuticals in wastewater and removal using a membrane bioreactor. Analytical and Bioanalytical Chemistry 387: 1365–1377.

Radjenovic, J., M. Petrovic and D. Barcelo. 2009. Fate and distribution of pharmaceuticals in wastewater and sewage sludge of the conventional activated sludge (CAS) and advanced membrane bioreactor (MBR) treatment. Water Research 43: 831–841.

Rivera-Utrilla, J., M.A. Sanchez-Polo, G .Prados-Joya and R. Ocampo-Perez. 2013. Pharmaceuticals as emerging contaminants and their removal from water. A review. Chemosphere 93: 1268–1287.

Sabirova, J.S., L.F.F. Cloetens, L. Vanhaecke, I. Forrez, W. Verstraete and N. Boon. 2008. Manganese-oxidizing bacteria mediate the degradation of 17 alpha-ethinylestradiol. Microbial Biotechnology 1: 507–512.

Sahar, E., I. David, Y. Gelman, H. Chikurel, A. Aharoni, R. Messalem and A. Brenner. 2011a. The use of RO to remove emerging micropollutants following CAS/UF or MBR treatment of municipal wastewater. Desalination 273: 142–147.

Sahar, E., M. Ernst, M. Godehardt, A. Hein, J. Herr, C. Kazner, T. Melin, H. Cikurel, A. Aharoni, R. Messalem, A. Brenner and M. Jekel. 2011b. Comparison of two treatments for the removal of selected organic micropollutants and bulk organic matter: conventional activated sludge followed by ultrafiltration versus membrane bioreactor. Water Science and Technology 63: 733–740.

Sahar, E., R. Messalem, H. Cikurel, A. Aharoni, A. Brenner, M. Godehardt, M. Jekerl and M. Ernst. 2011c. Fate of antibiotics in activated sludge followed by ultrafiltration (CAS-UF) and in a membrane bioreactor (MBR). Water Research 45: 4827–4836.

Santos, L., A.N. Araujo, A. Fachini, A. Pena, C. Deleruse-Matos and M. Montenegro. 2010. Ecotoxicological aspects related to the presence of pharmaceuticals in the aquatic environment. Journal of Hazardous Materials 175: 45–95.

Santos, L., M. Gros, S. Rodriguez-Mozaz, C. Delerue-Matos, A. Pena, D. Barcelo and M. Montenegro. 2013. Contribution of hospital effluents to the load of pharmaceuticals in urban wastewaters: Identification of ecologically relevant pharmaceuticals. Science of the Total Environment 461: 302–316.

Sauve, S. and M. Desrosiers. 2014. A review of what is an emerging contaminant. Chemistry Central Journal 8: 7.

Schmitt, H., K. Stoob, G. Hamscher, E. Smit and W. Seinen. 2006. Tetracyclines and tetracycline resistance in agricultural soils: Microcosm and field studies. Microbial Ecology 51: 267–276.

Schroeder, H.F., J.L. Tambosi, R.F. Sena, R.F.P.M. MOreira, H.J. Jose and J. Pinnekamp. 2012. The removal and degradation of pharmaceutical compounds during membrane bioreactor treatment. Water Science and Technology 65: 833–839.

Shariati, F.P., M.R. Mehrnia, B.M. Salmasi, M. Heran, C. Wisniewski and M.H. Sarrafzadeh. 2010. Membrane bioreactor for treatment of pharmaceutical wastewater containing acetaminophen. Desalination 250: 798–800.

Shen, L., X. Yuan, W. Shen, N. He., Y. Wang, H. Lu and Y. Lu. 2014. Positive impact of biofilm on reducing the permeation of ampicillin through membrane for membrane bioreactor. Chemosphere 97: 34–39.

Stephenson, T., K. Brindle, S. Judd and B. Jefferson. 2007. Membrane Bioreactors for Wastewater Treatment. Volume 6, IWA Publishing, London, UK.

Subbiah, M., S.M. Mitchell, J.L. Ullman and D.R. Call. 2011. Beta-lactams and florfenicol antibiotics remain bioactive in soils while ciprofloxacin, neomycin, and tetracycline are neutralized. Applied and Environmental Microbiology 77: 7255–7260.

Sustrasic, M. 2009. Wastewater treatment: Understanding the activated sludge process. Chemical Engineering Progress 105: 26–29.

Talesnik, D. 2016. Antibiotic use, resistance threaten global health. NIH Record, LXVIII, 1–12.

Tambosi, J.L., R.F. De Sena, M. Favier, W. Gebhardt, H.J. Jose, H.F. Schroeder and R.D.F. Peralta Muniz Moreira. 2010. Removal of pharmaceutical compounds in membrane bioreactors (MBR) applying submerged membranes. Desalination 261: 148–156.

Taskan, B., O. Hanay, E. Taskan, M. Erdem and H. Hasar. 2016. Hydrogen-based membrane biofilm reactor for tetracycline removal: biodegradation, transformation products, and microbial community. Environmental Science and Pollution Research 23: 21703–21711.

Tiwari, B., B. Sellamurthu, Y. Ouarda, P. Drogui, R.D. Tyagi and G. Buelna.. 2017. Review on fate and mechanism of removal of pharmaceutical pollutants from wastewater using biological approach. Bioresource Technology 224: 1–12.

Tuc, D.Q., M.G. Elodie, L. Pierre, A. Fabrice, T. Marie-Jeanne, B. Martine, E. Joelle and C. Marc. 2017. Fate of antibiotics from hospital and domestic sources in a sewage network. Science of the Total Environment 575: 758–766.

van der Roest, H.F., A.G. van Bentem and D.P. Lawrence. 2002. MBR-technology in municipal wastewater treatment: challenging the traditional treatment technologies. Water Science and Technology 46: 273–280.

Verlicchi, P., M. Al Aukidy and E. Zambello. 2015. What have we learned from worldwide experiences on the management and treatment of hospital effluent? An overview and a discussion on perspectives. Science of the Total Environment 514: 467–491.

Volkmann, H., T. Schwartz, P. Bischoff, S. Kirchen and U. Obst. 2004. Detection of clinically relevant antibiotic-resistance genes in municipal wastewater using real-time PCR (TaqMan). Journal of Microbiological Methods 56: 277–286.

Wang, J., Y. Wei, K. Li, Y. Cheng, M. Li and J. Xu. 2014. Fate of organic pollutants in a pilot-scale membrane bioreactor-nanofiltration membrane system at high water yield in antibiotic wastewater treatment. Water Science and Technology 69: 876–881.

Wang, J., K. Li, Y. Wei, Y. Cheng, D. Wei and M. Li. 2015. Performance and fate of organics in a pilot MBR-NF for treating antibiotic production wastewater with recycling NF concentrate. Chemosphere 121: 92–100.

Wang, P., Y.-L. He and C.-H. Huang. 2011. Reactions of tetracycline antibiotics with chlorine dioxide and free chlorine. Water Research 45: 1838–1846.

Wang, X.-C., J.-M. Shen, Z.-L. Chen, X. Zhao and H. Xu. 2016. Removal of pharmaceuticals from synthetic wastewater in an aerobic granular sludge membrane bioreactor and determination of the bioreactor microbial diversity. Applied Microbiology and Biotechnology 100: 8213–8223.

Watkinson, A.J., E.J. Murby and S.D. Costanzo. 2007. Removal of antibiotics in conventional and advanced wastewater treatment: Implications for environmental discharge and wastewater recycling. Water Research 41: 4164–4176.

WHO. 2016. Global Action Plan on Antimicrobial Resistance. World Health Organization, Geneva, Switzerland.

Wijekoon, K.C., F.I. Hai, J. Kang, W.E. Price, W. Guo, H.H. Ngo and L.D. Nghiem. 2013. The fate of pharmaceuticals, steroid hormones, phytoestrogens, UV-filters and pesticides during MBR treatment. Bioresource Technology 144: 247–254.

Wu, N., M. Qiao, B. Zhang, W.-D. Cheng and Y.-G. Zhu. 2010. Abundance and Diversity of tetracycline resistance genes in soils adjacent to representative swine feedlots in China. Environmental Science & Technology 44: 6933–6939.

Xia, S., R. Jia, F. Feng, K. Xie, H. Li, D. Jing and X. Xu. 2012. Effect of solids retention time on antibiotics removal performance and microbial communities in an A/O-MBR process. Bioresource Technology 106: 36–43.

Xia, Z., W. Xiao-Chun, C. Zhong-Lin, X. Hao and Z. Qing-Fang. 2015. Microbial community structure and pharmaceuticals and personal care products removal in a membrane bioreactor seeded with aerobic granular sludge. Applied Microbiology and Biotechnology 99: 425–33.

Xu, J., Y. Xu, H.M. Wang, C.S. Guo, H.Y. Qiu, Y. He, Y. Zhang, X.C. Li and W. Meng. 2015. Occurrence of antibiotics and antibiotic resistance genes in a sewage treatment plant and its effluent-receiving river. Chemosphere 119: 1379–1385.

Yoo, R., J. Kim, P.L. Mccarty and J. Bae. 2012. Anaerobic treatment of municipal wastewater with a staged anaerobic fluidized membrane bioreactor (SAF-MBR) system. Bioresource Technology 120: 133–139.

Yuan, X., Z. Qiang, W. Ben, B. Zhu and J. Qu. 2015. Distribution, mass load and environmental impact of multiple-class pharmaceuticals in conventional and upgraded municipal wastewater treatment plants in East China. Environmental Science-Processes & Impacts 17: 596–605.

Zhang, S., H.-L. Song, X.-L. Yang, K.-Y. Yang and X.-Y. Wang. 2016. Effect of electrical stimulation on the fate of sulfamethoxazole and tetracycline with their corresponding resistance genes in three-dimensional biofilm-electrode reactors. Chemosphere 164: 113–119.

Zhang, S., H.L. Song, X.L. Yang, X.Z. Long, X. Liu and T.Q. Chen. 2017. Behavior of tetracycline and sulfamethoxazole and their corresponding resistance genes in three-dimensional biofilm-electrode reactors with low current. Journal of Environmental Science and Health, Part A-Toxic/Hazardous Substances & Environmental Engineering 52: 333–340.

Zhang, T., M. Zhang, X.X. Zhang and H.H. Fang. 2009a. Tetracycline resistance genes and tetracycline resistant lactose-fermenting enterobacteriaceae in activated sludge of sewage treatment plants. Environmental Science & Technology 43: 3455–3460.

Zhang, Y., H. Zhu, U. Szewzyk and S.U. Szewzyk. 2015. Removal of pharmaceuticals in aerated biofilters with manganese feeding. Water Research 72: 218–226.

Zhang, Y.Z., C.M. Ma, F. Ye, Y. Kong and H. Li. 2009b. The treatment of wastewater of paper mill with integrated membrane process. Desalination 236: 349–356.

Zhao, X., Z.-L. Chen, X.-C. Wang, J.-M. Shen and H. Xu. 2014. PPCPs removal by aerobic granular sludge membrane bioreactor. Applied Microbiology and Biotechnology 98: 9843–9848.

Zhu, Y.-G., T.A. Johnson, J.-Q. Su, M. Qiao, G.-X. Guo, R.D. Stedtfeld, S.A. Hashsham and J.M. Tiedje. 2013. Diverse and abundant antibiotic resistance genes in Chinese swine farms. Proceedings of the National Academy of Sciences of the United States of America 110: 3435–3440.

Zupanc, M., T. Kosjek, M. Petkovsek, M. Dular, B. Kompare, B. Sirok, Z. Blazeka and E. Health. 2013. Removal of pharmaceuticals from wastewater by biological processes, hydrodynamic cavitation and UV treatment. Ultrasonics Sonochemistry 20: 1104–1112.

CHAPTER 9

Bioremediation Approaches for Persistent Organic Pollutants Using Microbial Biofilms

Sudhir K. Shukla,[1,2] *Neelam Kungwani*[3] *and T. Subba Rao*[1,2,*]

1. Introduction

The term bioremediation is coined from two words *bios* means life and *remediate* means a correction measure to clean up the environment. In general, bioremediation can be defined as *a unique microbial/ biological process that hastens the biodegradation of a pollutant by optimizing the environmental conditions. As such, it is an ecologically safe and economically viable method for remediation of contaminated environments.* In other words, bioremediation is the use of microorganisms (e.g., bacteria and fungi), plants/algae (termed phytoremediation), or enzymes to realize the effective removal of hazardous pollutants from contaminated soil, sediments and water. Bioremediation can target a variety of media (wastewater, groundwater, soil/sludge, air) with multiple possibilities, for example, complete mineralization of organic compounds, and immobilization of non-degradable contaminants (Figure 1). Many studies have shown that bioremediation has received worldwide attention for cleaning most pollutant types, such as pesticides, polycyclic aromatic hydrocarbons and organic solvents. Of late, the concomitant biotechnological processes are also gaining importance for treatment of radioactive waste and military wastes. Despite having high prospective of this technology, the applications are still scarce due to limited understanding of the metabolic degradation pathways. Microorganisms like bacteria, fungi and algae take part in bioremediation and there are many reviews and journal publications in support of this remediation technology (Abdulsalam et al. 2011, Adams et al. 2015, Agamuthu and Dadrasnia 2013, Gavrilescu and Chisti 2005, Sasek and Cajthaml 2005, Singh et al. 2006, Vidali 2001).

[1] Biofouling & Thermal Ecology Section, Water and Steam Chemistry Division, BARC Facilities, Kalpakkam-603102, India.
[2] Homi Bhabha National Institute, Mumbai 400094, India.
[3] Biotechnology Faculty, Government Science College, K K Shastri Education campus, Ahmedabad-380008, India.
* Corresponding author: subbarao@igcar.gov.in

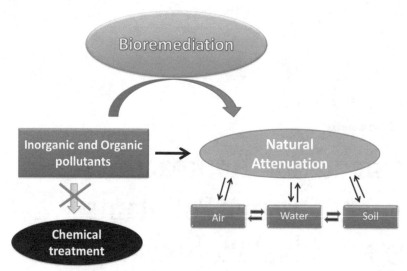

Figure 1. Natural attenuation and bioremediation are widely accepted environment cleanup procedures (Shukla et al. 2014).

Bioremediation techniques are currently being used to mitigate highly toxic metals, chemicals, effluents and pollutants. If the toxic pollutants are accumulated into water bodies, they can cause the death of fishes and marine animals. Algal blooms are generally formed once the water body becomes eutrophic which have a significant effect on the ecosystem productivity. They block the sunlight and reduce dissolved oxygen levels in the water, resulting in the death of aquatic biota. Algal blooms in drinking water sources makes the water unfit for human consumption as the algal growth give colour and bad odour to the water. Oil spills causes similar effects making the water unfit for aquatic biota growth and cause their death. A large number of marine biota was lost due to oil spills, thus causing a disturbance in food chains and affecting the functioning of the ecosystem. Fertilizers and agrochemicals are commonly added to such polluted water bodies to increase the metabolic rate of the microbiota which degrades the pollutant (Adams et al. 2015, Al-Sulaimani et al. 2011, Atlas and Bragg 2009, Cohen 2002).

According to the polluted site to be remediated, there are two common bioremediation methods, *in-situ* and *ex-situ*. In the case of *in situ* treatment, it is often at the site of contamination, while *ex situ* treatment is away from the site of contamination. *Ex situ* bioremediation is generally applied to media readily accessible aboveground (e.g., in treatment cells/soil piles or bioreactors). Some authors have described various types of bioremediation processes (Hamzah et al. 2013, Kumar et al. 2006, Orji et al. 2012). However, it will be worthwhile to determine what exactly it means *in situ* and *ex situ* remediation and use the same to describe the different types of technologies. Bioremediation can easily be observed by keeping a check on the redox potential, pH, temperature, oxygen content and residue concentrations from catabolic activity. Based on the type of toxicant/pollutant, the bioremediation process may be mediated naturally or by augmenting the intrinsic potential. In *ex situ* bioremediation process, microorganisms which are more efficient when compared to that of indigenous ones, or sometimes genetically modified microorganism can be used to scale-up the bioprocess, e.g., bacteria like *Deinococcus radiodurans*, was successfully modified to demonstrate the reduction of mercury and in the degradation of aromatic hydrocarbons like toluene (Brim et al. 2006).

2. Persistent organic pollutants

Many hydrocarbons such as benzene, toluene, ethyl benzene, xylene (BTEX) and polyaromatic hydrocarbons (PAHs) are persistent organic pollutants (POPs) that are released into the environment majorly by petroleum industry activities. Apart from petroleum industry, chemical and pharmaceutical industries also contribute a number of new hydrocarbon molecules that are being discovered and synthesized. Persistence of these organic pollutants and emerging contamination is of increasing concern. These chemicals are abundant in the human habitat owing to their application in industry, agriculture, therapeutic products and households (Doll and Frimmel 2004). A number of natural or anthropogenic organic complexes are there in our biosphere, ~ 16 million molecular species are known of which roughly 40,000 are in the eye of attention due to their daily use. The detail structure, application and toxicity of some of these organic chemicals are illustrated in Table 1.

2.1 Polycyclic aromatic hydrocarbons

Polycyclic aromatic hydrocarbons consist of carbon and hydrogen atoms along with N, S and O atoms. These organic molecules do have two or more fused benzene rings which may have linear, angular or cluster arrangements (Cerniglia and Heitkamp 1989, Varjani and Upasani 2017, Wilson and Jones 1993) (Table 1). PAHs are ubiquitous and are natural pollutants of air, water and soil. They are a class of multi-phase compounds that are released into the environment naturally and by anthropogenic means. Crude oil and coal deposits are natural repositories of PAHs and they are mainly in the refinery process of crude oil and incomplete combustion of fossil fuels. PAHs are one of the most significant classes of organic compounds that have caused rising concern regarding their harmful effects to humans and other living organisms. High molecular weight are PAHs more carcinogenic, mutagenic and teratogenic as compared to low molecular weight PAHs (Cerniglia 1992, Okere and Semple 2012, Tiwari et al. 2016). On the basis of genotoxicity and abundant use, the U.S. EPA has listed as many as 16 PAH compounds as priority pollutants (Perelo 2010).

Presence of dense π-electrons in ring structure of PAH molecule makes it less accessible to nucleophilic attack, thus enhancing its biochemical persistence as compared to that of any other hydrocarbons present in the environment (Haritash and Kaushik 2009). They are highly hydrophobic in nature and tend to adsorb onto the surface of soil (or sediments in a marine environment), therefore less available for the microbial metabolic activity. Increase in the molecular weight increases the hydrophobicity of PAHs as well as reduces the water solubility. Hence, when the molecular weight of PAHs is higher, it becomes poorly bioavailable and recalcitrant to microbial degradation when compared with low molecular weight PAHs (Ghosal et al. 2016, Louvado et al. 2015, Okere and Semple 2012).

2.2 Nitro-aromatic compounds

Nitro-aromatic compounds consist of a minimum of one nitro group ($-NO_2$) attached to an aromatic ring (see Table 1). These compounds consist of an imperative group of chemical which are widely used in various industries at present. Nitro-aromatic compounds are released to the environment by both synthetic and natural means; however, the major contribution is through synthetic means. The unique chemistry of the nitro group has led to the use of several nitro-aromatic compounds as agrochemicals, textile, and chemicals in pharmaceutical industry (Su et al. 2012). They are highly resistant to biodegradation because of their recalcitrant character. They are toxic to both humans and animals; the reported manifestations in human are sensitization of skin, immuno-toxicity, and methaemoglobinemia. Nitro-aromatic compounds are also common toxic pollutants observed in many aquatic ecosystems (Katritzky et al. 2003).

2.3 Organo-chlorine compounds

Organo-chlorines include a class of organic compounds containing chlorine ($-Cl$) atoms bound by covalent bonds to the carbon atom. Organo-chlorines have diverse structures and applications, they

Table 1. Organic Pollutants: Structure, Toxicity and Source.

	Pollutants	Structure	Toxicity/health effect	Use/source
PAHs	Pyrene		Toxic to kidneys and liver, carcinogenic	Dye and dye precursor, combustion
	Napthalene		Carcinogenic, haemolytic anemia	Industrial application
	Phenanthrene		Carcinogenic, dye	Coal tar, pesticides
	Benzo(a)pyrene		Mutagenic, carcinogen	Industries, coal tar, automobile exhaust fumes
Nitroaromatic compounds	2,4,6-trinitrotoluene (TNT)		Skin irritation, carcinogenic, immune modulator	Explosive, pesticides
	Hexanitrobenzene		Skin irritation, carcinogenic, fertility	Highly explosive
Organochlorines	DDT(dichlorodiphenyl-richloroethane)		Xenoestrogenic, endocrine disruptor	Insecticide
	2,4-D (2,4-Dichlorophenox-yacetic acid)		Neurotoxic, carcinogenic	Pesticide
Phthalates	di(2-ethylhexyl) phthalate		Endocrine disruptor	Plasticizer, dielectric fluid
Azo dye	Azobenzene		Carcinogenic	Dyeing

are mainly synthetic compounds that have both aliphatic chains and aromatic rings. Aliphatic organo-chlorides, e.g., vinyl chloride is used to make polymers, whereas compounds like chloromethane are common solvents used in dry cleaning and textile industry (see Table 1). Aromatic organo-chlorines like DDT, endosulfan, chlordane, aldrin, dieldrin, endrin and mirex are used as pesticides. They are designated in the list of persistent organic pollutants (POPs) due to their toxicity and environmental persistence. Organo-chlorines like *Polychlorinated biphenyls* (*PCBs*) are used as coolant and insulators in transformers. Tons of commercial PCBs (e.g., Aroclors) are present in the terrestrial and aquatic environment (Salvo et al. 2012). Commercial PCBs are frequently used as plasticizers, adhesives, lubricants and hydraulic fluids (Perelo 2010).

2.4 Phthalates

Phthalate esters are a class of chemical compounds mainly used as plasticizers for polyvinyl chloride (PVC) or to a lesser extent other resins in different industrial activities (see Table 1) (Liu et al. 2013). Phthalates are 'semi-volatile' chemicals that are broadly used in the manufacture of plastics, insect repellents, synthetic fibres, lubricant and cosmetics (Liang et al. 2008). Derived from oil, around 2 million tonnes of phthalates are produced across the world each year, and more than 20 different types are in common use. Paper and plastic industries are major manufacturers of phthalates, often discharged into the environment at relatively high levels and hence cause environmental pollution. Being classified as endocrine disruptors or hormonal active agents (HAAs) and their potential impacts on public health, phthalates have evoked great interest among the research community in the past decade. Phthalates like butyl benzyl phthalate are reported to cause eczema and rhinitis in children. Butyl benzyl phthalate is also classified as a possible class-C human carcinogen (under the 1986 U.S. EPA guidelines) (Matsumoto et al. 2008).

2.5 Azo dyes

The wastewater from textile industries has been designated as a major pollutant of water bodies as well as ground water resources. The effluent is characterized to have intense colour due to the presence of azo dyes, high suspended solids and high chemical oxygen demand emanating from dyes, cellulose fibres, surfactants, detergents and solvents (Lotito et al. 2012). Among all the above mentioned pollutants, azo dyes constitute ~ 60–70% of all the dye stuff which are highly persistent and extremely recalcitrant (Saratale et al. 2011). Azo dyes have the functional azo group R-N=N-R', where R and R' can be either aryl or alkyl (see Table 1) (Puvaneswari et al. 2006). Azo dyes and their transformed products are recalcitrant and mutagenic, and are shown to be associated with toxicity to aquatic flora and fauna due to reduction in photosynthesis in water bodies (Pearce et al. 2003). A survey by the National Institute for Occupational Safety and Health reported the carcinogenicity of azo dyes (e.g., Benzidine) based on animal experiments (Mittal et al. 2013). Around 2000, different azo-dyes are presently used and annually, over 7×10^5 tons of these compounds are produced globally (Zollinger 2003). Azo dyes are not very acquiescent towards oxidative biodegradation via oxygenases due to the presence of strong electron withdrawing azo groups (Hong and Gu 2010). However, azo groups can be microbiologically reduced to the corresponding amines with concurrent removal of colour (Hong and Gu 2010, Tomei et al. 2016).

3. Bioremediation strategies for organic pollutants

Among the various bioremediation technologies, *in situ* bioremediation (ISB) is widely applied in the treatment of subsurface pollution. ISB can be applied in the unsaturated zone (e.g., bioventing) or in saturated soils and groundwater. ISB technology was originally developed as a less costly, more effective alternative to the standard pump and treat methods used to clean up aquifers and soils contaminated with organic chemicals (e.g., fuel hydrocarbons, chlorinated solvents). This process has been expanded for the degradation of explosives (e.g., TNT), inorganics (e.g., nitrate) and toxic metals (e.g., chromium). One way to categorize ISB is by the type of metabolism involved. The two main categories of metabolism are aerobic and anaerobic. The target metabolism for an ISB system will generally depend on the contaminants of concern. Some contaminants (e.g., fuel hydrocarbons) are degraded via an aerobic pathway, some anaerobically (e.g., carbon tetrachloride), and some contaminants can be biodegraded under either aerobic or anaerobic conditions (e.g., trichloroethane). In accelerated ISB, substrate or nutrients are added to an aquifer to stimulate the growth of a targeted indigenous consortium of bacteria. Usually the enriched cultures of bacteria that are highly efficient at degrading a particular contaminant can be introduced into the aquifer, which is termed bioaugmentation (Paul et al. 2005). Accelerated ISB is used where it is desired to increase the rate of contaminant biotransformation, which may be limited by lack of required nutrients and electron donor/acceptor. Aerobic ISB may only require the addition of oxygen as the electron acceptor, while anaerobic ISB generally requires the addition of an electron donor (e.g., lactate, benzoate) and potentially an electron acceptor (e.g., nitrate, sulfate). Among the various bioprocesses used for *in situ* remediation, the common methods are, viz., bio-venting, biosorption, bio-augmentation, co-metabolism and supplemented by myco-remediation and phyto-remediation.

3.1 Merits and limitations of in situ bioremediation

3.1.1 Merits of in situ Bioremediation

- Complete degradation (mineralization) of organic contaminants to harmless substances (e.g., carbon dioxide and water).
- Accelerated ISB can provide volumetric treatment for both dissolved and sorbed contaminants. The time required to treat subsurface pollution will be faster than pump and treat process.
- ISB has lower risk of human exposure to the contaminated media and lower equipment/operating costs.
- The areal zone of treatment can be larger when compared to other technologies since the treatment process moves with the plume and can reach inaccessible areas.
- In ISB technology, typically no secondary waste is generated.

3.1.2 Limitations of in situ Bioremediation

- *In situ* bioremediation usually requires an acclimatized population of microorganisms. Some contaminants may be recalcitrant and cannot be biodegraded or completely mineralized.
- If biotransformation of the toxicant stops at an intermediate compound, it may be more toxic and environmentally mobile than the parent compound.
- When applied in low-permeability or heterogeneous sub-surface systems, the addition of nutrients, electron donor/acceptor into the injection wells may get clogged due to dense growth of bacteria.
- Heavy metals and toxic concentrations of organic compounds may inhibit the activity of indigenous microorganisms.

3.2 Bioventing

Bioventing is an *in situ* remediation technology which utilizes indigenous microorganisms for degrading organic pollutants adsorbed to the soil particles in the unsaturated soil zone. In the bioventing process, the activity of the indigenous bacteria is enhanced by inducing air flow inside the unsaturated zone (using injection wells or extraction) and, when required, is augmented with the addition of nutrients.

In the bioventing process, aerobically biodegradable chemicals can be treated easily. This process was proven to be very effective in remediating petroleum products such as gasoline, jet fuels, kerosene and diesel. Bioventing is often used at sites which are contaminated with medium range molecular weight hydrocarbons (i.e., diesel and jet fuel), since lighter products such as gasoline tend to volatilize readily and can be removed more rapidly by means of soil aeration which is referred as vapour extraction. High molecular weight products (e.g., lubricating oils) take a longer time to biodegrade. Bioventing is one of the most attractive options for reducing or eliminating toxic pollutants in soil and water. Air sparging forces the compressed air into the toxicant saturated soil, which enhances the biodegradation process. In this biologically enhanced process, volatilization is reduced and biodegradation is maximized. The advantage is that by minimizing the volatilization, the process eliminates the treatment of obnoxious gases. Bioventing can also be called as a stepped up version of natural attenuation that happens without intervention, where the native bacterial population degrades the pollutants. Oxygen is the limiting factor since it indirectly limits the microbial degradation process. In an extended extraction process, well mediated bioventing is similar to Soil Vapor Extraction (SVP) (Verhagen et al. 2011). However, SVE removes constituents mainly through volatilization, and a venting system promotes biodegradation and reduces volatilization (by using lower air flow rates). In practice, some degree of volatilization and biodegradation ensues when either SVE or bioventing is used. It may be noted that the environment is maintained in such a manner so as to promote the growth of the indigenous microorganisms which hastens the biodegradation of the contaminants (Leahy and Erickson 1995a, b). Bioventing eliminates the need for excavation of the site, therefore reducing the transportation of polluted soil to remote places for processing or disposal. Thus, one can remediate an area with little or no influence to the geological composition, and no re-compaction or restoration of the location is required. The technology is simple, cost effective and has minimal site disturbance, also reducing the pollutant concentrations to a safe and acceptable limit (Seo et al. 2009).

3.3 Bio-augmentation and bio-stimulation

These are the two basic forms of bioremediation presently being practiced: the microbiological approach (*bio-augmentation*) or the microbial ecological approach (*bio-stimulation*). Bio-augmentation is the addition of microorganisms with the capability of degradation of the pollutants to the contaminated soil, whereas bio-stimulation is addition of nutrients and other supplementary components to the native microbial population to induce the growth of specific microbes at a hastened rate. The bio-augmentation approach involves addition of specialized populations of a particular microbial species at high cell densities into a polluted site to enhance the rate of contaminant biodegradation. On the other hand, in bio-stimulation approach, emphasis is placed on identifying and adjusting certain physical and chemical factors (such as soil temperature, pH, moisture and nutrient content) that may be impeding the rate of degradation of the pollutant by the indigenous microorganism in the affected site (Thomassin-Lacroix et al. 2002).

Successful bio-augmentation treatments depend on the use of the right inoculum consisting of a variety of microbial strains or microbial consortia that have been well adapted to the site to be remediated. The process will be successful only when the microbial species or consortium would be able to compete with indigenous microorganisms, survive predation, and perform in the presence of various hostile abiotic factors. In bio-augmentation process, addition of oil-degrading microorganisms to supplement the indigenous populations has been projected as an alternate strategy for the remediation of oil contaminated sites (Head et al. 2006). The rationale for this methodology is that indigenous microbial populations may not be capable of degrading the wide range of organic pollutants present in the complex concoction such as petroleum (Seo et al. 2009). Other conditions under which bio-augmentation is chosen are when the indigenous hydrocarbon-degrading population is low in number and there is a need to speed up the bioremediation process. In this procedure, microbial seeding may reduce the lag period to start the remediation process. For successful application of this approach in the field, the seed microorganisms must be able to degrade most of the hydrocarbon constituents and should maintain their genetic stability and cell viability during storage/preservation. The microbes should also survive and move through the pores of the sediment to degrade the contaminants. Different microbial species have different enzymatic abilities and preferences for the degradation of hydrocarbon compounds. Factors affecting proliferation of microorganisms used for bio-augmentation include the chemical structure and concentration of the pollutants, the availability of the contaminant to the microorganisms, and the nature and the size of the microbial population, and therefore the surrounding physical environment should also be considered when screening for microorganisms. Some microorganisms degrade linear, branched, or cyclic compounds. Others are capable of degrading mono- or poly-nuclear aromatics, while some others can degrade both alkanes and aromatic hydrocarbons (Thomassin-Lacroix et al. 2002).

Bio-stimulation involves the modification of the environment to stimulate autochthonous bacteria which are capable of biodegrading the pollutant. This can be done by addition of limiting nutrients and electron acceptors, such as phosphorus, nitrogen, oxygen, or organic carbon, which are otherwise available in low quantities that constrain the microbial activity. Bio-stimulation can be considered as an appropriate remediation technique for removing petroleum pollutants from the soil and requires the evaluation of both the intrinsic degradation capacities of the autochthonous microbiota and the environmental parameters involved in the kinetics of the *in situ* process. The process was tested for hydrocarbon biodegradation in the soil, i.e., limited by many factors, including nutrients, pH, temperature, moisture, oxygen, soil properties and the concentration of the contaminant (Al-Sulaimani et al. 2011). According to Agamuthu and Dadrasnia (2013), addition of nutrients, oxygen or other electron donors and acceptors to the contaminated site should increase the population or activity of naturally occurring microorganisms available for remediation. The primary advantage of bio-stimulation is that the remediation process is undertaken by the native microorganisms that are well-suited to thrive in the subsurface environment, and are well distributed spatially. The addition of nutrients may also promote the growth of other heterotrophic microorganisms that are not the innate degraders of the pollutant, thereby creating a competition between the resident microbiota (Adams et al. 2015).

3.4 Co-metabolism

Co-metabolism is defined as the transformation of a hazardous waste to environmentally benign products indirectly as a consequence of the metabolism of another chemical that the bacterium uses as a source of carbon and energy. An aspect of co-metabolism that is of concern is the production of partially oxidized end products that may not be readily degraded by the indigenous microbes because of the shorter adaptation time or lag phase. In another process, co-metabolism is an aerobic process in which chlorinated ethenes are degraded as a result of unusual biochemical interactions that yield no benefit to the bacteria. Alvarez-Cohen and McCarty (1991) reported that tetra-chloroethylene (TCE) is degraded under aerobic conditions by methanotrophic bacteria in soils rich with methane and oxygen. Further studies revealed that the methane monooxygenase enzyme was responsible for catalysing the oxidation of TCE. In this oxidation reaction metabolic enzymes are used. But they do not contribute to any energy generation in return. According to Ely et al. (1997), co-metabolism is not a sustainable process under stagnant conditions because of substrate concentration and enzyme inhibition and inactivation. Competitive inhibition occurs when enzymes co-metabolize chlorinated solvents to their natural substrates, ultimately depleting the bacteria of energy molecules. The TCE oxidation yields by-products for TCE-epoxide which may result in the inactivation of the oxygenase activity. Thus, the process has its limitations when the enhanced conditions of bioremediation are in the process (Kim et al. 2002). In recent times, co-metabolism is gaining prominence as an emergent bioremediation processes.

3.5 Biosorption

Biosorption can be defined as a passive and active process wherein the toxic pollutants are adsorbed by microbial biomass through various types of interactions/reactions (Hetzer et al. 2006, Huang et al. 2010). Microbial biosorption is one of the green technologies for bioremediation of toxic metals. Biosorption (sorption of metal ions from solutions by live or dried biomass) offers an alternative to the remediation of industrial effluents as well as the recovery of metals contained in other media. Biosorbents are prepared from naturally abundant and waste biomass. Due to the high uptake capability and economical source of the raw material, biosorption is a prospective technique. It has been demonstrated that both living and non-living biomass may be utilized in biosorption processes, as they often exhibit a marked tolerance towards persistent organic pollutants and other hostile conditions. One of the major advantages of this process is the treatment of bulk volumes of effluents with low concentrations of contaminants (Gavrilescu 2004, Hetzer et al. 2006). Moreover, many non-biodegradable organic compounds that are discharged into the environment tend to "bioaccumulate" via various food chains. Biosorption offers a promising alternative to remove such persistent organic pollutants from the waste waters. Among persistent pollutants, dyes, phenolics and pesticides are of great concern due to their extreme toxicity and persistency. Many studies have shown the biosorption of these types of hazardous organics by various live and dead biomasses (Aksu 2005). Zhou and Banks (1991) first reported the adsorption of humic acid by dead *Rhizopus arrhizus* from the raw water. This study was quickly followed by the study by Hu (1992), which showed efficient removal of 11 azo dyes using *Aeromonas* biomass. In another study of Zhou and Banks (1993), a biomaterial, chitin/chitosan was shown to be a major active component of *R. arrhizus* for humic acid adsorption. Since then, a number of studies have shown the efficacy of biosorption process for removal of organic pollutants from water wastes.

3.6 Myco-remediation

Bioremediation process in which fungi are used to remediate the toxic pollutants from the environment is termed myco-remediation. Fungi feed on dead and decaying organic matter as natural decomposers, hence, they are called saprophytes. Fungi aids in the degradation of oil, aromatic compounds, hydrocarbons, and petroleum products, which pollute soil and water. Fungal hyphae secrete acids and enzymes which decomposes lignin and cellulose. Fungal mycelium uses micro-filtration process for the removal of toxic wastes (Sasek and Cajthaml 2005). The use of fungi in decomposing toxic pollutants *in situ* is well documented particularly in case of 2, 4, 6-trinitrotoluene (TNT) (Pointing 2001, Rhodes 2012, Sasek and Cajthaml 2005). The other important study is treating diesel oil contaminated soil by inoculating with oyster mushrooms (*Pleurotusostreatus*) mycelia. The mycelia have converted 95% of the poly aromatic hydrocarbons to non-toxic compounds within 30 days. It seems that the naturally present community of microbes act in concert with the fungi to decompose the contaminants, finally resulting in complete mineralization. Wood-degrading fungi are extremely effective in decomposing toxic aromatic pollutants from petroleum and also chlorine containing resilient pesticides (Rhodes 2012). The prospects of using fungi, principally white-rot fungi, for cleaning contaminated soil are surveyed. White-rot fungi are useful in degrading a wide range of organic compounds since they release extra-cellular lignin degrading enzymes. The enzymes include lignin-peroxidases, and hydrogen peroxide producing enzymes (de Gonzalo et al. 2016). The degradation processes can be augmented by adding carbon sources such as sawdust and straw at the polluted sites.

Myco-filtration is a procedure that utilizes fungal mycelia to filter and remove toxic pollutants from contaminated water. Although edible mushrooms are used for myco-remediation, the prospect of whether they would be safe to eat afterward (Kulshreshtha et al. 2014) is questionable. However, this depends on the nature of the pollutant-if they are heavy/toxic metals, it will be a problem (if they are absorbed and concentrated into the mushroom), while hydrocarbon contaminants might be decomposed or mineralized completely without any toxicity. In either case, the benefit is that the land that is contaminated and unfit for agricultural purpose can be cleaned and used for cultivating crops (Rhodes 2012, Sasek and Cajthaml 2005).

3.7 Phytoremediation

Natural plants and transgenic plants have the ability to accrue some toxic pollutants in their basal parts and this process of accumulation is called as phytoremediation (Zhang et al. 2009). Phyto-extraction is also referred as phyto-accumulation or phyto-mining. The process involves the removal of contaminants (metals and certain organic compounds) from the environment by direct uptake into the plant tissue. The plants and some macro algae can complex the contaminants into their basal system and retain them, thus effectively remove pollutants from the soil or water environment. Normally, all plants cannot readily accumulate heavy metals and organic toxicants. The process involves the planting of one or more plant species that are hyper-accumulators of a particular pollutant. Phyto-remediation is appropriate to apply wherever the soil or static water bodies are chronically polluted. Phytoremediation has been used to remediate soil-water environment contaminated with heavy metals, radionuclides, chlorinated solvents (TCE, PCE) and petroleum hydrocarbons (BTEX) (Seo et al. 2009, Zhang et al. 2011).

Plants can stimulate microbial (bacteria and fungi) bioactivity in the root zone/rhizosphere by the excretion of bio-enhancing compounds. Typically, there are few 100 times higher microbial population in the rhizosphere compared to the bulk soil. The plant-excreted root exudates provide a carbon and nitrogen source for the soil bacteria. The excreted compounds commonly include amino acids, carbohydrates, polysaccharides, flavonoids and phenols. Specifically, it has been shown that flavonoids can support the growth of PCB-degrading bacteria. Also, the phenols excreted by some plants such as *Asparagus* can stimulate PCB-degrading bacteria and inhibit the growth of other microbes. There has been some promising research into plant-based remediation of explosives contaminated soils and groundwater (Rhodes 2012). Poplar trees have been shown to rapidly uptake TNT from hydroponic solutions. TNT within the plant was transformed to metabolites such as amino-2, 6-dinitrotoluenes and 2, 4, diamino-6-nitrotoluene in the roots and stems; however, there was no complete mineralization of TNT. When the TNT was present in the soil, it was less bioavailable for uptake due to sorption competition by the microbiota present in the milieu (Thompson et al. 1998). Treatment of TNT and RDX in wetland systems was also investigated, wherein it was showed that the enzyme nitro-reductase can rapidly breakdown TNT to simpler compounds (Oh et al. 2001).

In addition to precipitation and binding of aromatic compounds, other intracellular plant enzymes can degrade and transform organic compounds. For example, the nitro reductase enzyme in plants rapidly degrades nitro-aromatics. In laboratory studies, it was shown that hairy root cultures can degrade TNT, and also, some common aquatic plants. Also, poplar trees have been demonstrated to have the ability to dealkylate and hydrolyze atrazine and the metabolites are incorporated into plant roots and leaves (Hughes et al. 1996).

4. Factors influencing the bioremediation of organic pollutants

There are many factors which can directly or indirectly affect the efficacy of the bioremediation process (Figure 2). Most of the organic pollutants are highly hydrophobic compounds. Poor solubility in water makes it difficult for a living system to utilize it as a source of energy. However, several bacterial genera have been shown to efficiently degrade PAHs. Number of environmental factors such as temperature, pH, presence of electron acceptor and water activity together with intrinsic factors of the microbes involved, such as quorum sensing, plasmid and chemotaxis can influence the degradation of organic contaminants. Understanding of these factors will be useful for designing a suitable bioremediation strategy.

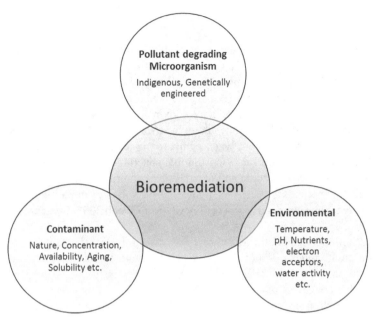

Figure 2. Factors affecting the bioremediation of persistent organic pollutants.

4.1 Nature of the contaminant

Bioavailability is one of the most critical factors in bioremediation of persistent organic compounds (Mueller et al. 1996). It is defined as the effect of physicochemical and microbiological factors on the rate of biodegradation of a pollutant. Since many of the persistent organic pollutants are hydrophobic molecules having low water solubility, they are resistant to biological, chemical and photolytic breakdown, hence have a very low bioavailability (Semple et al. 2003). Solubility is inversely proportional to the molecular weight of PAHs and thus its availability for metabolism by the microbial cell diminishes with the increase in the molecular weight (Fewson 1988). Another factor which is known as persistence, also determines the bioavailability of contaminants. PAHs undergo rapid sorption to soil constituents such as mineral surfaces (e.g., clay) and organic matter (e.g., humic acids). The longer contact time of PAHs with soil results in irreversible sorption, which directly influences the lower chemical and biological extractability of the PAHs. This phenomenon is called '*ageing*' of the contaminant (Hatzinger and Alexander 1995).

4.2 Environmental factors

4.2.1 pH

The pH of the contaminated site is one of the critical factors that greatly impact the growth and metabolism of the microorganisms. Therefore, it is very essential to monitor and regulate it as per the requirement to favour the growth of *Persistent Organic Pollutants* (POP) degrading microbes. On the other hand, it should also be noted that the pollutant may also be influenced by the pH of contaminated sites, which can turn the microenvironment acidic or alkaline, or the contaminated sites itself may not have the optimal pH for bioremediation. Such conditions may not provide an optimum micro milieu for the indigenous POP degrading microorganisms. Therefore, pH of such sites should be adjusted by the addition of lime to maintain favourable conditions for the growth of microorganisms and bioremediation.

4.2.2 Oxygen

Bioremediation of PAHs can take place under both aerobic and anaerobic conditions depending on the microbes involved in the process. The presence or absence of oxygen determines the pathway of PAHs degradation. These pathways initiate the biodegradation of PAHs by introducing atoms of molecular oxygen into the aromatic nucleus. These reactions are catalyzed by a multi-component dioxygenase, which consists of a reductase, a ferredoxin and an iron–sulfur protein (Harayama et al. 1992, Haritash and Kaushik 2009). Previous studies have shown that in general, aerobic biodegradation of PAHs is significantly more efficient when compared to anaerobic biodegradation (Rockne and Strand 1998). However, few reports also showed that anaerobic PAHs degradation could be just as efficient as that of those under aerobic and denitrifying conditions (McNally et al. 1998). Aerobic *in situ* bioremediation methods are easier as indigenous microbial communities can be stimulated by using oxidising agents like hydrogen peroxide (Bewley and Webb 2001, Coates et al. 1999, Pardieck et al. 1992).

In general, shortcomings of anaerobic bioremediation dominate the benefits of the process. For example, (1) all environments do not contain anaerobes with PAHs degrading capability (Coates et al. 1997), (2) Imposing the reducing conditions may significantly change the geochemistry of the soil. In an anaerobic condition, other electron acceptors such as nitrate, ferric iron and sulphate are reduced during respiration (Stumm and Morgan 1981). The reduction to ferrous iron and the release of phosphate from iron–phosphate complexes are toxic processes to the environment. (3) Reducing conditions also enhance the pH which in turn results in the solubilisation of carbonate minerals and the release of trace metals (Ponnamperuma 1972). Therefore, thorough understandings on implications of the anaerobic bioremediation process are needed before it can be used for *in situ* bioremediation.

4.2.3 Temperature

The solubility of PAHs increases with an increase in temperature, which in turn increases the bioavailability of the organic pollutants (Margesin and Schinner 2001). Therefore, temperature is a major governing factor that determines the microbial degradation ability of PAHs *in situ*. In general, contaminated sites may not be at the optimum temperature throughout the year for the bioremediation process. On the other hand, oxygen concentration in water goes down at higher temperature, which in turn reduces the metabolic activity of aerobic microorganisms. Therefore, most studies tend to use mesophilic temperatures to achieve higher bioavailability but low enough to retain substantial microbial metabolic activity (Lau et al. 2003).

4.3 Biological factors

The potential of microbes to degrade organic pollutants depends on a lot of factors, such as presence of specific genes (Table 3), ability to form biofilm, and many biofilm associated functions such as quorum sensing, horizontal gene transfer, production of biosurfactant, etc. Although the presence of specific genes is a crucial factor, but other factors such as biofilm production and its associated feature are of high importance as studies have shown that same microbe can be more efficient if employed in biofilm phase to degrade PAHs when compared to planktonic phase.

5. Role of biofilm and its associated functions in bioremediation

5.1 Biofilm formation

Microbial biofilms are surface-attached microbial communities wherein the microbial cells reside in self-produced extra-cellular polymeric substances (EPS) in a cooperative manner and benefit each other by forming different ecological niches. The properties of these microbial communities are governed partially by its structure, diffusion of nutrients, and physiological activity of the cells (Costerton et al. 1995). Biofilm cells remain in the physiological dormant state; therefore, biofilms are resilient to a wide variety of environmental stresses and show high tolerance to physical, chemical, and biological stressors. This characteristic also makes them a potential candidate for the development of bioremediation process to remove toxic pollutants.

The biofilm development is a sequential process governed by cellular, surface-related, macromolecules and environmental factors (Costerton et al. 1995). In the first step, the substratum is conditioned by many inorganic and organic molecules secreted by microbes themselves, thereby providing a favourable place for cell settlement. In the second step, microbial cells attach to the surface; firstly, in a reversible manner and then irreversibly. In the third step, attached cells multiply and secrete EPS to grow in form of microcolonies, then transformation into a mature biofilm with more complex architecture. In the last step, a part of the biofilm gets dispersed due to certain environmental cues and changes in the microenvironments within the biofilm. Detached or dispersed cells can attach and form biofilms on new surfaces (see Figure 3). Biofilm matrix components such as polysaccharides (Wingender et al. 2001), proteins (Shukla and Rao 2013a, b), lipids and nucleic acids (Whitchurch et al. 2002) play a critical role in the mechanical stability of biofilms.

The implication of biofilms for the bioremediation of contaminated settings with organic pollutant represents a promising approach. Biofilms are a rich source of cells and essential biomolecules, which improves bioremediation via both adsorption and catabolic approaches. EPS, cell signaling,

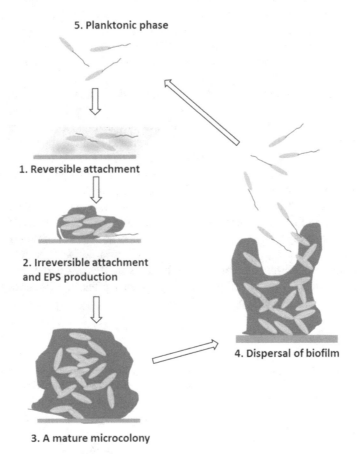

Figure 3. Different stages of microbial biofilm formation and its disassembly. (1) bacterial settlement and reversible attachment, (2) irreversible phase of adhesion with exopolymer production, (3) biofilm formation with a consortium of bacteria, (4) mature biofilm and dispersion stage, i.e., disperal phase for resettlement, (5) dispersed bacteria for settlement on new surface.

Table 2. Biofilm regulatory components and their importance in bioremediation.

Biofilm component	Functions	Use in bioremediation	References
Extracellular polymeric substances	• Biofilm structure and stability • Source of nutrient during starvation	• Mineralization, bioaccumulation and solubilization of organic and inorganic compounds	(Pan et al. 2010, Zhang et al. 2013)
Quorum sensing	• Biofilm formation, gene expression and cell signaling	• Assist in pollutant catabolic gene expression • Increase in QS gene expression or presence of exogenous autoinducer increases degradation kinetics of aromatic compounds	(Mangwani et al. 2016a, Yong and Zhong 2010)
Horizontal gene transfer	• Transfer of gene among population	• Bioaugmentation • Catabolism of organic pollutant	(Dejonghe et al. 2000, Goris et al. 2003)
Extracellular enzymes	• EPS formation, • Nutrient availability during starvation	• Degradation of organic compounds • Oxidoreductase and hydrolyase in EPS involved in the degradation of organic compounds	(Gianfreda and Rao 2004, Jia et al. 2011)

gene transfer, and exogenous enzymes are some of the important assets of bacterial biofilms useful from the bioremediation perspective (Table 2). EPS is a major component of biofilm, which plays a significant role in enhancing the bioremediation process. The production of EPS in biofilm provides protection to the cells from environment, desiccation, predation, erosion and toxic chemicals (Xavier and Foster 2007). The EPS component of biofilm is composed of ionic, hydrophilic and hydrophobic moieties. As a result, it has the capacity to bind with any class of organic compounds. For instance, Eriksson et al. (2002) have reported bacterial colonization and biofilm formation over pyrene crystals. In another study, Dasgupta et al. (2013) observed proficient growth of *Pseudomonas* biofilms near hydrocarbon oil-water interface. These properties further enhanced mineralization of many hydrophobic compounds, which makes the degradation process efficient and easy for a bacterial cell. The degradation by biofilm of organic pollutant is chiefly due to enhanced solubility in EPS matrix and high cell density. There are many studies which support a strong positive correlation between the EPS content of biofilm and degradation of hydrophobic pollutants by biofilm cells (Mangwani et al. 2016b, Shukla et al. 2014). In nature, bacterial cells utilize organic compounds as a source of energy by secreting EPS thereby increasing degradation and biofilm formation (Perumbakkam et al. 2006). Thus, EPS provides stability and constitutes an important part of biofilm framework. Many times EPS trap pollutants such as herbicides, petroleum hydrocarbon, etc., which act as a source of nutrients for biofilm cells during starvation (Wolfaardt et al. 1995). Apart from these factors, the biofilm architecture along with its composition significantly affect the PAHs degradation ability (Mangwani et al. 2012, 2014a, 2016b).

5.2 *Quorum sensing*

Biofilm formation is often regulated by density-sensing mechanism via autoinducers known as Quorum Sensing (QS). Acyl-Homserine Lactone (AHLs) based QS is predominant in Gram-negative bacteria which plays important role in biofilm formation and organic compound degradation. Increase in the degradation of aromatic hydrocarbons in the presence of exogenous AHLs has been also reported (Mangwani et al. 2012, 2016b). QS is known to be one of the major factors influencing biofilm architecture (Mangwani et al. 2012, Nadell et al. 2008); therefore, it is proposed that QS-mediated bioremediation process of organic pollutants can be employed and developed. QS-molecules are known to be involved in biofilm formation and EPS production thus significantly impacting the bioremediation of organic pollutants. For instance, in a study by Yong and Zhong (2010), an increase in biodegradation of phenol by *Pseudomonas aeruginosa* upon exogenous addition of AHLs was reported. The contribution of QS to hexadecane degradation by *Acinetobacter* sp. was also described (Kang and Park 2010).

Moreover, expression of AHLs gene also increases with increase in the hydrophobicity of compounds (Mangwani et al. 2015). Thus, it can be concluded that QS also adds value towards bioremediation potential of biofilms. For the metabolism of organic compounds, many intracellular and extracellular enzymes are involved in the degradation process. Many enzymes belonging to the groups dioxygenase, hydrolyase, decarboxylase dehydrogenase, and transferase are involved in the degradation processes (Shukla et al. 2014). Bacterial biofilms represent alternate approach to enzyme immobilization process used in bioremediation technology. Many extracellular enzymes are naturally immobilized within biofilm-EPS matrix, which makes chain for start-up reaction involved in the degradation process. Jia et al. (2011) reported presence of extracellular oxidoreductase and hydrolyase in EPS which were involved in the degradation of pyrene in soil. Extracellular lipase or elastase influences biofilm architecture and EPS composition (Tielen et al. 2010). Most of the extracellular enzymes within biofilm are involved in the degradation of biopolymers, which are both water soluble and water insoluble. Many times organic particles trapped inside EPS are also degraded by these enzymes. Thus, it can be concluded that extracellular enzymes from biofilm act as self-purification agents relatively more in soils and sediments than in water. These processes form the base of wastewater treatment using biofilms or flocs to clean organic substances (Flemming and Wingender 2010).

5.3 Chemotaxis

Chemotaxis is the movement of organisms in response to a nutrient source or a chemical gradient (Paul et al. 2005). Cells with chemotatic capabilities can sense xenobiotic chemicals adsorbed to soil particles and swim towards them, thereby overcoming the mass-transfer limitations in bioremediation process. Under conditions of limited carbon or energy sources, chemotaxis helps bacteria to find the optimum conditions for growth after exposure to such chemicals. Chemotaxis also governs the biofilm formation in several microorganisms (Pratt and Kolter 1999). It guides bacteria to travel towards hydrophobic pollutants which act as carbon source and is followed by surface attachment. The flagellum has been reported to be critical locomotory organ for adhesion to surfaces and also facilitates the initiation of biofilm formation (Pratt and Kolter 1999). Chemotaxis and motility are required for a bacterium to develop a biofilm and move along the surface, grow and form micro-colonies (Nicolella et al. 2000). Thus, bioavailability and biodegradation of organic contaminants can be significantly improved by exploiting chemotactic response of bacteria. PAHs (e.g., naphthalene) are reported to be chemo-attractant for bacteria (Law and Aitken 2003).

5.4 Horizontal gene transfer

Horizontal Gene Transfer (HGT) or lateral transfer of genetic material between existing bacteria is a very common process in a biofilm community. The HGT processes occur in biofilm generally by conjugation and transformation. Moreover, DNA release and transformation processes are very common in a biofilm, which adds new avenues for augmentation of metabolic potential of a strain with an adaptable catabolic potential (Figure 4). The catabolic gene encoding for degradation of PAHs (or xenobiotic) is often encoded on plasmid or transposon (Springael et al. 2002) (Table 3). Plasmids are important structural and functional component of biofilm, stabilized by plasmid coded mechanisms. They are often lost by rapidly growing planktonic culture. In terms of that, the plasmid stability is high in biofilms because of the quiescent nature of cells in the biofilm (Madsen et al. 2012). Consequently, biofilm community is often a good reservoir of active plasmid. The natural transformation potential of biofilm is 600-fold more as compared to planktonic cells (Molin and

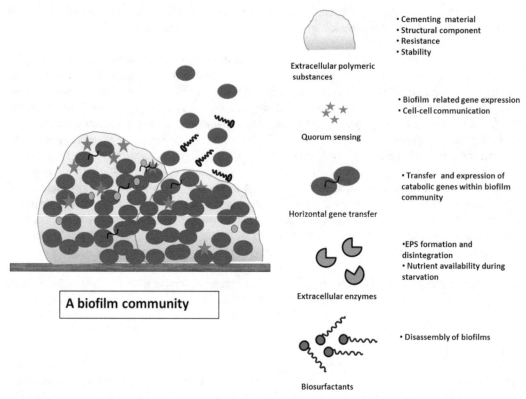

Figure 4. A schematic diagram to explain biofilm associated components, which play critical roles in the biological removal of organic pollutants.

Color version at the end of the book

Tolker-Nielsen 2003). Therefore, horizontal transfers of those catabolic plasmids in biofilms provide a transformed population that has potential to degrade several compounds. The auto-gene relocation among cells is promoted by the dense and packed structure of the biofilms. Owing to this, HGT is more helpful as compared to traditional bio-augmentation (*in situ* method) for bioremediation.

5.5 Biosurfactant production

Another characteristic of biofilm that aids in the bioremediation of organic pollutants is the ability to produce biosurfactants. Some microorganisms also produce surfactants known as biosurfactants during the dispersion phase of the biofilms (Figure 4). As described earlier, with ageing bioavailability of PAHs decreases as these molecules strongly bind to soil particles. PAHs can be released from the surface of minerals and organic matter with the help of surface-active agents or surfactants, which are amphipathic in nature and contain a hydrophobic and a hydrophilic moiety. Thus, these biosurfactant molecules can assist in desorption of PAHs from the soil which provides access to the hydrophilic microbial cell by solubilising the hydrophobic PAH molecules (Makkar and Rockne 2003). One major advantage of using biosurfactant is that they are less toxic to indigenous microbes and do not produce micelles, which can encapsulate contaminant PAHs and prevent the microbial access (Makkar and Rockne 2003). Rhamno-lipids are bio-surfactants that have been found in *Pseudomonas aeruginosa* biofilm matrix; they display a variety of surface activities and have been reported to assist in micro-colony formation of water channels in mature biofilm as well as in biofilm dispersion (Boles et al.

Table 3. Examples of organic compound degradation regulatory genes and key metabolites in bacteria.

Pollutants	Organism	Key metabolites	Regulator genes	References
m-Alkane (C10-20)	*Acinetobacter* sp. DSM 17874	Methanol, Formic acid	*alkMaMb*	(Singh et al. 2012)
Naphthalene	*Pseudomonas putia* AK5	Salicylate, gentisate	*nah, sgp*	(Izmalkova et al. 2013)
Pyrene	*Mycobacterium vanbaalenii* PYR-1	o-phthalate, β-ketoadipate	*nidA*	(Khan et al. 2001) (Kim et al. 2007)
PCB	*Pseudomonas aeruginosa* JP-11	2-hydroxy-6-oxo-6-phenylhexa-2, 4-dienoate, 4-cyclohexadiene-1-carboxylic acid	bphA	(Chakraborty and Das 2016)
Methyl red	*Bacillus subtilis*	N,N'dimethyl-p-phenyle-nediamine, 2-aminobenzoic acid	*azoR1*	(Leelakriangsak and Borisut 2012)
2,4-D	*Ralstonia eutropha* JMP134	2,4-dichlorophenol, 3,5-dichlorocatechol	*tfd*	(Kitagawa et al. 2002) (Plumeier et al. 2002)
Ibuprofen	Sphingomonas Ibu-2	isobutylcatechol	ipfABDEF	(Murdoch and Hay 2013)
4-chloronitrobenzene	*Pseudomonas putida* ZWL73	2-amino-5-chlorophenol	chbAB	(Xiao et al. 2006)
Carbofuran	*Achromobacter* sp. WM111	Carbofuran 7-phenol, methylamine	mcd	(Topp et al. 1993)

2004). Therefore, biosurfactant may be highly useful for bioremediation of oil recovery and oil spills, as they can successfully disperse hydrophobic substances and make them more bioavailable.

6. Application of biofilms in bioremediation

Bioremediation of hazardous organic pollutants using biofilm has gained lot of prominence in past few decades. Biofilm mediated bioremediation is a safe, cost effective and environmental friendly approach. Moreover, biofilms are also resistant to a variety of environmental stress and toxic chemicals (Nisha et al. 2015). Their use for the bioremediation of petrochemicals, pesticides, textile wastewater, pharmaceuticals and polychlorinated biphenyls are well studied (Table 4).

6.1 Petrochemicals (Aliphatic and Aromatic hydrocarbons)

Petroleum hydrocarbons are major source of energy for both industrial and domestic use. They are one of the major contaminant of soil and water. The bioremediation of soil-water environment contaminated with petrochemicals has been extensively studied using planktonic microorganisms. Some of them (see Tables 3 and 4) do have the potential to degrade a variety of hydrocarbons (Das and Chandran 2010). However, most of the bioremediation operations takes place under harsh operational conditions and produce some toxic intermediates which can reduce the efficacy of the bioprocess. In terms of efficacy, biofilms have been found to be more effective for the degradation of variety of petroleum hydrocarbons. The degradation of PAHs is higher during biofilm mode of growth as compared to planktonic culture, which is also supported by numerous studies (Mangwani et al. 2014b, 2016b). In a study by Mangwani et al. (2016b), bacterial biofilms of *Pseudomonas*, *Stenotrophomonas* and *Alcaligenes* showed higher degradation of phenanthrene and pyrene as compared to planktonic

Table 4. Biofilm mediated organic pollutant degradation.

Pollutant	Experimental conditions	Organism	Degradation efficiency	References
Benzo [a] pyrene	Biofilm grown over polytetrafluoroethylene filters	*Pseudomonas*, *Bacillus*, and *Microbacterium* species	44%	(Folwell et al. 2016)
Phenanthrene	Biofilm grown over glass beads	*Stenotrophomonas acidaminiphila* NCW702	71.1%	(Mangwani et al. 2014b)
Pyrene	Biofilm grown over glass beads	*Stenotrophomonas acidaminiphila* NCW702	40.2%	(Mangwani et al. 2014b)
Naphthalene	Biofilm grown over glass tubes	*Pseudomonas stutzeri* T102	70%	(Shimada et al. 2012)
n-Alkanes	Rotating biological contactors	*Prototheca zopfii*	65%	(Yamaguchi et al. 1999)
Crude oil	Natural biofilm	*Pseudomonas* spp.	40%	(Dasgupta et al. 2013)
Acid Orange 10, 14	Rotating drum biofilm reactor	*Methylosinus trichosporium*	60%	(Zhang et al. 1995)
Everzol Turquoise Blue G	Laboratory-scale activated sludge unit	*Coriolus versicolor*	82%	(Kapdan and Kargi 2002)
2,4-D	Granular activated-carbon biofilm reactor	Mixed culture of herbicide-degrading bacteria	100%	(Oh et al. 1994)
Polychlorinated biphenyls (PCBs) mixed PCB contamination	Bioactive granular activated carbon	*Dehalobium chlorocoercia* DF1 and *Burkholderia xenovorans strain* LB400 together with activated carbon	56%	(Payne et al. 2011)
Toluene	Hollow-fibre membrane biofilm reactor	*Pseudomonas putida*	98%	(Kumar et al. 2010)
Para-chloronitrobenzene (p-CNB)	Hollow-fibre membrane biofilm reactor	Consortium	95.7%	(Li et al. 2014)
Methylaminoantipyrine	Continuous flow biofilm reactor	Consortium	100%	(Pieper et al. 2010)
Ibuprofen	Oxic biofilm reactor	Consortium	35%	(Zwiener and Frimmel 2003)
Carbofuran	Natural river biofilms	Consortium	59.4%	(Tien et al. 2013)
Methomyl	Natural river biofilms	Consortium	99%	(Tien et al. 2013)
Pentachlorophenol	Biofilm over non porous glass beads	*Arthrobacter strain* ATCC 33790	90%	(Edgehill 1996)
Linuron		Consortium	85%	(Horemans et al. 2014)

cultures of same organisms. Moreover, the accumulation of metabolic intermediates and enzyme activity is also higher during biofilm mode of growth (Mangwani et al. 2014b). The degradation of high molecular weight PAHs is mainly dependent on the bioavailability. As a result, biodegradation of PAHs is often limited when planktonic cultures are used. In contrast, biofilms are more efficient for the degradation of high molecular weight-PAHs (Folwell et al. 2016).

6.2 Polychlorinated biphenyls

Polychlorinated biphenyls (PCBs) are toxic and are persistent pollutants of concern worldwide. They are highly resistant to microbial degradation and have low water solubility. Since they are lipophilic, they can pass through food chain and show bioaccumulation tendency (Zhang et al. 2013). Even, the metabolic intermediates produced during PCB degradation can affect biofilm community structure and functioning. Macedo et al. (2005) monitored growth of PCB degrading biofilm grown over Aroclor 1242, for 31 days. The young biofilm showed diverse microbial community with species of genera *Herbaspirillum* and *Bradyrhizobium* as dominant members. The PCB degradation was initially between 5–45% in 10 days, which gradually declined. During second stage, PCB degradation was very poor due to metabolic inhibition and increase in cell damage within the biofilm. The ratio of damaged cells reached up to 70%, with changes in microbial diversity and cell attachment and detachment process. In another study, Borja et al. (2006) investigated PCB bioremediation, using mixed culture PCB acclimatized bacterial biofilm developed on modified cement particles. Further, Acroclor 1260 degradation was studied in a three-phase fluidized-bed reactor operated in batch mode. Initial PCB removal was $80 \pm 2.38\%$ from medium, which gradually increased to $92 \pm 2.48\%$ in one day, from 8th batch onwards. The study supports that exposure of biofilm to pollutant can significantly increase bioremediation over time. Some recent studies also support the use of biofilms for PCB bioremediation (Edwards and Birthe 2013). Approximately 84.4% degradation of PCB under aerobic condition and 70% under anaerobic conditions have been observed by using moving-bed biofilm bioreactor by Dong et al. (2015).

6.3 Pharmaceuticals

The presence of pharmaceutical compounds in terrestrial and aquatic environment rose tremendously in recent times (Pieper et al. 2010). Many of these compounds are persistent in nature and have not been eliminated completely by common effluent treatment process. The manifestation of these chemicals in aquatic ecosystems exposes organisms to mixture of pharmaceutical chemicals. The presence of pharmaceutical agents has been reported to cause feminization of male fish and inhibition of molting in crustaceans (Kidd et al. 2007, Rodríguez et al. 2007, Sumpter 1998). The adaptation of microorganism to pharmaceuticals is slow, but exposure and mode of operation can affect rate of degradation. In a study by Pieper et al. (2010), methylaminoantipyrine and phenazone degradation was studied using natural biofilms derived from river water in continuous and batch mode, where 100% degradation of methylaminoantipyrine within 12 h was achieved. Phenazone was not easily degradable in continuous flow biofilm bioreactor. But slow degradation was observed in biofilm reactor operated in batch mode. Thus, it can be concluded that mode of operation can affect the biofilm-mediated degradation process. As compared to that of activated sludge process, the removal rate per unit biomass was higher for the biofilm culture. The higher removal rate using suspended biofilm for pharmaceutical substances like mefenamic acid, gemfibrozil, ibuprofen, ketoprofen, naproxen, diclofenac and clofibric acid was observed in some studies (Falås et al. 2012). Biofilms are also efficient in removing pharmaceutical compounds from wastewater. More than 90% recovery of pharmaceuticals and endocrine disruptors has been reported in river biofilms (Huerta et al. 2016). Thus, biofilms could be an important cubicle for accumulation and transformation of pharmaceutical compounds in natural water bodies.

6.4 Synthetic dyes

Textile industries are one of the foremost contributors towards environmental pollution. These industries discharge untreated effluent rich in complex and poorly degradable synthetic dyes (Engade and Gupta 2015). Synthetic dyes are stable to light, temperature, and microbial attack, thus persistent in the environment. The exoneration of dye is unpleasant from aesthetic as well as biological point of view. These dyes reduce the penetration of sunlight, which in turn affects primary productivity of water bodies. As a result, contamination of waters with dyes is harmful to aquatic and terrestrial life (Sarnaik and Kanekar 1995). In a study by Ribeiro and de Aragão Umbuzeiro (2014), an aquatic organism *Girardia tigrina* was shown to be vulnerable upon exposure to azo dyes. Thus, discharge of dyes causes major health problem to respective ecosystems due to their toxic impacts on receiving waters (Kumari and Naraian 2016). Furthermore, dye wastewater contains large amount of suspended solids, fluctuating pH, and is highly coloured (Subramonian and Wu 2014, Wu et al. 2013).

Standard methods like ozonation, adsorption, chemical precipitation, flocculation, photolysis and ion pair extraction are presently offered to treat dye-containing effluents (Grant and Buchanan 2000). Although biological methods are very effective for the bioremediation of synthetic dyes, but some of the biological processes face challenges while scale-up (Kumari and Naraian 2016). Initial degradation of synthetic dyes such as azo dyes requires anaerobic conditions for the cleavage of azo bond, resulting in the formation of colourless aromatic amines. Further degradation of amines is not efficient under anaerobic condition and subsequent degradation takes place effectively under aerobic condition. The mineralization of aromatic amines occurs through hydroxylation and the ring-fission mechanism under aerobic conditions (Mohan et al. 2013). Most of the available dyes' remediation processes do not support complete degradation rather only offer decolonization of textile wastewaters. Thus, biofilms are effective means when it comes to bioremediation of azo dye contaminated waters. Biofilm is a highly stratified and non-homogenous structure, which is usually characterized by the co-existing aerobic and anaerobic zones. The anaerobic and aerobic zones can be used to breakdown azo bond and mineralization of recalcitrant intermediates, respectively. The productivity of biofilm is reported to be higher than suspension culture with dye removal efficiency of 93.14% and 84.20%, respectively, at 350 mg dye/l (Zhang et al. 1995). In another study, Coughlin et al. (2002) reported complete mineralization of azo dye acid Orange 7 in a rotating drum bioreactor (RDBR). The biofilm community was able to reduce azo bond under aerobic condition, whereas intermediate sulfanilic acid and 1-amino-2-naphthol were completely utilized without exogenous carbon and nitrogen source.

6.5 Pesticides

The presence of pesticides in the environment is of great concern, due to their toxic effects on animals and human. Most of them show bioaccumulation and biomagnification due to their persistent nature in natural settings. Similar to other organic pollutants, pesticides are highly resistant to degradation due to their high molecular mass and lyophilic nature. Owing to that, biofilms are suitable candidates for the biodegradation of pesticides. The EPS matrix is very much efficient in accumulating pesticides from the surrounding, whereas cells in biofilm can breakdown pesticides (Verhagen et al. 2011). Feng et al. (2015) reported that pesticides, i.e., cypermethrin and chlorpyrifos were removed from water using biofilm. Microbial degradation and adsorption to EPS matrix are found to be key processes for removing the pesticides. In another study, Chen et al. (2007) used Fenton-coagulation process along with biological treatment using biofilm to reduce chemical oxygen demand (COD) in pesticide contaminated wastewater. In Fenton-coagulation process, wastewater is subjected to biological oxidation using moving-bed biofilm reactor (MBBR) and > 85% COD removal efficiency can be achieved in this process. Biofilms are also effective for the remediation of phenylurea herbicide linuron. In a recent study, Marrón-Montiel et al. (2014) reported 100% linuron removal within 30 days using packed-bed biofilm channel reactor.

7. Limitations of biofilm based bioremediation

Bioremediation is preferred over the chemical treatment due to involvement of natural processes and is economical. However, biofilm based bioremediation processes still have certain limitations

(Heitzer and Sayler 1993). Bioremediation processes are relatively slower as compared to the chemical treatment processes for pollutant degradation. Major limiting points are as below:

- Bioremediation may be slow or ineffective when essential nutrients for supporting microbial growth are limited *in situ*, particularly when the pollutant concentration is very high.
- One of the most limiting factors is the reliability of bioremediation process when dealing with heavy contamination of pollutants. Being a biologically dependent process, there are always some tolerance limits up to which the processes or microorganism can sustain, beyond which it is bound to fail. The strategy of bioremediation is ideally useful in situations when the level of pollution is relatively low or does not require immediate restoration or where chemical treatment is not ideal.
- Not all organic chemicals are amenable to biodegradation, particularly the man-made unnatural recalcitrant compounds such as plastics and certain halogenated aromatic compounds.
- The metabolic by-products generated in the pollutants degradation process are to be checked for their toxicity.
- Bioavailability of pollutant to microbes also determines the efficiency of bioremediation and pollutants that are not enclosed by other materials such as clay which is more amenable to biodegradation.
- Biofilm reactors may not be used with rapidly growing microorganisms where the reactor capacity is dependent on oxygen diffusion.

8. Conclusion

Till date, many bioremediation strategies have been developed and adopted to remove many persistent organic pollutants. Each strategy has its own advantages and disadvantages. Major advantages of developing biofilm-based bioremediation process are the reusability of bacterial biomass and it is a low cost process as compared to artificial whole cell or enzyme immobilization. However, very low bioavailability and the dormant state of the cells in biofilms is the disadvantage. The biofilm cells can respond to the environmental pollutants by chemotaxis or quorum sensing, which, in turn, can elicit variable physiological responses in bacteria such as EPS and biosurfactant production, and thus yield higher efficiency in organic pollutant degradation. There are many factors and limitations which determine the efficiency of the bioremediation process. Apart from the bioavailability of the pollutant, other parameters also need to be considered such as temperature, aerobic/anaerobic conditions, levels of nutrients and co-substrates, presence of toxic intermediate by-products or co-contaminant and the physiological potential of microorganisms. Therefore, it is important to gain a better understanding of the metabolic cooperation among the microbiota within the biofilm community. The structural and functional *in situ* studies of microbial communities contaminated with PAHs using community fingerprinting and environmental genomic techniques are valuable. Ample technical expertise and interdisciplinary approach from different fields such as environmental microbiology, civil engineering and soil science are also required for successfully implementing biofilm mediated bioremediation. In addition, further research is required to develop potential anaerobic technologies that can be deployed to remediate subsurface sites such as marine/river bed sediments.

References

Abdulsalam, S., I. Bugaje, S. Adefila and S. Ibrahim. 2011. Comparison of biostimulation and bioaugmentation for remediation of soil contaminated with spent motor oil. International Journal of Environmental Science & Technology 8: 187–194.

Adams, G.O., P.T. Fufeyin, S.E. Okoro and I. Ehinomen. 2015. Bioremediation, biostimulation and bioaugmention: a review. International Journal of Environmental Bioremediation and Biodegradation 3: 28–39.

Agamuthu, P. and A. Dadrasnia. 2013. Potential of biowastes to remediate diesel fuel contaminated soil. Global NEST Journal 15(4): 474–484. https://doi.org/10.30955/gnj.001031.

Aksu, Z. 2005. Application of biosorption for the removal of organic pollutants: A review. Process Biochemistry 40: 997–1026.

Al-Sulaimani, H., S. Joshi, Y. Al-Wahaibi, S. Al-Bahry, A. Elshafie and A. Al-Bemani. 2011. Microbial biotechnology for enhancing oil recovery: current developments and future prospects. Biotechnology, Bioinformatics and Bioengineering 1: 147–158.

Alvarez-Cohen, L. and P.L. McCarty. 1991. A cometabolic biotransformation model for halogenated aliphatic compounds exhibiting product toxicity. Environmental Science and Technology 25(8): 1381–1387.

Atlas, R. and J. Bragg. 2009. Bioremediation of marine oil spills: when and when not–the Exxon Valdez experience. Microbial Biotechnology 2: 213–221.

Bewley, R.J. and G. Webb. 2001. *In situ* bioremediation of groundwater contaminated with phenols, BTEX and PAHs using nitrate as electron acceptor. Land Contamination & Reclamation 9: 335–347.

Boles, B.R., M. Thoendel and P.K. Singh. 2004. Self-generated diversity produces "insurance effects" in biofilm communities. Proceedings of the National Academy of Sciences of the United States of America 101: 16630–16635.

Borja, J., J. Auresenia and S. Gallardo. 2006. Biodegradation of polychlorinated biphenyls using biofilm grown with biphenyl as carbon source in fluidized bed reactor. Chemosphere 64: 555–559.

Brim, H., J.P. Osborne, H.M. Kostandarithes, J.K. Fredrickson, L.P. Wackett and M.J. Daly. 2006. *Deinococcus radiodurans* engineered for complete toluene degradation facilitates Cr (VI) reduction. Microbiology 152: 2469–2477.

Cerniglia, C.E. and M.A. Heitkamp. 1989. Microbial degradation of polycyclic aromatic hydrocarbons (PAH) in the aquatic environment. Metabolism of Polycyclic Aromatic Hydrocarbons in the Aquatic Environment. CRC Press, Inc., Boca Raton, Fla, 41–68.

Cerniglia, C.E. 1992. Biodegradation of polycyclic aromatic hydrocarbons. Biodegradation 3: 351–368.

Chakraborty, J. and S. Das. 2016. Characterization of the metabolic pathway and catabolic gene expression in biphenyl degrading marine bacterium *Pseudomonas aeruginosa* JP-11. Chemosphere 144: 1706–1714.

Chen, S., D. Sun and J.-S. Chung. 2007. Treatment of pesticide wastewater by moving-bed biofilm reactor combined with Fenton-coagulation pretreatment. Journal of Hazardous Materials 144: 577–584.

Coates, J.D., J. Woodward, J. Allen, P. Philp and D.R. Lovley. 1997. Anaerobic degradation of polycyclic aromatic hydrocarbons and alkanes in petroleum-contaminated marine harbor sediments. Applied and Environmental Microbiology 63: 3589–3593.

Coates, J.D., U. Michaelidou, R.A. Bruce, S.M. O'Connor, J.N. Crespi and L. Achenbach. 1999. Ubiquity and diversity of dissimilatory (per) chlorate-reducing bacteria. Applied and Environmental Microbiology 65: 5234–5241.

Cohen, Y. 2002. Bioremediation of oil by marine microbial mats. International Microbiology 5: 189–193.

Costerton, J.W., Z. Lewandowski, D.E. Caldwell, D.R. Korber and H.M. Lappin-Scott. 1995. Microbial biofilms. Annual Reviews in Microbiology 49: 711–745.

Coughlin, M.F., B.K. Kinkle and P.L. Bishop. 2002. Degradation of acid orange 7 in an aerobic biofilm. Chemosphere 46: 11–19.

Das, N. and P. Chandran. 2010. Microbial degradation of petroleum hydrocarbon contaminants: an overview. Biotechnology Research International 2011, Article ID 941810, 13 pages. http://dx.doi.org/10.4061/2011/941810.

Dasgupta, D., R. Ghosh and T.K. Sengupta. 2013. Biofilm-mediated enhanced crude oil degradation by newly isolated *Pseudomonas* species. ISRN biotechnology 2013, Article ID 250749, 13 pages. http://dx.doi.org/10.5402/2013/250749.

de Gonzalo, G., D.I. Colpa, M.H. Habib and M.W. Fraaije. 2016. Bacterial enzymes involved in lignin degradation. Journal of Biotechnology 236: 110–119.

Dejonghe, W., J. Goris, S. El Fantroussi, M. Höfte, P. De Vos, W. Verstraete and E.M. Top. 2000. Effect of dissemination of 2, 4-dichlorophenoxyacetic acid (2, 4-D) degradation plasmids on 2, 4-D degradation and on bacterial community structure in two different soil horizons. Applied and Environmental Microbiology 66: 3297–3304.

Doll, T.E. and F.H. Frimmel. 2004. Development of easy and reproducible immobilization techniques using TiO_2 for photocatalytic degradation of aquatic pollutants. CLEAN–Soil, Air, Water 32: 201–213.

Dong, B., H.Y. Chen, Y. Yang, Q.B. He and X.H. Dai. 2015. Biodegradation of polychlorinated biphenyls using a moving-bed biofilm reactor. CLEAN–Soil, Air, Water 43: 1078–1083.

Edgehill, R.U. 1996. Degradation of pentachlorophenol (PCP) by Arthrobacter strain ATCC 33790 in biofilm culture. Water Research 30: 357–363.

Edwards, S.J. and V.K. Birthe. 2013. Applications of biofilms in bioremediation and biotransformation of persistent organic pollutants, pharmaceuticals/personal care products, and heavy metals. Applied Microbiology and Biotechnology 97: 9909–9921.

Ely, R.L., K.J. Williamson, M.R. Hyman and D.J. Arp. 1997. Cometabolism of chlorinated solvents by nitrifying bacteria: kinetics, substrate interactions, toxicity effects, and bacterial response. Biotechnology and Bioengineering 54: 520–534.

Engade, K. and S. Gupta. 2007. Adsorption of synthetic dye and dyes from a textile effluent by dead microbial mass. Journal of Industrial Control Pollution 23(1): 145–150.

Eriksson, M., G. Dalhammar and W.W. Mohn. 2002. Bacterial growth and biofilm production on pyrene. FEMS Microbiology Ecology 40: 21–27.

Falås, P., A. Baillon-Dhumez, H.R. Andersen, A. Ledin and J. la Cour Jansen. 2012. Suspended biofilm carrier and activated sludge removal of acidic pharmaceuticals. Water Research 46: 1167–1175.

Feng, L., G. Yang, L. Zhu, X. Xu, F. Gao, J. Mu and Y. Xu. 2015. Enhancement removal of endocrine-disrupting pesticides and nitrogen removal in a biofilm reactor coupling of biodegradable phragmites communis and elastic filler for polluted source water treatment. Bioresource Technology 187: 331–337.

Fewson, C.A. 1988. Biodegradation of xenobiotic and other persistent compounds: the causes of recalcitrance. Trends in Biotechnology 6: 148–153.

Flemming, H.-C. and J. Wingender. 2010. The biofilm matrix. Nature Reviews Microbiology 8: 623–633.

Folwell, B.D., T.J. McGenity and C. Whitby. 2016. Biofilm and planktonic bacterial and fungal communities transforming high-molecular-weight polycyclic aromatic hydrocarbons. Applied and Environmental Microbiology 82: 2288–2299.

Gavrilescu, M. 2004. Removal of heavy metals from the environment by biosorption. Engineering in Life Sciences 4: 219–232.

Gavrilescu, M. and Y. Chisti. 2005. Biotechnology—a sustainable alternative for chemical industry. Biotechnology Advances 23: 471–499.

Ghosal, D., S. Ghosh, T.K. Dutta and Y. Ahn. 2016. Current state of knowledge in microbial degradation of polycyclic aromatic hydrocarbons (PAHs): a review. Frontiers in Microbiology 7: 1369, doi: 10.3389/fmicb.2016.01369.

Gianfreda, L. and M.A. Rao. 2004. Potential of extra cellular enzymes in remediation of polluted soils: A review. Enzyme and Microbial Technology 35: 339–354.

Goris, J., N. Boon, L. Lebbe, W. Verstraete and P. De Vos. 2003. Diversity of activated sludge bacteria receiving the 3-chloroaniline-degradative plasmid pC1gfp. FEMS Microbiology Ecology 46: 221–230.

Grant, J. and I.D. Buchanan. 2000. Colour removal from pulp mill effluents using immobilized horseradish peroxidase. http://sfmn.ualberta.ca/Portals/88/Documents/Publications/PR_2000-8.pdf?ver=2015-12-17-091207-910.

Hamzah, A., C.-W. Phan, N.F. Abu Bakar and K.-K. Wong. 2013. Biodegradation of crude oil by constructed bacterial consortia and the constituent single bacteria isolated from Malaysia. Bioremediation Journal 17: 1–10.

Harayama, S., M. Kok and E. Neidle. 1992. Functional and evolutionary relationships among diverse oxygenases. Annual Reviews in Microbiology 46: 565–601.

Haritash, A. and C. Kaushik. 2009. Biodegradation aspects of polycyclic aromatic hydrocarbons (PAHs): a review. Journal of Hazardous Materials 169: 1–15.

Hatzinger, P.B. and M. Alexander. 1995. Effect of aging of chemicals in soil on their biodegradability and extractability. Environmental Science and Technology 29: 537–545.

Head, I.M., D.M. Jones and W.F. Röling. 2006. Marine microorganisms make a meal of oil. Nature Reviews Microbiology 4: 173–182.

Heitzer, A. and G.S. Sayler. 1993. Monitoring the efficacy of bioremediation. Trends in Biotechnology 11: 334–343.

Hetzer, A., C.J. Daughney and H.W. Morgan. 2006. Cadmium ion biosorption by the thermophilic bacteria *Geobacillus stearothermophilus* and *G. thermocatenulatus*. Applied and Environmental Microbiology 72: 4020–4027.

Hong, Y.-G. and J.-D. Gu. 2010. Physiology and biochemistry of reduction of azo compounds by Shewanella strains relevant to electron transport chain. Applied Microbiology and Biotechnology 88: 637–643.

Horemans, B., J. Vandermaesen, P. Breugelmans, J. Hofkens, E. Smolders and D. Springael. 2014. The quantity and quality of dissolved organic matter as supplementary carbon source impacts the pesticide-degrading activity of a triple-species bacterial biofilm. Applied Microbiology and Biotechnology 98: 931–943.

Hu, T. 1992. Sorption of reactive dyes by *Aeromonas* biomass. Water Science and Technology 26: 357–366.

Huang, A., M. Teplitski, B. Rathinasabapathi and L. Ma. 2010. Characterization of arsenic-resistant bacteria from the rhizosphere of arsenic hyperaccumulator *Pteris vittata*. Canadian Journal of Microbiology 56: 236–246.

Huerta, B., S. Rodriguez-Mozaz, C. Nannou, L. Nakis, A. Ruhi, V. Acuña, S. Sabater and D. Barcelo. 2016. Determination of a broad spectrum of pharmaceuticals and endocrine disruptors in biofilm from a waste water treatment plant-impacted river. Science of the Total Environment 540: 241–249.

Hughes, J.B., J. Shanks, M. Vanderford, J. Lauritzen and R. Bhadra. 1996. Transformation of TNT by aquatic plants and plant tissue cultures. Environmental Science and Technology 31: 266–271.

Izmalkova, T.Y., O.I. Sazonova, M.O. Nagornih, S.L. Sokolov, I.A. Kosheleva and A.M. Boronin. 2013. The organization of naphthalene degradation genes in *Pseudomonas putida* strain AK5. Research in Microbiology 164: 244–253.

Jia, C., P. Li, X. Li, P. Tai, W. Liu and Z. Gong,. 2011. Degradation of pyrene in soils by extracellular polymeric substances (EPS) extracted from liquid cultures. Process Biochemistry 46: 1627–1631.

Kang, Y.S. and W. Park. 2010. Contribution of quorum-sensing system to hexadecane degradation and biofilm formation in Acinetobacter sp. strain DR1. Journal of Applied Microbiology 109: 1650–1659.

Kapdan, I.K. and F. Kargi. 2002. Simultaneous biodegradation and adsorption of textile dyestuff in an activated sludge unit. Process Biochemistry 37: 973–981.

Katritzky, A.R., P. Oliferenko, A. Oliferenko, A. Lomaka and M. Karelson. 2003. Nitrobenzene toxicity: QSAR correlations and mechanistic interpretations. Journal of Physical Organic Chemistry 16: 811–817.

Khan, A.A., R.-F. Wang, W.-W. Cao, D.R. Doerge, D. Wennerstrom and C.E. Cerniglia. 2001. Molecular cloning, nucleotide sequence, and expression of genes encoding a polycyclic aromatic ring dioxygenase from *Mycobacterium* sp. strain PYR-1. Applied and Environmental Microbiology 67: 3577–3585.

Kidd, K.A., P.J. Blanchfield, K.H. Mills, V.P. Palace, R.E. Evans, J.M. Lazorchak and R.W. Flick. 2007. Collapse of a fish population after exposure to a synthetic estrogen. Proceedings of the National Academy of Sciences of the United States of America 104: 8897–8901.

Kim, S.-J., O. Kweon, R.C. Jones, J.P. Freeman, R.D. Edmondson and C.E. Cerniglia. 2007. Complete and integrated pyrene degradation pathway in *Mycobacterium vanbaalenii* PYR-1 based on systems biology. Journal of Bacteriology 189: 464–472.

Kim, Y., D.J. Arp and L. Semprini. 2002. Kinetic and inhibition studies for the aerobic cometabolism of 1, 1, 1-trichloroethane, 1, 1-dichloroethylene, and 1, 1-dichloroethane by a butane-grown mixed culture. Biotechnology and Bioengineering 80: 498–508.

Kitagawa, W., S. Takami, K. Miyauchi, E. Masai, Y. Kamagata, J.M. Tiedje and M. Fukuda. 2002. Novel 2, 4-dichlorophenoxyacetic acid degradation genes from oligotrophic *Bradyrhizobium* sp. strain HW13 isolated from a pristine environment. Journal of Bacteriology 184: 509–518.

Kulshreshtha, S., N. Mathur and P. Bhatnagar. 2014. Mushroom as a product and their role in mycoremediation. AMB Express 4: 29.

Kumar, A., X. Yuan, S. Ergas, J. Dewulf and H. van Langenhove. 2010. Model of a polyethylene microporous hollow-fiber membrane biofilm reactor inoculated with *Pseudomonas putida* strain Tol 1A for gaseous toluene removal. Bioresource Technology 101: 2180–2184.

Kumar, K., S.S. Devi, K. Krishnamurthi, S. Gampawar, N. Mishra, G. Pandya and T. Chakrabarti. 2006. Decolorisation, biodegradation and detoxification of benzidine based azo dye. Bioresource Technology 97: 407–413.

Kumari, S. and R. Naraian. 2016. Decolorization of synthetic brilliant green carpet industry dye through fungal co-culture technology. Journal of Environmental Management 180: 172–179.

Lau, K., Y. Tsang and S.W. Chiu. 2003. Use of spent mushroom compost to bioremediate PAH-contaminated samples. Chemosphere 52: 1539–1546.

Law, A.M. and M.D. Aitken. 2003. Bacterial chemotaxis to naphthalene desorbing from a nonaqueous liquid. Applied and Environmental Microbiology 69: 5968–5973.

Leahy, M.C. and G.P. Erickson. 1995a. Bioventing reduces clean-up costs. Hydrocarbon surface charge on waste sludge biomass. Water SA 20(73-76): 73–76.

Leahy, M.C. and G.P. Erickson. 1995b. Bioventing reduces soil cleanup costs: Hydrocarbon Processing Hydrocarbon Process 74: 63–66.

Leelakriangsak, M. and S. Borisut. 2012. Characterization of the decolorizing activity of azo dyes by *Bacillus subtilis* azoreductase AzoR1. Songklanakarin Journal of Science and Technology 34: 509–516.

Li, H., Z. Zhang, X. Xu, J. Liang and S. Xia. 2014. Bioreduction of para-chloronitrobenzene in a hydrogen-based hollow-fiber membrane biofilm reactor: effects of nitrate and sulfate. Biodegradation 25: 205–215.

Liang, P., J. Xu and Q. Li. 2008. Application of dispersive liquid–liquid microextraction and high-performance liquid chromatography for the determination of three phthalate esters in water samples. Analytica Chimica Acta 609: 53–58.

Liu, Y., Z. Chen and J. Shen. 2013. Occurrence and removal characteristics of phthalate esters from typical water sources in northeast China. Journal of Analytical Methods in Chemistry 2013, Article ID 419349, 8 pages, http://dx.doi.org/10.1155/2013/419349.

Lotito, A.M., U. Fratino, A. Mancini, G. Bergna and C. Di Iaconi. 2012. Effective aerobic granular sludge treatment of a real dyeing textile wastewater. International Biodeterioration and Biodegradation 69: 62–68.

Louvado, A., N. Gomes, M. Simões, A. Almeida, D. Cleary and A. Cunha. 2015. Polycyclic aromatic hydrocarbons in deep sea sediments: Microbe–pollutant interactions in a remote environment. Science of the Total Environment 526: 312–328.

Macedo, A.J., U. Kuhlicke, T.R. Neu, K.N. Timmis and W.-R. Abraham. 2005. Three stages of a biofilm community developing at the liquid-liquid interface between polychlorinated biphenyls and water. Applied and Environmental Microbiology 71: 7301–7309.

Madsen, J.S., M. Burmølle, L.H. Hansen and S.J. Sørensen. 2012. The interconnection between biofilm formation and horizontal gene transfer. FEMS Immunology and Medical Microbiology 65: 183–195.

Makkar, R.S. and K.J. Rockne. 2003. Comparison of synthetic surfactants and biosurfactants in enhancing biodegradation of polycyclic aromatic hydrocarbons. Environmental Toxicology and Chemistry 22: 2280–2292.

Mangwani, N., H.R. Dash, A. Chauhan and S. Das. 2012. Bacterial quorum sensing: functional features and potential applications in biotechnology. Journal of Molecular Microbiology and Biotechnology 22: 215–227.

Mangwani, N., S.K. Shukla, T.S. Rao and S. Das. 2014a. Calcium-mediated modulation of *Pseudomonas mendocina* NR802 biofilm influences the phenanthrene degradation. Colloids and Surfaces B: Biointerfaces 114: 301–309.

Mangwani, N., S.K. Shukla, S. Kumari, T.S. Rao and S. Das. 2014b. Characterization of *Stenotrophomonas acidaminiphila* NCW-702 biofilm for implication in the degradation of polycyclic aromatic hydrocarbons. Journal of Applied Microbiology 117: 1012–1024.

Mangwani, N., S. Kumari and S. Das. 2015. Involvement of quorum sensing genes in biofilm development and degradation of polycyclic aromatic hydrocarbons by a marine bacterium *Pseudomonas aeruginosa* N6P6. Applied Microbiology and Biotechnology 99: 10283–10297.

Mangwani, N., S. Kumari and S. Das. 2016a. Effect of synthetic N-acylhomoserine lactones on cell–cell interactions in marine *Pseudomonas* and biofilm mediated degradation of polycyclic aromatic hydrocarbons. Chemical Engineering Journal 302: 172–186.

Mangwani, N., S.K. Shukla, S. Kumari, S. Das and T.S. Rao. 2016b. Effect of biofilm parameters and extracellular polymeric substance composition on polycyclic aromatic hydrocarbon degradation. RSC Advances 6: 57540–57551.

Margesin, R. and F. Schinner. 2001. Biodegradation and bioremediation of hydrocarbons in extreme environments. Applied Microbiology and Biotechnology 56: 650–663.

Marrón-Montiel, E., N. Ruiz-Ordaz, J. Galíndez-Mayer, S. Gonzalez-Cuna, F.S. Tepole and H. Poggi-Varaldo. 2014. Biodegradation of the herbicide linuron in a plug-flow packed-bed biofilm channel equipped with top aeration modules. Environmental Engineering and Management Journal 13: 1939–1944.

Matsumoto, M., M. Hirata-Koizumi and M. Ema. 2008. Potential adverse effects of phthalic acid esters on human health: a review of recent studies on reproduction. Regulatory Toxicology and Pharmacology 50: 37–49.

McNally, D.L., J.R. Mihelcic and D.R. Lueking. 1998. Biodegradation of three-and four-ring polycyclic aromatic hydrocarbons under aerobic and denitrifying conditions. Environmental Science and Technology 32: 2633–2639.

Mittal, A., V. Thakur and V. Gajbe. 2013. Adsorptive removal of toxic azo dye Amido Black 10 B by hen feather. Environmental Science and Pollution Research 20: 260–269.

Mohan, S.V., C.N. Reddy, A.N. Kumar and J.A. Modestra. 2013. Relative performance of biofilm configuration over suspended growth operation on azo dye based wastewater treatment in periodic discontinuous batch mode operation. Bioresource Technology 147: 424–433.

Molin, S. and T. Tolker-Nielsen. 2003. Gene transfer occurs with enhanced efficiency in biofilms and induces enhanced stabilisation of the biofilm structure. Current Opinion in Biotechnology 14: 255–261, doi:S0958166903000363 [pii].

Mueller, J.G., C. Cerniglia and P.H. Pritchard. 1996. Bioremediation of environments contaminated by polycyclic aromatic hydrocarbons. Biotechnology Research Series 6: 125–194.

Murdoch, R.W. and A.G. Hay. 2013. Genetic and chemical characterization of ibuprofen degradation by *Sphingomonas* Ibu-2. Microbiology 159: 621–632.

Nadell, C.D., J.B. Xavier, S.A. Levin and K.R. Foster. 2008. The evolution of quorum sensing in bacterial biofilms. PLoS Biology 6: e14. https://doi.org/10.1371/journal.pbio.0060014.

Nicolella, C., M.C.M. van Loosdrecht and S.J. Heijnen. 2000. Particle-based biofilm reactor technology. Trends in Biotechnology 18: 312–320.

Nisha, K.N., V. Devi, P. Varalakshmi and B. Ashokkumar. 2015. Biodegradation and utilization of dimethylformamide by biofilm forming *Paracoccus* sp. strains MKU1 and MKU2. Bioresource Technology 188: 9–13.

Oh, B.-T., G. Sarath and P. Shea. 2001. TNT nitroreductase from a *Pseudomonas aeruginosa* strain isolated from TNT-contaminated soil. Soil Biology and Biochemistry 33: 875–881.

Okere, U. and K. Semple. 2012. Biodegradation of PAHs in 'pristine' soils from different climatic regions. Journal of Bioremediation and Biodegradation S1: 006. doi: 10.4172/2155-6199.S1-006.

Orji, F.A., A.A. Ibiene and E.N. Dike. 2012. Laboratory scale bioremediation of petroleum hydrocarbon—polluted mangrove swamps in the Niger Delta using cow dung. Malaysian Journal of Microbiology 8: 219–228.

Pan, X., J. Liu and D. Zhang. 2010. Binding of phenanthrene to extracellular polymeric substances (EPS) from aerobic activated sludge: A fluorescence study. Colloids and Surfaces B: Biointerfaces 80: 103–106.

Pardieck, D.L., E.J. Bouwer and A.T. Stone. 1992. Hydrogen peroxide use to increase oxidant capacity for *in situ* bioremediation of contaminated soils and aquifers: A review. Journal of Contaminant Hydrology 9: 221–242.

Paul, D., G. Pandey, J. Pandey and R.K. Jain. 2005. Accessing microbial diversity for bioremediation and environmental restoration. Trends in Biotechnology 23: 135–142.

Payne, R.B., H.D. May and K.R. Sowers. 2011. Enhanced reductive dechlorination of polychlorinated biphenyl impacted sediment by bioaugmentation with a dehalorespiring bacterium. Environmental Science and Technology 45: 8772–8779.

Pearce, C., J. Lloyd and J. Guthrie. 2003. The removal of colour from textile wastewater using whole bacterial cells: a review. Dyes and Pigments 58: 179–196.

Perelo, L.W. 2010. *In situ* and bioremediation of organic pollutants in aquatic sediments. Journal of Hazardous Materials 177: 81–89.

Perumbakkam, S., T.F. Hess and R.L. Crawford. 2006. A bioremediation approach using natural transformation in pure-culture and mixed-population biofilms. Biodegradation 17: 545–557.

Pieper, C., D. Risse, B. Schmidt, B. Braun, U. Szewzyk and W. Rotard. 2010. Investigation of the microbial degradation of phenazone-type drugs and their metabolites by natural biofilms derived from river water using liquid chromatography/tandem mass spectrometry (LC-MS/MS). Water Research 44: 4559–4569.

Plumeier, I., D. Pérez-Pantoja, S. Heim, B. González and D.H. Pieper. 2002. Importance of different tfd genes for degradation of chloroaromatics by *Ralstonia eutropha* JMP134. Journal of Bacteriology 184: 4054–4064.

Pointing, S. 2001. Feasibility of bioremediation by white-rot fungi. Applied Microbiology and Biotechnology 57: 20–33.

Ponnamperuma, F. 1972. The chemistry of submerged soils. Advances in Agronomy 24: 29–96.

Pratt, L.A. and R. Kolter. 1999. Genetic analyses of bacterial biofilm formation. Current Opinion in Microbiology 2: 598–603.

Puvaneswari, N., J. Muthukrishnan and P. Gunasekaran. 2006. Toxicity assessment and microbial degradation of azo dyes. Indian Journal of Experimental Biology 44(8): 618–626.

Rhodes, C.J. 2012. Feeding and healing the world: through regenerative agriculture and permaculture. Science Progress 95(4): 345–446.

Ribeiro, A.R. and G. de Aragão Umbuzeiro. 2014. Effects of a textile azo dye on mortality, regeneration, and reproductive performance of the planarian, Girardia tigrina. Environmental Sciences Europe 26: 22.

Rockne, K.J. and S.E. Strand. 1998. Biodegradation of bicyclic and polycyclic aromatic hydrocarbons in anaerobic enrichments. Environmental Science and Technology 32: 3962–3967.

Rodríguez, E.M., D.A. Medesani and M. Fingerman. 2007. Endocrine disruption in crustaceans due to pollutants: a review. Comparative Biochemistry and Physiology Part A: Molecular & Integrative Physiology 146: 661–671.

Salvo, L.M., A.C. Bainy, E.C. Ventura, M.R. Marques, J.R.M. Silva, C. Klemz and H.C. Silva de Assis. 2012. Assessment of the sublethal toxicity of organochlorine pesticide endosulfan in juvenile common carp (*Cyprinus carpio*). Journal of Environmental Science and Health, Part A 47: 1652–1658.

Saratale, R.G., G.D. Saratale, J.-S. Chang and S. Govindwar. 2011. Bacterial decolorization and degradation of azo dyes: a review. Journal of the Taiwan Institute of Chemical Engineers 42: 138–157.

Sarnaik, S. and P. Kanekar. 1995. Bioremediation of colour of methyl violet and phenol from a dye-industry waste effluent using *Pseudomonas* spp. isolated from factory soil. Journal of Applied Microbiology 79: 459–469.

Sasek, V. and T. Cajthaml. 2005. Mycoremediation: current state and perspectives. International Journal of Medicinal Mushrooms 7.

Semple, K.T., A. Morriss and G. Paton. 2003. Bioavailability of hydrophobic organic contaminants in soils: fundamental concepts and techniques for analysis. European Journal of Soil Science 54: 809–818.

Seo, J.-S., Y.-S. Keum and Q.X. Li. 2009. Bacterial degradation of aromatic compounds. International Journal of Environmental Research and Public Health 6: 278–309.

Shimada, K., Y. Itoh, K. Washio and M. Morikawa. 2012. Efficacy of forming biofilms by naphthalene degrading *Pseudomonas stutzeri* T102 toward bioremediation technology and its molecular mechanisms. Chemosphere 87: 226–233.

Shukla, S.K. and T.S. Rao. 2013a. Effect of calcium on Staphylococcus aureus biofilm architecture: A confocal laser scanning microscopic study. Colloids and Surfaces B: Biointerfaces 103: 448–54.

Shukla, S.K. and T.S. Rao. 2013b. Dispersal of Bap-mediated *Staphylococcus aureus* biofilm by proteinase K. The Journal of Antibiotics 66: 55–60.

Shukla, S.K., N. Mangwani, T.S. Rao and S. Das. 2014. 8–Biofilm-mediated bioremediation of polycyclic aromatic hydrocarbons. Microbial Biodegradation and Bioremediation, 203–232. https://doi.org/10.1016/B978-0-12-800021-2.00008-X.

Singh, R., D. Paul and R.K. Jain. 2006. Biofilms: implications in bioremediation. Trends in Microbiology 14: 389–397.

Singh, S., B. Kumari and S. Mishra. 2012. Microbial Degradation of Alkanes. Microbial Degradation of Xenobiotics. Springer, pp. 439–469.

Springael, D., K. Peys, A. Ryngaert, S.V. Roy, L. Hooyberghs, R. Ravatn, M. Heyndrickx, J.R.V.D. Meer, C. Vandecasteele and M. Mergeay. 2002. Community shifts in a seeded 3-chlorobenzoate degrading membrane biofilm reactor: indications for involvement of in situ horizontal transfer of the clc-element from inoculum to contaminant bacteria. Environmental Microbiology 4: 70–80.

Stumm, W. and J. Morgan. 1981. Aquatic Chemistry, 780 pp. J. Wiley & Sons.

Su, L., X. Zhang, X. Yuan, Y. Zhao, D. Zhang and W. Qin. 2012. Evaluation of joint toxicity of nitroaromatic compounds and copper to *Photobacterium phosphoreum* and QSAR analysis. Journal of Hazardous Materials 241: 450–455.

Subramonian, W. and T.Y. Wu. 2014. Effect of enhancers and inhibitors on photocatalytic sunlight treatment of methylene blue. Water, Air, & Soil Pollution 22: 1922.

Sumpter, J.P. 1998. Xenoendocrine disrupters—environmental impacts. Toxicology Letters 102: 337–342.

Thomassin-Lacroix, E., M. Eriksson, K. Reimer and W. Mohn. 2002. Biostimulation and bioaugmentation for on-site treatment of weathered diesel fuel in Arctic soil. Applied Microbiology and Biotechnology 59: 551–556.

Thompson, P.L., L.A. Ramer and J.L. Schnoor. 1998. Uptake and transformation of TNT by hybrid poplar trees. Environmental Science and Technology 32: 975–980.

Tielen, P., F. Rosenau, S. Wilhelm, K.-E. Jaeger, H.-C. Flemming and J. Wingender. 2010. Extracellular enzymes affect biofilm formation of mucoid *Pseudomonas aeruginosa*. Microbiology 156: 2239–2252.

Tien, C.-J., M.-C. Lin, W.-H. Chiu and C.S. Chen. 2013. Biodegradation of carbamate pesticides by natural river biofilms in different seasons and their effects on biofilm community structure. Environmental Pollution 179: 95–104.

Tiwari, B., N. Manickam, S. Kumari and A. Tiwari. 2016. Biodegradation and dissolution of polyaromatic hydrocarbons by *Stenotrophomonas* sp. Bioresource Technology 216: 1102–1105.

Tomei, M.C., J.S. Pascual and D.M. Angelucci. 2016. Analysing performance of real textile wastewater bio-decolourization under different reaction environments. Journal of Cleaner Production 129: 468–477.

Topp, E., R.S. Hanson, D. Ringelberg, D. White and R. Wheatcroft. 1993. Isolation and characterization of an N-methylcarbamate insecticide-degrading methylotrophic bacterium. Applied and Environmental Microbiology 59: 3339–3349.

Varjani, S.J. and V.N. Upasani. 2017. A new look on factors affecting microbial degradation of petroleum hydrocarbon pollutants. International Biodeterioration and Biodegradation 120: 71–83.

Verhagen, P., L. De Gelder, S. Hoefman, P. De Vos and N. Boon. 2011. Planktonic versus biofilm catabolic communities: importance of the biofilm for species selection and pesticide degradation. Applied and Environmental Microbiology 77: 4728–4735.

Vidali, M. 2001. Bioremediation. an overview. Pure and Applied Chemistry 73: 1163–1172.

Whitchurch, C.B., T. Tolker-Nielsen, P.C. Ragas and J.S. Mattick. 2002. Extracellular DNA required for bacterial biofilm formation. Science 295: 1487, doi:10.1126/science.295.5559.1487.

Wilson, S.C. and K.C. Jones. 1993. Bioremediation of soil contaminated with polynuclear aromatic hydrocarbons (PAHs): a review. Environmental Pollution 81: 229–249.

Wingender, J., M. Strathmann, A. Rode, A. Leis and H.-C. Flemming. 2001. Isolation and biochemical characterization of extracellular polymeric substances from *Pseudomonas aeruginosa*. Methods Enzymology 336: 302–14, doi:S0076-6879(01)36597-7 [pii].

Wolfaardt, G., J. Lawrence, R. Robarts and D. Caldwell. 1995. Bioaccumulation of the herbicide Diclofop in extracellular polymers and its utilization by a biofilm community during starvation. Applied and Environmental Microbiology 61: 152–158.

Wu, T.Y., A.W. Mohammad, S.L. Lim, P.N. Lim and J.X.W. Hay. 2013. Recent advances in the reuse of wastewaters for promoting sustainable development. Wastewater Reuse and Management. Springer, pp. 47–103.

Xavier, J.B. and K.R. Foster. 2007. Cooperation and conflict in microbial biofilms. Proceedings of the National Academy of Sciences of United States of America 104: 876–881.

Xiao, Y., J.-F. Wu, H. Liu, S.-J. Wang, S.-J. Liu and N.-Y. Zhou. 2006. Characterization of genes involved in the initial reactions of 4-chloronitrobenzene degradation in *Pseudomonas putida* ZWL73. Applied Microbiology and Biotechnology 73: 166–171.

Yamaguchi, T., M. Ishida and T. Suzuki. 1999. Biodegradation of hydrocarbons by *Prototheca zopfii* in rotating biological contactors. Process Biochemistry 35: 403–409.

Yong, Y.-C. and J.-J. Zhong. 2010. N-Acylated homoserine lactone production and involvement in the biodegradation of aromatics by an environmental isolate of *Pseudomonas aeruginosa*. Process Biochemistry 45: 1944–1948.

Zhang, H., Z. Dang, L. Zheng and X. Yi. 2009. Remediation of soil co-contaminated with pyrene and cadmium by growing maize (Zea mays L.). International Journal of Environmental Science and Technology 6: 249–258.

Zhang, M., Q. Zhao and Z. Ye. 2011. Organic pollutants removal from 2, 4, 6-trinitrotoluene (TNT) red water using low cost activated coke. Journal of Environmental Sciences 23: 1962–1969.

Zhang, T.C., Y. Fu, P.L. Bishop, M. Kupferle, S. FitzGerald, H.H. Jiang and C. Harmer. 1995. Transport and biodegradation of toxic organics in biofilms. Journal of Hazardous Materials 41: 267–285.

Zhang, X., F. Li, T. Liu, C. Peng, D. Duan, C. Xu, S. Zhu and J. Shi. 2013. The influence of polychlorinated biphenyls contamination on soil protein expression. ISRN Soil Science 2013, Article ID 126391, 6 pages, http://dx.doi.org/10.1155/2013/126391.

Zhou, J. and C. Banks. 1991. Removal of humic acid fractions by *Rhizopus arrhizus*: Uptake and kinetic studies. Environmental Technology 12: 859–869.

Zhou, J. and C. Banks. 1993. Mechanism of humic acid colour removal from natural waters by fungal biomass biosorption. Chemosphere 27: 607–620.

Zollinger, H. 2003. Color Chemistry: Syntheses, Properties, and Applications of Organic Dyes and Pigments. 3rd Ed. John Wiley & Sons.

Zwiener, C. and F. Frimmel. 2003. Short-term tests with a pilot sewage plant and biofilm reactors for the biological degradation of the pharmaceutical compounds clofibric acid, ibuprofen, and diclofenac. Science of the Total Environment 309: 201–211.

CHAPTER 10

Bioprocesses for the Treatment of Volatile Organic Compounds

Theresa Ugochukwu Ukwamedua,[1] *Eric D. van Hullebusch,*[2]
M. Estefanía López,[3] *Manivannan Sethurajan*[1] *and Eldon R. Rene*[1,*]

1. Introduction

Volatile organic compounds (VOCs) are emitted from various industries such as chemical production, petroleum refining and processing and they cause adverse effects on human health and the environment (Singh et al. 2006). According to Probhat et al. (2011), VOCs are not easily removed from the environment; hence, they tend to accumulate over a long period of time in the ambient atmosphere depending on the prevailing environmental conditions. Different methods have been developed for its removal from point-source emissions, but the use of biological techniques has continued to gain more popularity.

Volatile organic compounds (VOCs) are organic chemicals that evaporate rapidly at room temperature because of their high vapor pressure and low boiling point (Zhang et al. 2017a). Dutta et al. (2017) stated that most VOC emanates from natural and man-made sources such as the combustion of fossil fuels, chemical production, industry processes, coating of equipment and emissions from vegetation (natural source), etc. The emission of VOC is a concern because they play a role in triggering photochemical reactions in the atmosphere, contributing to the formation of ozone and also organic aerosols found in air particles (USEPA 2014). According to Mudliar et al. (2010), VOC emissions can pose a risk to human health and the environment. Example of VOC include xylene, tetrachloroethylene, phenols, toluene, benzene, ethylene, formaldehyde and methanol, etc. Considering the incessant emission of VOC from several industries and the health and environmental impacts associated with it, several agencies such as the European Environmental Agency (EEA 2016) have continued to make and implement strict regulations to minimize and eliminate the emission of

[1] Department of Environmental Engineering and Water Technology, UNESCO-IHE Institute for Water Education, 2601DA Delft, The Netherlands.
[2] Institut de Physique du Globe de Paris, Sorbonne Paris Cité, Université Paris Diderot, UMR 7154, CNRS, F-75005 Paris, France.
[3] IES García Lorca, Fernando de Herrera, 11207 Algeciras, Cádiz, Spain.
* Corresponding author: e.raj@un-ihe.org

these compounds (Zhang et al. 2017b, Rene et al. 2009, 2011). On the other hand, industries have continued to develop effective removal pollution control technologies based on physical, chemical and biological techniques for VOC removal. The physical and chemical techniques are classified as recovery or destruction based methods. According to Devinny et al. (1998), biological techniques are widely promoted because they are inexpensive and environmentally friendly for removing VOC from the gas-phase.

The fate of VOCs in the environment depends on the following factors: (i) weather conditions, (ii) species correlations, (iii) daytime and night-time abatement mechanisms, (iv) secondary pollutant transformations, (v) air-mass origins and (vi) source apportionments (Song et al. 2019). According to Sahu et al. (2017), biogenic VOC emissions are the largest natural contributor of VOCs in the atmosphere and anthropogenic sources are mostly related to urban and industrial VOC contributions. Considering all these factors, variations in the VOC concentration and composition could be expected depending on the prevailing climatic/environmental conditions. Besides, depending on the source of the VOC and the mechanism of VOC release, the chemical reactivity of the VOC and its interaction with sunlight will also change (Zhang et al. 2017b). For example, these photochemical reactions include the reactions with hydroxyl radicals (OH°), O_3 and NO_3 radicals, as well as the oxidation reaction of peroxide radicals (RO_2) and NO (Cheng et al. 2013, Lyu et al. 2016). Hu et al. (2019) studied the effect of atmospheric photochemical reactions on particulate matter pollution in France and observed that VOC oxidation is an important factor that contributes to the formation of secondary organic aerosols, while the oxidation of SO_2 and NO_x contributes to secondary inorganic aerosols. Some of the well-known sources of VOC emission include the petrochemical, paint and varnish, pharmaceutical, pulp and paper industry, gas storage tanks, petroleum filling stations, fossil fuel volatilization, biomass combustion and exhausts from gasoline and diesel operated vehicles (López et al. 2013, Yang et al. 2013, Zheng et al. 2014, López et al. 2017).

2. Physico-chemical techniques

The physico-chemical techniques include incineration, condensation, absorption, adsorption, and catalytic oxidation. As an example, for the removal of VOC present in the liquid phase, it requires the adsorption of VOCs by contact with activated carbon (Zhang et al. 2017a). The pollutants are adsorbed to the media (activated carbon) by physical or chemisorption and the removal of VOC is achieved. Adsorption based technologies have been recognized as an efficient and economical control strategy because it has the potential to recover and reuse both the adsorbent and the adsorbate (Zhang et al. 2017a,b). Specific properties of the adsorbents used for VOC removal and recovery include: (i) large specific surface area, (ii) highly porous structure, (iii) high adsorption capacity, (iv) ability to regenerate and reuse during multiple cycles of operation. Berenjian et al. (2012) stated that the physico-chemical techniques also involve the treatment of VOCs through oxidation by thermal, internal combustion engine (ICE) and ultraviolet oxidation (UV). The by-products of this process include CO_2, H_2O and HCl. The operational process for thermal oxidation is carried out at high temperature with supporting oxidation agents. The removal is mostly done at a contact time of one second. The ICE works in a similar manner, although its operating condition requires high inlet concentration of the VOCs because they are used as energy source for combustion. Regarding the UV (catalytic) oxidation method, a catalyst and oxidizer is used to speed up the oxidation process (Benitez 1993), and the removal takes place at the surface of the catalyst by adsorption including organic compounds and oxygen reactions (Berenjian et al. 2012). From a VOC recovery view point, adsorption, condensation, absorption, and membrane separation are commonly used at the industrial scale. Concerning adsorption techniques for recovering VOC, finding the optimal porous solid adsorbent with good process stability is very important for practical applications. Other emerging technologies can also be combined with adsorption processes, such as temperature swing adsorption (YSA), pressure swing adsorption (PSA) and electro thermal temperature swing adsorption (ETSA) (Swetha et al. 2017).

3. Biological techniques

Besides the physical and chemical techniques, the biological techniques are another techniques that involve biocatalysts such as bacteria, fungi, yeast or a mixture of microorganisms for treating the contaminants (Kennes and Thalasso 1998). Rene et al. (2011) explained that this technology is widely promoted and has continued to develop because it's cost effective, reliable, easy to operate and most importantly, it's a cleaner technology compared to the physico-chemical techniques. Kennes et al. (2009) explained that biological treatments are carried out in different types of bioreactors, which includes biofilter, bioscrubber and biotrickling filter. They are the most common biological method and have similar removal mechanisms but are configured differently in terms of the use of microorganisms, different packing materials, and it can remove different types of contaminants from the gaseous phase.

Biological waste gas systems have proven to be effective for the treatment of gas streams containing low concentrations of VOC at gas flow in the range of 60 to 150,000 m³/h. The different waste gas treatment systems commonly used in practice are: (i) biofilters, (ii) biotrickling filters, and (iii) bioscrubbers. They are differentiated by the flow of the polluted gas and the trickling water phase, where there exists a continuous flow of liquid only in the case of the biotrickling filter and the bioscrubber. In the case of the biofilter, water is intermittently added (sprinkled) on the filter bed in order to prevent drying and loss of microbial activity. In these bioreactor configurations, transfer of pollutants occur from the gas phase to the biofilm phase and the subsequent biodegradation of the pollutant to CO_2, H_2O, end-products and biomass. In the biofilter and the biotrickling filter, the biofilm is grown on the surface of an appropriate packing material (natural or inert).

3.1 Bioscrubbers

In bioscrubbers, two bioreactors are required for the treatment of contaminant (Figure 1). The contaminants are treated in two stages where absorption and bioreaction takes place (Groenestijn and Hesselink 1993). The treatment of contaminants is carried out by passing the gas phase pollutant

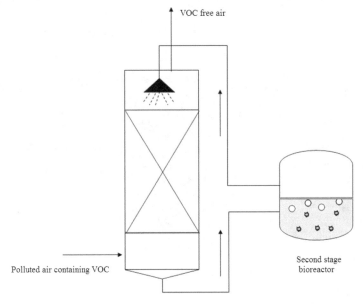

Figure 1. Schematic of a bioscrubber showing a first state absorption tower and a second stage bioreactor for VOC removal from polluted air.

into the reactor, and the gas is dissolved and transferred to the top of the column of the reactor, while the liquid containing the dissolved contaminants is transferred to the packing bed for treatment. However, in this technology, it is difficult to control and manage the excessive residue and liquid waste released. Besides, there are also problems such as poor surface area, difficult to start-up, and it incurs expensive operations and maintenance cost (van Groenestijn and Hesselink 1993). Recently, anaerobic bioscrubbers have been successfully used as a pre-treatment step to treat emissions from a printing press air emission and when combined with an expanded granular sludge bed reactor, it was used to produce biogas. According to Bravo et al. (2017), bioscrubbers are more versatile for handling fluctuating loads of hydrophilic VOCs, at capacities ranging from 3000–4000 m³/m².h. In that study, the authors ascertained the anaerobic removal of VOC mixtures from a printing press containing mainly ethanol, ethyl acetate and 1-ethoxy-2-propanol and smaller amounts of ethyl acetate, 1-propanol, 2-propanol, 1 methoxy-2-propanol, and 3-ethoxy-1-propanol. Although the bioscrubber was operated under wide fluctuations in the VOC emissions, with interruptions during nights, weekends and temperature oscillations, stable conversion of alcohols, esters, and glycol ethers to enriched methane biogas was demonstrated. One of the frequently encountered problem with bioscrubber is the problems associated with the disposal of excess sludge/effluent and a drop in the treatment efficiency if the waste gas contains hydrophobic pollutants.

3.2 Biotrickling filter

This biological removal technique is carried out in one reactor configuration, where the absorption and biodegradation process of the contaminant takes place (Rene et al. 2011) (Figure 2). According to Soccol et al. (2003), the contaminated gas is fed into the bioreactor, and continuous release of nutrient-rich liquid is carried out for improved microbial activity in the biofilm where the contaminants are degraded. In this removal process, the liquid is continuously re-circulated. According to Wu et al. (2018), the BTF system offers the following advantages: (i) low operating and capital costs, (ii) lower pressure drop during long-term operation, (iii) durability of the packing materials, (iv) good pH and process control, (v) capacity to remove the acidic by-products formed, and (vi) possibility to recover valuable chemicals such as volatile fatty acids when the BTF is operated anaerobically

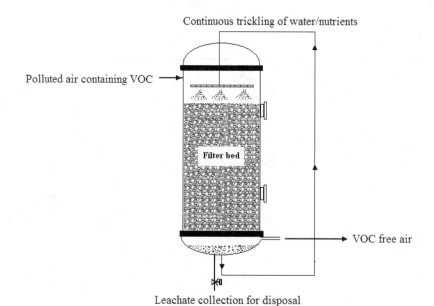

Figure 2. Schematic of a biotrickling filter showing a continuous trickling of the water phase and VOC removal from polluted air.

(Eregowda et al. 2019a,b). Nevertheless, the construction, operation and maintenance costs of BTFs are much higher than a conventional biofilter (Mudliar et al. 2010).

BTFs are usually inoculated with mixed biocatalysts that contain bacteria, fungi, yeasts and molds. Nevertheless, biofilm attachment and growth on the packing material depends on the following factors: (i) pH, (ii) temperature, (iii) concentration of the pollutant, (iv) gas flow rate, (v) presence of inhibiting carbon source, (vi) oxygen concentration, (vii) nutrient and trace element concentration, (viii) predators and (ix) surface property of the packing material (Eregowda et al. 2019a,b, Kasperczyk et al. 2019, López et al. 2013, 2017). Nevertheless, if the waste gas contains easily biodegradable pollutants, the rate of bacterial growth will be fast, leading to excess biomass formation on the surface of the packing material. The accumulation of biomass in BTF may lead to anaerobic zones, formation of localized water pits, increased pressure drop, clogging, channeling and finally a decrease in the reactor's performance (Eregowda et al. 2019a,b, Khanongnuch et al. 2019, López et al. 2013, 2017).

3.3 Biofilters

In biofilters, the humidified contaminated gas is fed into the bioreactor containing the packed bed characterized with good pore volume and surrounded by the contaminant degrading microbes (Ramirez-Lopez et al. 2010) (Figure 3). The contaminant elimination occurs in two distinct phases such as sorption and biodegradation. According to Soccol et al. (2003), the contaminant in the gaseous phase is fed into the reactor, and is transported to the aqueous phase where the biodegradation is carried out by microorganisms present in the biofilm. Adler (2001) highlighted that the mechanism governing the transport of the contaminated gas to the liquid phase for degradation includes, among others, contaminant adsorption on the filter material, contaminant dissolution in the liquid phase and the degradation of contaminant by the microorganisms. As the organic contaminant is dissolved in the biofilm, the organic or inorganic molecules from the contaminant serve as a carbon and energy source for the microbes. Thereafter, the contaminant is eliminated from the system. Mudliar et al. (2010) confirmed that more than 600 industries in Europe involved in chemical processing use biofilter to remove VOCs. Further research by Devinny et al. (1998) and Shareefdeen and Singh (2005) showed that 60 hazardous air pollutant (HAP) out of 189 have been successfully treated using biofilters.

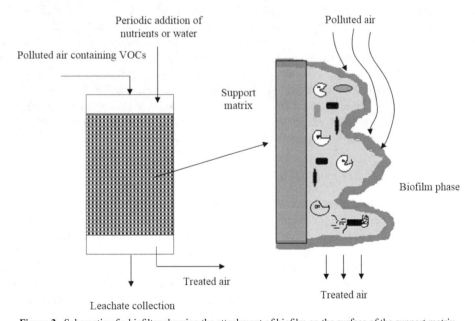

Figure 3. Schematic of a biofilter showing the attachment of biofilm on the surface of the support matrix.

Although Mudliar et al. (2010) explained the necessity for improvement of the biofilter with respect to efficient removal of highly concentrated contaminants, the elimination of clogging problems and flexible control of moisture and pH in the biofilter is still an important issue. According to Kumar et al. (2013), for the effective removal of contaminants to occur, the choice of media for the filter bed is a vital aspect in the removal process as well as the proper management of the factors that influence the operation of the biofilter. As a result, different studies have continued to explore different biological materials for the elimination of contaminants. Example of some of the media that have been explored include compost, wood chips and perlite. Table 1 shows the important results obtained from previous studies on the removal of organic compound with different filter materials.

Table 1. Previous studies on the biodegradation of methanol using different biofilter materials (Modified from Ramirez-Lopez et al. 2010).

Medium	Pollutant	EBRT (s)	Removal efficiency (%)	Removal rate (g/m³.h)	References
Diatomaceous earth and ceramic	Methanol	180	95	–	Arulneyam and Swaminathan (2003)
Lava rock	Methanol	91	98	–	Prado et al. (2005)
Ceramic spheres	Methanol	65	95	–	Avalos Ramirez et al. (2008)
Compost and wood chips	Methanol		95	70	Johnson and Deshusses (1997)
Wood chips and compost	Methanol	–	90-95	250	Mohseni and Allen (2000)

3.3.1 Compost as a biofilter material for bioremediation

Compost is a biological material with unique physico-chemical properties. Some of its characteristics include high surface area, good water retention capacity, availability of nutrients and stable microbial activities (biofilm) that facilitate the degradation of pollutants (Zilli et al. 2005). Research done by Johnson and Deshusses (1997) and Mohseni and Allen (2000) have shown that 95% and 98% removal of methanol can be achieved using compost as the biofilter media. Similarly, Arulneyam and Swaminathan (2000) reported a maximum elimination capacity (EC) of 195 g/m³.h for ethanol removal using compost and perlite. However, Zilli et al. (2005) explained the limitation of compost regarding biofilm thickness which inhibits the microbial processes, consequently reducing the removal efficiency of the contaminants. Kumar et al. (2013) reported other hindrances encountered when using compost as a biofilter media; for example, channeling, compaction, pH regulation and drying out as a result of liquid migration. Furthermore, the authors suggested mechanical assistance for buffering when using compost as the biofilter media. For example, the addition of wood chips or polystyrene beads will provide structural stability to the compost biofilter (Johnson and Deshusses 1997, Mohseni and Allen 2000) and polystyrene (Arulneyam and Swaminathan 2003).

3.3.2 Biochar as a biofilter material for bioremediation

Biochar is the result of the bioconversion of biomass feedstock. It is cost effective and easy to produce (Chintala et al. 2013). According to the research by Bartocci et al. (2017), biochar has been found to have unique characteristics that favors its use for bioremediation purposes. It has large surface area, high porosity and several surface-rich functional groups such as carboxylic, alcohol and hydroxyl group in the surface of biochar (Tang et al. 2013). Zhang et al. (2012) further explained that in aqueous solution, this biological material can adsorb nutrients including phosphorus and nitrogen. According to Zhang et al. (2012), activation of biochar for effective removal of the contaminant is possible. Furthermore, this activation produces higher yield at low temperatures resulting in reduced burn-off of the char. Another potential of biochar as a bioremediation media is the ability to control the migration of the adsorbed contaminant (Chintala et al. 2013), which is an additional advantage compared to some of the limitations of biofilters (Mudliar et al. 2010). Further studies on this topic should focus on assessing the mechanism of pollutant removal (adsorption + biodegradation) using hybrid phenomenological models that consider the adsorptive capacity of the biochar and the elimination capacity of the biofilm.

4. Factors affecting the biofiltration process

Biofiltration process is controlled by a number of factors; the most commonly considered parameters include pH, temperature, source of inoculum, moisture content, nutrient supply, oxygen, empty bed residence time (EBRT), inlet loading rate (ILR) and resilience to shock loads.

4.1 pH

According to Kennes et al. (2009), the pH of the biofilter is one of the major factors that influences the microbial activities. Tang et al. (2009) stated that microorganisms that carry out activities in acidic solution, i.e., with low pH, could be hindered in medium (slightly acidic) to high (alkaline) pH. Most microbial activities have been reported to occur within a pH of 5.0 (Brennan et al. 1996). Notwithstanding, Adler et al. (2014) reported that some of these microbes that survive in acidic solution such as the ethanol degraders (*Acetobacter*) are mostly active and productive in the pH range of 4.0 to 6.0. Ramirez-Lopez et al. (2010) reported methanol removal within a pH value of 6.5 to 8.5 with removal efficiency as high as 96%. Kennes et al. (2009) recommended the optimal pH as 6.0 to 7.0 for pollutant containing C, H and O atoms in order to have optimal microbial activity in the biofilter. Another study by Berenjian et al. (2012) showed that biofilters can also operate with high microbial activity within a pH range of 7.0 to 8.0.

4.2 Temperature

Microbial activity depends on the temperature of the biofilter; therefore, microorganisms should be selected based on the operating temperature of the biofilter and according to local conditions. Soccol et al. (2003) stated that temperature differences from the inlet air and the produced temperature from exothermic reactions also influences the temperature of the biofilter. At higher inlet loading rate (ILR), temperature is increased due to exothermic bioreaction (Kennes et al. 2009). Lim (2005) reported methanol removal at a maximum biofilter temperature of 25°C, resulting in a removal efficiency of 96%. Therefore, temperature of biofilters within a range of 20 to 30°C are suitable for mesophilic microorganisms to achieve optimal microbial activity. It is also important to note that temperature less than 20°C can affect the microbial growth (Kennes et al. 2009).

4.3 Nutrient supply

The pollutants fed into the biofilter are the major source of carbon and energy for the microbes. Other nutrients such as nitrogen, phosphorus, potassium, sulphur and trace metals are released by the biofilter material depending on the material selected (Soccol et al. 2003). According to Anit and Artuz (2002), a good biofilter material should be able to retain nutrients and supply them when needed to the biofilm that has been formed by the microbes. However, when these nutrients are in limited supply, it is essential that a suitable growth nutrient medium should be prepared to supply to the microbial community present in the biofilter. Compost is another nutrient rich material that has been successfully utilized to remove pollutants, including organic compounds. However, studies have shown inefficiencies as a result of deterioration (Morgenroth et al. 1996). Soccol et al. (2003) reported that nutrient supply is an important aspect for the biodegradation process; thus, the media selected should be able to provide high retention and capacity to supply nutrients to the microbes. The amount of nutrient supplied should be proportional to the concentration of the organic compound in order to achieve for an efficient biodegradation process. Example of nutrients used in biofilters includes, KH_2PO_4, KNO_3, $(NH_4)_2SO_4$, NH_4Cl, NH_4HCO_3, etc. Considering the type of nutrients required for microbial growth, it is essential to maximize its availability to the microbes.

4.4 Biocatalysts

The major factor influencing the efficient and quick removal of the contaminant is the microorganisms. According to Kumar et al. (2011), microorganisms majorly found in biofilters are aerobic and heterotrophic organisms and they are responsible for the biodegradation of the contaminants. Example of heterotrophs are bacteria and fungi and these microorganisms can be prepared in several ways including the application of pure bacterial culture (Reichert et al. 1997), isolation and characterization (Shareefdeen et al. 1993), and the preparation of mixed microbial population (Cox and Deshusses 1999). Kennes et al. (2009) suggested that for easily biodegradable contaminants such as methanol, natural release of microorganisms indigenous to the filter media can effectively carry out the degradation, while for pollutant that are difficult to degrade, inoculating specialized bacteria culture is suitable to degrade them. Hence, a mixed bacteria culture of any sludge is suitable for pollutant that are moderately difficult to degrade. Soccol et al. (2003) also acknowledged that the major microbial activity commences after the biofilter is allowed to adjust to its microbial ecology, and the most active microbes consume the transferred contaminant almost immediately. Example of organism that easily degrade one carbon compound without C-C bond such as methanol are methylotrophs, and this group of organisms is capable of utilizing methanol as their sole carbon and energy source and they can be found in mixed culture (Hanson and Hanson 1996). The micro kinetics of the degradation process are generally investigated and modeled with pure cultures of suspended microorganisms. However, in a continuously operated biofilter, during long term operation heterogeneous mixed culture of microorganisms is present rather than monocultures. Some of the predominant microorganisms in biofilters treating specific volatile compounds are given in Table 2.

Table 2. Frequently identified microorganisms in a biofilter.

Microorganism	Substrate
Pseudomonas sp.	Butyraldehyde, diethylamine and benzene, toluene, ethylbenzene, xylene (BTEX)
Coryneformic bacterium	2-Ethyl hexanol
Bacillus sp.	Thiophenol
Alcaligenes sp.	Indole
Pseudomonas sp.	Chlorobenzene
Methylo bacterium	Dichloromethane
Mycobacterium	Vinyl chloride
Exophiala, Aspergillus, Phanerochaete, Cladosporium, Paecilomyces, Trichoderma, and *Trametes* sp.	Wide variety of hydrophobic VOCs such as n-hexane, BTEX, α-pinene
Pseudomonas, Acinetobacter, Proteus, Aspergillus and *Fusarium* sp.	1, 1-dimethylhydrazine

4.5 Moisture content

The moisture content of the filter bed is vital for the optimal performance of the biofilter and efficient removal of the contaminants. Kennes and Veiga (2001) stated that an optimal moisture content in the range of 40 to 60% is enough to get the biofilter operating at efficient capacity. The moisture availability enables easy dilution of contaminant from gas to liquid phase so that the microorganism can degrade the contaminant with ease (Kennes et al. 2009). The microorganisms are most effective in carrying out their activities in a moist environment. However, excessive moisture availability for the microbes limits the transfer of oxygen and non-polar contaminants to the biofilm; consequently, channelling and pressure drop may occur (Schroeder 2002).

4.6 Oxygen availability

Another important factor for the biofilter is the availability of oxygen. Considering the activities of the aerobic microbes in the biofilm, sufficient oxygen is required by the microorganisms to carry out biodegradation that will help to eliminate the contaminant. According to Kennes and Veiga (2002), the liquid phase is responsible for transferring the oxygen and the substrate from the gaseous phase to the biofilm, where microbial activities are carried out. Aerobic microorganisms depend majorly on the oxygen availability to facilitate elimination of contaminant through microbial metabolism.

4.7 Empty bed residence time (EBRT)

EBRT is important for the effective removal of the pollutant because the removal efficiency depends on how much contact time is provided for the pollutants to stay inside the biofilter. Studies by Arulneyam and Swaminathan (2003) showed that high contact time increases the efficiency of pollutant removal. However, Kennes et al. (2009) argued that when the maximum elimination is achieved, increase in contact time may not be necessary because the same removal yield will be recorded. Even though methanol is readily biodegradable (Zhao et al. 2007), there is limited information on the EBRT required to attain optimal removal using biochar and compost as biofilter material.

4.8 Inlet loading rate (ILR)

This is the amount of contaminant that is constantly fed into the biofilter. It is expressed as the ratio of the gas flow rate and inlet concentration of the contaminant to the volume of the filter bed. According to Vergara-Fernandez et al. (2008), the removal efficiency of the biofilter depends on the loading rate of the contaminant; thus, high loading rate of the contaminant that is more than the rate at which the biodegradation occurs can reduce the removal efficiency. This was confirmed by the continuous reduction in removal efficiency of methanol when the loading rate was more than 280 g/m^3.h (Prado et al. 2005). Furthermore, Kennes and Veiga (2002) and Luvsanjamba et al. (2007) stated that higher loading rate usually requires effective, optimal working condition and improved performance of the filter bed. However, research has shown that there is increase in the development of innovative bioreactors that will be capable of handling high loading rates of the pollutants with improved removal efficiency (Rene et al. 2012). These bioreactors are expected not to be hindered by limitations associated with high loading rate such as clogging, limitation of oxygen, channeling and pH regulation (Soccol et al. 2003).

4.9 Shock loading conditions

This occurs when a sudden, yet unexpected, increase of pollutant is fed into the biofilter. Shock loading is capable of reducing the performance of the bioreactor, consequently affecting the biodegradation of pollutant by the microbes (Kennes and Veiga 2001, Jin et al. 2007). High and sudden loading of contaminant could overwhelm the microbes, thereby inhibiting the microbial activities and the degradation process. Conversely, low or absence of pollutant fed into the biofilter can result in low microbial activities (Rene et al. 2012). Soccol et al. (2003) reported that microbes adapt to biodegradation environment when the pollutants are gradually fed into the biofilter and not when a sudden increase or a very low concentration of pollutant is applied.

4.10 Performance parameters

Removal efficiency (RE)

Removal efficiency is evaluated to ascertain the effective removal of the contaminant (VOCs) by the biofilter. This is given by the percentage of the ratio of the amount of pollutant removed to the initial concentration of pollutant fed into the biofilter (Eq. 1).

$$RE = \frac{c_i - c_o}{c_i} \times 100 \tag{1}$$

where,

C_i = the inlet concentration of the pollutant, g/m³

C_o = the outlet concentration of pollutant, g/m³

Elimination capacity (EC)

The amount of pollutant removed per unit volume of the filter bed is defined as the elimination capacity (Eq. 2). The elimination capacity depends on the rate at which the pollutants diffuses into the biofilter and the corresponding rate at which it is degraded in the biofilm. The result of this interaction depends largely on stable and effective operating condition of the biofilm (Rene et al. 2009). According to Rene et al. (2009), the maximum elimination capacity is achieved at the critical loading rate of the pollutant. Under this condition, the biofilter is able to tolerate biodegradation process without inhibition.

$$EC = \frac{Q(c_i - c_o)}{V} \tag{2}$$

where,

Q = gas flow rate, m³/h

C_i = inlet concentration of the pollutant, g/m³

C_o = outlet concentration of the pollutant, g/m³

V = Filter bed volume, m³

5. Case-studies related to the removal of methanol and other pollutants from the pulp and paper industry

5.1 Properties of methanol and its emission

Methanol is a hydrophilic VOC, in the group of aliphatic alcohol, colorless and flammable. This volatile organic compound consists of 1 carbon, 4 hydrogen and 1 oxygen atom (EBTP 2011). This aliphatic alcohol has a freezing point of –97°C and a boiling point of 65°C, and is miscible with water and other organic solvents such as alcohol and ethers. It is most widely used in the pulp and paper industries. Heitz et al. (2008) stated that the methanol emission in Canada and the USA was estimated to be 17.7 and 71.6 kilotons in 2005, respectively. The pulp and paper industries accounted for 65% of these emissions.

The production of methanol includes biochemical and thermochemical process. Methanol is produced from fossil fuels and biological materials and can occur naturally in some plants as well as anaerobic metabolism of some microbes (EBTP 2011). Other biochemical processes include the bioconversion of methanotrophic bacteria such as *Methylococcus capsulatus*. Hwang et al. (2014) mentioned that methanol is also produced from methane oxidation. Figure 4 shows the biological sources of methanol production.

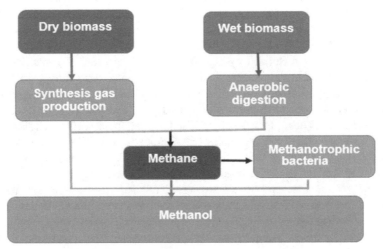

Figure 4. Biological production of methanol (adapted from EBTP 2011).

Thermochemical conversion process is another source of methanol production. In this process, fossil fuel feedstock produces methanol from synthesis gas such as CO, H_2 and CO_2 and methanol is produced through an exothermic reaction. The synthesis gas reaction can be illustrated according to Eqs. (3) to (5):

$$CO + 2H_2 \quad \rightarrow \quad CH_3OH \tag{3}$$

$$CO_2 + 3H_2 \quad \rightarrow \quad CH_3OH + H_2O \tag{4}$$

$$CO_2 + H_2 \quad \rightarrow \quad CO + H_2O \tag{5}$$

Understanding the biodegradation pathway of methanol is essential to manage the effluent and most importantly, to ensure that the intermediate product is not more toxic or hazardous than the initial contaminants (Rosenberg et al. 2014). Biodegradation of methanol is achieved through microbial activities that depend on carbon and energy for growth (Kennes and Thalasso 1998, Waweru et al. 2000). Aerobic bacteria degrade methanol by utilizing the carbon in the organic compound, giving a by-product of CO_2, H_2O, biomass and heat (Rene et al. 2011), as shown in Eq. (6):

$$\text{Organic pollutant} + O_2 \rightarrow CO_2 + H_2O + \text{Heat} + \text{Biomass} \tag{6}$$

5.2 Wastewater treatment in the pulp and paper industry

The pulp and paper making industry is a major consumer of fresh water, as a result generating large amount of wastewater of diverse characteristics (Azimvand and Mirshokraie 2016). Studies carried out by (Tharshanapriya et al. 2017) explain that this wastewater has significant negative impact on the environment and poses a serious threat to human health. According to Savant et al. (2006), the pulp and paper industry produces the third largest amount of wastewater after primary metals and chemicals industries.

The production of pulp and paper produce different environmental waste such as wastewater, waste gases and solid wastes (Azimvand and Mirshokraie 2016). During production of pulp and paper, large amount of water is required for the processing, cleaning and auxiliary purposes. Water is recirculated several times within the process (as high as ten passes) (Mladenov and Pelovski 2010). As a result, the production process is intense and requires industrial wastewater treatment. Currently, there are different and cost effective techniques for the treatment of the wastewater from pulp and paper production, such as physico-chemical, biological, and integrated treatment processes (Goswami et al. 2019).

This wastewater is characterized by organic and inorganic contaminants that emanate from tannins, lignins, resins, and chlorine compounds, while the major contaminants include COD, TSS, nitrogen compounds, and adsorbable organic halides (Kumar et al. 2014). Physico-chemical technique is employed to eliminate suspended solids, colloidal particles, toxic compounds, floating matters, and colors from wastewaters. These processes include sedimentation, ultra-filtration (Bhattacharjee et al. 2007), flotation (Kamali et al. 2019), screening (El-Ashtoukhy et al. 2009), coagulation, flocculation (Wong et al. 2006), ozonation and electrolysis (Kishimoto et al. 2010). These processes are capable of emitting GHG directly and indirectly as a result of their energy requirement. Thus, in choosing physicochemical treatment, several factors are considered, such as the type of contaminants in the wastewaters and the desired removal efficiencies. However, the removal efficiencies of the COD and BOD are usually inadequate under all observed conditions (Ashrafi et al. 2015). Therefore, these processes should be combined with biological processes to achieve acceptable results (Kumar et al. 2015).

Aerobic and anaerobic biological wastewater treatment process are mostly used in WWTP's to eliminate contaminants in wastewater. Aerobic processes are preferred because they are easier to operate and are more cost effective (Ashrafi et al. 2015). Although there are various aerobic technologies, activated sludge (AS) and aerated lagoons are commonly used in the pulp-and-paper industry (Pokhrel and Viraraghavan 2004). The application of anaerobic process in the pulp-and-paper industry is commonly used for the secondary treatment of wastewater treatment in the pulp and paper industry because it is more efficient in treating high strength organic effluents (Azimvand and Mirshokraie 2016).

However, the sulphur content in the wastewaters is the main disadvantage for application of anaerobic systems because one of the end products is hydrogen sulphide in the anaerobic biodegradation in the presence of sulphate (Lettinga et al. 1991). Both aerobic and anaerobic processes have certain disadvantages which include the high sludge production of aerobic processes and sensitivity of anaerobic bacteria to toxic materials. Ashrafi et al. (2015) showed that high sulphur content of chemical pulping wastewater had a detrimental effect on the contaminant removal capacity of anaerobic processes, especially at low pH values. In order to take advantage of these treatment processes, an integrated biological treatment process is required (aerobic and anaerobic or physicochemical and biological processes), and this process should be carried out under favorable environmental condition to achieve the desired result (Kumar et al. 2015).

5.3 Waste gas treatment in the pulp and paper industry

Pulp, paper and wood-related industries produce toxic air-pollutants like H_2S, α-pinene and methanol, which appear at different stages of unit operations. There are only a few reports that have focused on the removal of air emissions containing H_2S and VOCs from the pulp and paper industry. When biological waste gas treatment systems are used to treat such complex emissions, the pH of the biofilm was shown to drop when H_2S was converted to sulphuric acid, which in turn hindered the activity of the microbes that were degrading the VOC. Recently, one-stage and two-stage bioreactors were developed and successfully tested for the removal of gas-phase methanol, α-pinene and H_2S, either as stand-alone pollutants or as mixtures (Figures 5 and 6) (López et al. 2013, 2017). In the two-stage bioreactor, the first stage BTF was inoculated with a mixture of an autotrophic H_2S degrading culture

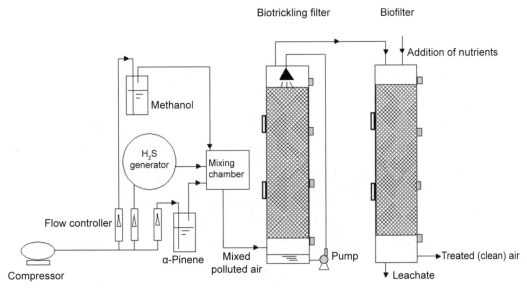

Figure 5. Two-stage bioreactor (BTF + BF) used for the combined treatment of emissions containing methanol, α-pinene and H₂S.

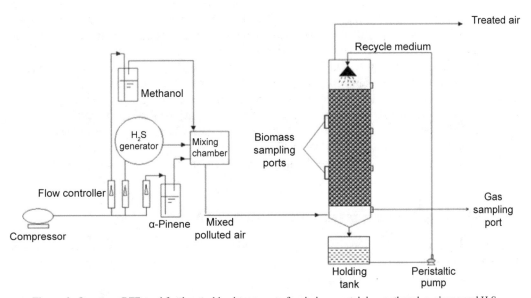

Figure 6. One-stage BTF used for the combined treatment of emissions containing methanol, α-pinene and H₂S.

and an acid-tolerant methanol degrading yeast (*Candida boidinii*), while an *Ophiostoma stenoceras* sp., was used to inoculate the second-stage BF (Figure 5). In the second experiment, with the one-stage BTF, a mixture of the above mentioned consortium was used and a *Rhodococcus* strain was added. The BTFs were packed with pall rings, while the BF was packed with a mixture of perlite and pall rings (Figure 6).

The empty bed residence times (EBRTs) used in the two-stage bioreactor were: 83.4, 41.7 and 27.8 s (BTF) and 146.4, 73.2 and 48.8 s (BF). In the one-stage BTF, EBRTs of 38 and 26 s were used. Concerning the key results achieved in these two bioreactor configurations (López et al. 2013, 2017), the performance of the two different bioreactor configuration can be compared in Figure 7. In the two-stage bioreactor, H₂S and methanol were better removed in the first-stage BTF with elimination

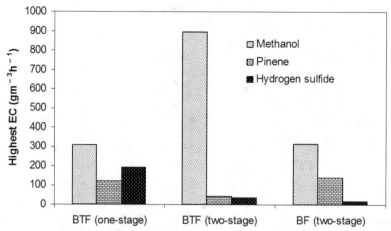

Figure 7. Highest elimination capacities reached for the three compounds in the one (BTF) and two-stage (BTF + BF) bioreactor configurations.

capacities (ECs) of 45 and 894 g/m^3.h, respectively, when compared to α-pinene (35 g/m^3.h). In the second-stage BF, the EC was 138 g/m^3.h for α-pinene, yet a high EC was observed for methanol (~ 315 g/m^3.h). In the one-stage BTF, the highest ECs were observed for methanol (302 g/m^3.h) followed by H$_2$S (126 g/m^3.h) and α-pinene (77 g/m^3.h). The behavior of the one-stage BTF was explored when maintaining the pH at a constant value (6.0 ± 0.3), leading to better removal of the hydrophobic pollutant (α-pinene) by the fungus, when compared to the other reactor configuration. It also helped to maintain a high activity for the surviving bacterial and yeast populations that removed methanol and H$_2$S. After long-term operation, the results from microbial community analysis (samples collected along the filter bed height) showed that the inoculated autotrophic H$_2$S-degrading culture was less diverse than the BTF samples. The low diversity of the inoculum can be explained by the emergence of a specific community able to degrade H$_2$S. A DGGE profile analysis of BTF samples after long-term operation suggests that two populations coming from the original autotrophic H$_2$S-degrading culture, as well as the inoculated *Rhodococcus* strain, are dominantly present within the BTF.

6. Outlook and future research directions

The following research directions are suggested in order to completely envisage aspects pertaining to the application of bioreactor technology (BT) for treating waste gases containing VOC mixtures from a wide variety of highly polluting industries:

 (i) An estimation of the bio-kinetics of the process and characterization of the biomass present in the bioreactor using actual industrial wastewater effluents.

 (ii) Transient state experiments should be performed under nitrogen and phosphorus limiting conditions in order to clearly understand the metabolic assimilation pathway of the microbial consortia.

(iii) Technologies for the treatment of greenhouse gas emissions from different process industries should be developed and integrated (for example: algae photobioreactor coupled to a BTF).

(iv) Development of hybrid mathematical and artificial intelligence based models to describe steady and transient-state bioreactor operation for better process control.

 (v) Perform cost-benefit analysis of bioreactor operation during steady and transient-state operations.

7. Conclusions

The application of biological processes for the removal of waste streams (solid, liquid and gas) has broadened due to the recently wide variety of biotechnological research. Different studies have been conducted on the successful biological removal of pollutant by considering factors such as microbial ecology in the bioprocesses, the movement pattern of the waste stream and various properties of the media and waste material in diverse bioreactor configurations. Although aerobic waste gas treatment systems are more common at the industrial scale, anaerobic/anoxic biotrickling filters should also be tested to recover valuable products such as volatile fatty acids and platform chemicals from waste gases. Due to the varying composition (concentration) and flow rate (transient state behavior) of industrial waste gas streams, the dominant microbial species should be identified and used for efficient biodegradation of the VOC mixtures in bioreactors.

Acknowledgements

TUU thanks the MSc fellowship received from Nuffic to pursue her MSc degree at UNESCO-IHE, Delft. The authors thank UNESCO-IHE for providing staff time and infrastructural support to start the new research line on biological waste-gas treatment and the application of biochars for waste to energy conversion.

References

Adler, P., L.J. Frey, A. Berger, C.J. Bolten, C. Erik and C. Wittmann. 2014. The key to acetate: Metabolic fluxes of acetic acid bacteria under cocoa pulp fermentation-simulating conditions. Applied Environmental Microbiology 80: 4702–4716.

Adler, S.F. 2001. Biofiltration-a Primer. Chemical Engineering Progress 97(4): 33–41.

Anit, S.B. and R.J. Artuz. 2002. Biofiltration of air (https://www.rpi.edu/dept/chem-eng/Biotech-Environ/MISC/biofilt/biofiltration.htm) (Accessed on January 2019).

Arulneyam, D. and T. Swaminathan. 2000. Biodegradation of ethanol vapour in a biofilter. Bioprocess Engineering 22(1): 63–67.

Arulneyam, D. and T. Swaminathan. 2003. Biodegradation of methanol vapor in a biofilter. Journal of Environmental Sciences 15(5): 691–696.

Ashrafi, O., L. Yerushalmi and F. Haghighat. 2015. Wastewater treatment in the pulp-and-paper industry: A review of treatment processes and the associated greenhouse gas emission. Journal of Environmental Management 158: 146–157.

Avalos Ramirez, A., S. Bénard, A. Giroir-Fendler, J.P. Jones and M. Heitz. 2008. Treatment of methanol vapours in biofilters packed with inert materials. Journal of Chemical Technology & Biotechnology 83(9): 1288–1297.

Azimvand, J. and S.A. Mirshokraie. 2016. Assessment of physico-chemical characteristics and treatment method of paper industry effluents: A review. International Research Journal of Applied and Basic Sciences 10(1): 32–43.

Bartocci, P., F.P. Vaccari, M. Valagussa, A. Pozzi, S. Baronti, F. Liberti, G. Bidini and F. Fantozzi. 2017. Effect of biochar on water retention in soil, a comparison between two forms: Powder and pellet. pp. 1732–1736. *In*: Proceedings of the 25th European Biomass Conference and Exhibition, Stockholm, Sweden.

Benitez, J. 1993. Process engineering and design for air pollution control. 1st Ed, Prentice Hall, Englewood Cliffs, pp. 466.

Berenjian, A., N. Chan and H.J. Malmiri. 2012. Volatile organic compounds removal methods: a review. American Journal of Biochemistry and Biotechnology 8(4): 220–229.

Bhattacharjee, S., S. Datta and C. Bhattacharjee. 2007. Improvement of wastewater quality parameters by sedimentation followed by tertiary treatments. Desalination 212(1-3): 92–102.

Brennan, B.M., M. Donlon and E. Bolton. 1996. Peat Biofiltration as an odour control technology for sulphur-based odours. Water and Environment Journal 10(3): 190–198.

Cheng, H.R., S.M. Saunders, H. Guo, P.K.K. Louie and F. Jiang. 2013. Photochemical trajectory modeling of ozone concentrations in Hong Kong. Environmental Pollution 180(3): 101–110.

Chintala, R., J. Mollinedo, T.E. Schumacher, S.K. Papiernik, D.D. Malo, D.E. Clay, S. Kumar and D.W. Gulbrandson. 2013. Nitrate sorption and desorption in biochars from fast pyrolysis. Microporous and Mesoporous Materials 179: 250–257.

Cox, H.H.J. and M.A. Deshusses. 1999. Chemical removal of biomass from waste air biotrickling filters: screening of chemicals of potential interest. Water Research 33(10): 2383–2391.

Devinny, J.S., M.A. Deshusses and T.S. Webster. 1998. Biofiltration for Air Pollution Control. Lewis Publishers, CRC Press, Boca Raton, USA.

Dutta, T., E. Kwon, S.S. Bhattacharya, B.H. Jeon, A. Deep, M. Uchimiya and K.H. Kim. 2017. Polycyclic aromatic hydrocarbons and volatile organic compounds in biochar and biochar-amended soil: a review. Gcb Bioenergy 9(6): 990–1004.

EBTP. 2011. European Biofuel Technology Platform. Methanol from biomass. http://www.etipbioenergy.eu/images/methanol-fact-sheet.pdf.

EEA. 2016. Air quality in Europe report. European Environmental Agency. https://www.eea.europa.eu/publications/air-quality-in-europe-2016.

El-Ashtoukhy, E.S., N.K. Amin and O. Abdelwahab. 2009. Treatment of paper mill effluents in a batch-stirred electrochemical tank reactor. Chemical Engineering Journal 146(2): 205–210.

Eregowda, T., L. Matanhike, E.R. Rene and P.N.L. Lens. 2019a. Performance of a biotrickling filter for the anaerobic utilization of gas-phase methanol coupled to thiosulphate reduction and resource recovery through volatile fatty acids production. Bioresource Technology 263: 591–600.

Eregowda, T., E.R. Rene and P.N.L. Lens. 2019b. Bioreduction of selenate in an anaerobic biotrickling filter using methanol as electron donor. Chemosphere 225: 406–413.

Goswami, L., R.V. Kumar, K. Pakshirajan and G. Pugazhenthi. 2019. A novel integrated biodegradation-microfiltration system for sustainable wastewater treatment and energy recovery. Journal of Hazardous Materials 365: 707–715.

Hanson, R.S. and T.E. Hanson. 1996. Methanotrophic bacteria. Microbiological Reviews 60(2): 439–471.

Hu, D., Y. Chen, V. Daële, M. Idir, C. Yu, J. Wang and A. Mellouki. 2019. Photochemical reaction playing a key role in particulate matter pollution over Central France: Insight from the aerosol optical properties. Science of the Total Environment 657: 1074–1084.

Hwang, I.Y., S.H. Lee, Y.S. Choi, S.J. Park, J.G. Na, I.S. Chang, C. Kim, H.C. Kim, Y.H. Kim, J.W. Lee and E.Y. Lee. 2014. Biocatalytic conversion of methane to methanol as a key step for development of methane-based biorefineries. Journal of Microbiology and Biotechnology 24(12): 1597–1605.

Jin, Y., L. Guo, M.C. Veiga and C. Kennes. 2007. Fungal biofiltration of α-pinene: Effects of temperature, relative humidity, and transient loads. Biotechnology and Bioengineering 96(3): 433–443.

Johnson, C.T. and M.A. Deshusses. 1997. Quantitative structure-activity relationships for VOC biodegradation in biofilters. pp. 175–180. *In*: Proceeding of the 4th International In situ and On site Bioremedation Symposium, New Orleans, USA.

Kamali, M., S.A. Alavi-Borazjani, Z. Khodaparast, M. Khalaj, A. Jahanshahi, E. Costa and I. Capela. 2019. Additive and additive-free treatment technologies for pulp and paper mill effluents: Advances, challenges and opportunities. Water Resources and Industry 21: 100109.

Kasperczyk, K., K. Urbaniec, K. Barbusinski, E.R. Rene and R.F. Colmenares-Quintero. 2019. Application of a compact trickle-bed bioreactor for the removal of odor and volatile organic compounds emitted from a wastewater treatment plant. Journal of Environmental Management 236: 413–419.

Kennes, C. and F. Thalasso. 1998. Waste gas biotreatment technology. Journal of Chemical Technology and Biotechnology 72(4): 303–319.

Kennes, C. and M.C. Veiga. 2001. Bioreactors for Waste Gas Treatment. Kluwer Academic Publishers, Dordrecht. The Netherlands.

Kennes, C. and M.C. Veiga. 2002. Inert filter media for the biofiltration of waste gases-characteristics and biomass control. Reviews in Environmental Science and Biotechnology 1(3): 201–214.

Kennes, C., E.R. Rene and M.C. Veiga. 2009. Bioprocesses for air pollution control. Journal of Chemical Technology and Biotechnology 84(10): 1419–1436.

Khanongnuch, R., F. Di Capua, A.-M. Lakaniemi, E.R. Rene and P.N.L. Lens. 2019. H$_2$S removal and microbial community composition in an anoxic biotrickling filter under autotrophic and mixotrophic conditions. Journal of Hazardous Materials 367: 397–406.

Kishimoto, N., T. Nakagawa, H. Okada and H. Mizutani. 2010. Treatment of paper and pulp mill wastewater by ozonation combined with electrolysis. Journal of Water and Environment Technology 8(2): 99–109.

Kumar, K.V., V. Sridevi, N. Harsha, M.V.V. Lakshmi and K. Rani. 2013. Biofiltration and its application in treatment of air and water pollutants-A review. International Journal of Applied or Innovative Engineering and Management 2(9): 226–231.

Kumar, S., T. Saha and S. Sharma. 2015. Treatment of pulp and paper mill effluents using novel biodegradable polymeric flocculants based on anionic polysaccharides: a new way to treat the waste water. International Research Journal of Engineering and Technology 2(4): 1–14.

Kumar, T.P., M.A. Rahul and B. Chandrajit. 2011. Biofiltration of volatile organic compounds (VOCs): An overview. Research Journal of Chemical Sciences 1(8): 83–92.

Kumar, V., P. Dhall, S. Naithani, A. Kumar and R. Kumar. 2014. Biological approach for the treatment of pulp and paper industry effluent in sequence batch reactor. Journal of Bioremediation and Biodegradation 5(3): 1–10.

Lettinga, G., J.A. Field, R. Sierra-Alvarez, J.B. van Lier and J. Rintala. 1991. Future perspectives for the anaerobic treatment of forest industry wastewaters. Water Science and Technology 24(3-4): 91–102.

Lim, K.H. 2005. The treatment of waste-air containing mixed solvent using a biofilter: 2. Treatment of waste-air containing ethanol and toluene in a biofilter. Korean Journal of Chemical Engineering 22(2): 228–233.

López, M.E., E.R. Rene, L. Malhautier, J. Rocher, S. Bayle, M.C. Veiga and C. Kennes. 2013. One-stage biotrickling filter for the removal of a mixture of volatile pollutants from air: Performance and microbial community analysis. Bioresource Technology 138: 245–252.

López, M.E., E.R. Rene, Z. Boger, M.C. Veiga and C. Kennes. 2017. Modelling the removal of volatile pollutants under transient conditions in a two-stage bioreactor using artificial neural networks. Journal of Hazardous Materials 324: 100–109.

Luvsanjamba, M., B. Sercu, S. Kertész and H. van Langenhove. 2007. Thermophilic biotrickling filtration of a mixture of isobutyraldehyde and 2-pentanone. Journal of Chemical Technology and Biotechnology 82(1): 74–80.

Lyu, X.P., N. Chen, H. Guo, W.H. Zhang, N. Wang, Y. Wang and M. Liu. 2016. Ambient volatile organic compounds and their effect on ozone production in Wuhan, central China. Science of the Total Environment 541: 200–209.

Mladenov, M. and Y. Pelovski. 2010. Utilization of wastes from pulp and paper industry. Journal of the University of Chemical Technology and Metallurgy 45(1): 33–38.

Mohseni, M. and D.G. Allen. 2000. Biofiltration of mixtures of hydrophilic and hydrophobic volatile organic compounds. Chemical Engineering Science 55(9): 1545–1558.

Morgenroth, E., E.D. Schroeder, D.P. Chang and K.M. Scow. 1996. Nutrient limitation in a compost biofilter degrading hexane. Journal of the Air & Waste Management Association 46(4): 300–308.

Mudliar, S., B. Giri, K. Padoley, D. Satpute, R. Dixit, P. Bhatt, R. Pandey, A. Juwarkar and A. Vaidya. 2010. Bioreactors for treatment of VOCs and odours-a review. Journal of Environmental Management 91(5): 1039–1054.

Pokhrel, D., and T. Viraraghavan. 2004. Treatment of pulp and paper mill wastewater-a review. Science of the Total Environment 333(1-3): 37–58.

Prado, Ó.J., M.C. Veiga and C. Kennes. 2005. Treatment of gas-phase methanol in conventional biofilters packed with lava rock. Water Research 39(11): 2385–2393.

Ramirez-Lopez, E.M., J. Corona-Hernandez, F.J. Avelar-Gonzalez, F. Omil and F. Thalasso. 2010. Biofiltration of methanol in an organic biofilter using peanut shells as medium. Bioresource Technology 101(1): 87–91.

Reichert, K., A. Lipski and K. Altendorf. 1997. Degradation of dimethyl disulphide and dimethyl sulphide by *Pseudonocardia* strains. pp. 269–272. *In*: Prins, W.L. and J. Van Ham (Eds.). Proc. Intl. Symp. Biological Gas Cleaning, Maastricht 1997. VDI Verlag, Düsseldorf, Germany.

Rene, E.R., M.E. López, D.V.S. Murthy and T. Swaminathan. 2009. Removal of xylene in gas-phase using compost-ceramic ball biofilter. International Journal of Physical Sciences 4(11): 638–644.

Rene, E.R., M. Montes, M.C. Veiga and C. Kennes. 2011. Novel bioreactors for waste gas treatment. pp. 121–170. *In*: E. Lichtfouse, J. Schwarzbauer and D. Robert (eds.). Environmental Chemistry for a Sustainable World Vol 2: Remediation of Air and Water Pollution, Springer Publishers.

Rene, E.R., B.T. Mohammad, M.C. Veiga and C. Kennes. 2012. Biodegradation of BTEX in a fungal biofilter: influence of operational parameters, effect of shock-loads and substrate stratification. Bioresource Technology 116: 204–213.

Rosenberg, E., E.F. DeLong, S. Lory, E. Stackebrandt and F. Thompson. 2014. The Prokaryotes: Gammaproteobacteria. Springer Publishers, Berlin, Germany.

Sahu, L.K., N. Tripathi and R. Yadav. 2017. Contribution of biogenic and photochemical sources to ambient VOCs during winter to summer transition at a semi-arid urban site in India. Environmental Pollution 229: 595–606.

Savant, D.V., R. Abdul-Rahman and D.R. Ranade. 2006. Anaerobic degradation of adsorbable organic halides (AOX) from pulp and paper industry wastewater. Bioresource Technology 97(9): 1092–1104.

Schroeder, E.D. 2002. Trends in application of gas-phase bioreactors. Reviews in Environmental Science and Biotechnology 1(1): 65–74.

Shareefdeen, Z., B.C. Baltzis, Y.S. Oh and R. Bartha. 1993. Biofiltration of methanol vapor. Biotechnology and Bioengineering 41(5): 512–524.

Shareefdeen, Z. and A. Singh. 2005. Biotechnology for Odor and Air Pollution Control. Springer Science & Business Media, Berlin, Heidelberg, Germany.

Singh, R.S., S.S. Agnihotri and S.N. Upadhyay. 2006. Removal of toluene vapour using agro-waste as biofilter media. Bioresource Technology 97(18): 2296–2301.

Soccol, C.R., A.L. Woiciechowski, L.P. Vandenberghe, M. Soares, G.K. Neto and V.T. Soccol. 2003. Biofiltration: an emerging technology. Indian Journal of Biotechnology 2: 396–410.

Song, M., X. Liu, Y. Xhang, M. Shao, K. Lu, Q. Tan, M. Feng and Y. Qu. 2019. Sources and abatement mechanisms of VOCs in southern China. Atmospheric Environment 201: 28–40.

Swetha, G., T. Gopi, C.S. Shekar, C. Ramakrishna, B. Saini and P.V.L. Rao. 2017. Combination of adsorption followed by ozone oxidation with pressure swing adsorption technology for the removal of VOCs from contaminated air streams. Chemical Engineering Research and Design 117: 725–732.

Tang, J., W. Zhu, R. Kookana and A. Katayama. 2013. Characteristics of biochar and its application in remediation of contaminated soil. Journal of Bioscience and Bioengineering 116(6): 653–659.

Tang, K., V. Baskaran and M. Nemati. 2009. Bacteria of the sulphur cycle: an overview of microbiology, biokinetics and their role in petroleum and mining industries. Biochemical Engineering Journal 44(1): 73–94.

Tharshanapriya, K., P. Sagadevan, K. Jayaramjayaraj, V. Bhuvaneshwari, S.N. Suresh, J. Pavithra, S. Sarah, B. Chandar Shekar and B. Ranjith Kumar. 2017. Health hazards of pulp and paper industrials workers. Indo American Journal of Pharmaceutical Research 7: 157–163.

van Groenestijn, J.W. and P.G. Hesselink. 1993. Biotechniques for air pollution control. Biodegradation 4(4): 283–301.

Vergara-Fernández, A.O., E.F. Quiroz, G.E. Aroca and N.A. Alarcón Pulido. 2008. Biological treatment of contaminated air with toluene in an airlift reactor. Electronic Journal of Biotechnology 11(4): 3–4.

Waweru, M., V. Herrygers, H. van Langenhove and W. Verstraete. 2000. Process engineering of biological waste gas purification. pp. 258–274. In: H.-J. Rehm, G. Reed, A. Pühler and P. Stadler (eds.). Biotechnology: Environmental Processes III, Volume 11, 2nd Edn, WILEY-VCH Verlag GmbH, Germany.

Wong, S.S., T.T. Teng, A.L. Ahmad, A. Zuhairi and G. Najafpour. 2006. Treatment of pulp and paper mill wastewater by polyacrylamide (PAM) in polymer induced flocculation. Journal of Hazardous Materials 135(1-3): 378–388.

Wu, H., H. Yan, Y. Quan, H. Zhao, N. Jiang and C. Yin. 2018. Recent progress and perspectives in biotrickling filters for VOCs and odorous gases treatment. Journal of Environmental Management 222: 409–419.

Yang, H., B. Zhu, J. Gao, Y. Li and L. Xia. 2013. Source apportionment of VOCs in the northern suburb of Nanjing in summer. Environmental Sciences 34(12): 4519–4528.

Zhang, M., B. Gao, Y. Yao, Y. Xue and M. Inyang. 2012. Synthesis of porous MgO-biochar nanocomposites for removal of phosphate and nitrate from aqueous solutions. Chemical Engineering Journal 210: 26–32.

Zhang, X., B. Gao, A.E. Creamer, C. Cao and Y. Li. 2017a. Adsorption of VOCs onto engineered carbon materials: A review. Journal of Hazardous Materials 338: 102–123.

Zhang, X., Z. Xue, H. Li, L. Yan, Y. Yang, Y. Wang, J. Duan, L. Li, F. Chai, M. Cheng and W. Zhang. 2017b. Ambient volatile organic compounds pollution in China. Journal of Environmental Sciences 55: 69–75.

Zheng, W., X. Bi, J. Wu, J. Feng, X. Fu, Y. Weng and Y. Zhu. 2014. Pollution characteristics and key reactive species of ambient VOCs in Ningbo City. Research of Environmental Sciences 27(12): 1411–1419.

Zilli, M., C. Guarino, D. Daffonchio, S. Borin and A. Converti. 2005. Laboratory-scale experiments with a powdered compost biofilter treating benzene-polluted air. Process Biochemistry 40(6): 2035–2043.

CHAPTER 11

Bio- and Phytoremediation of Persistent Organic Pollutants in Stormwater Containment Systems and Soil

Alisha Y. Chan and *Birthe V. Kjellerup**

1. Introduction

Persistent organic pollutants (POPs) are carbon based compounds that do not easily break down under natural environmental conditions. The POPs are semi-volatile, have low solubility in water, and can be highly toxic to both the ecosystem and humans (Fiedler 2002). These compounds originate from a variety of sources including pesticides for agricultural practices, industrial chemicals in hydraulic systems, and are also released as by-products from industrial production (Fiedler 2002). There are thousands of known POPs, but only twelve groups of POPs are recognized under the Stockholm Convention as having serious dangerous effects on humans and the environment (Table 1). Many new POPs have been found to be harmful since the listing in Table 1. The properties of POPs result in likely bioaccumulation in the food chain and long-range transport effects such as the grasshopper effect (Figure 1), where POPs are transported by volatilizing in warmer areas and condensing in colder areas (Fiedler 2002). In addition, POPs have a tendency to bioaccumulate in the food chain due to their chemical characteristics such as low water solubility and attraction towards lipids in organisms, making their presence especially harmful towards predator species, including humans (Jones and de Voogt 1999). Birds have historically been affected by POPs through the bioaccumulation of dichlorodiphenyltrichloroethane (DDT) as it was brought to the public's attention by Rachel Carson's book, "Silent Spring." There are also many health effects of POPs on humans such as cancer, negative reproductive effects, and neurotoxicity. Health effects on humans and organisms of the twelve groups of POPs under the Stockholm Convention are shown in Table 1.

Department of Civil and Environmental Engineering, University of Maryland at College Park, College Park MD, 20742, USA.
* Corresponding author: bvk@umd.edu

Table 1. **Twelve POPs under Stockholm Convention and their potential health effects** (Agency for Toxic Substances and Disease Registry, 1994, 1995, 1996, 1998, 2000a, 2002a, 2002b, 2007, 2014, 2015, Kamrin 1997, National Center for Biotechnology Information 2010, Stockholm Convention 2008).

POPs	Trade names	Source	Known potential health effects
Aldrin	Aldrex, Drinox, Octalene, Seedrin, Compound 118	Insecticides	Nervous system effects (headaches, dizziness, vomiting, involuntary muscle movements), kidney damage, destruction of own blood cells
Chlordane	Chlordan, Chlor-Kil, CD-68, Octachlor, Termi-Ded, Toxichlor, Topichlor, Velsicol 1068	Insecticides	Nervous system effects (convulsions, headaches, vision problems), digestive system effects (vomiting, stomach cramps, diarrhea, jaundice)
DDT	Genitox, Anofex, Detoxan, Neocid, Gesarol, Pentachlorin, Dicophane, Chlorophenothane	Insecticides	Nervous system (tremors, seizures, headache, nausea, dizziness), liver cancer, reproductive/development effects (decrease in lactation, increased chance of pre-term birth)
Dieldrin	Alvit, Dieldrix, Octalox, Quintox, Red Shield	Insecticides	Nervous system effects (headaches, dizziness, vomiting, involuntary muscle movements), kidney damage, destruction of own blood cells
Endrin	Endrex, Hexadrin	Pesticides	Nervous system (brain and spinal cord injury, convulsions, headache, dizziness, nervousness, confusion, nausea, vomiting), Birth defects (abnormal bone formation)
Heptachlor	Heptagran, Basaklor, Drinox, Soleptax, Termide, Velsicol 104	Pesticides	Liver damage, excitability, decrease in fertility, liver cancer, nervous system and immune system effects
Hexachlorobenzene	Anticarie, Bent-cure, Bent-No-more, Ceku C.B., Granero, No Bunt, Perchlorobenzene, Res-Q	Fungacides; Industrial chemical to control wheat bunt; By-products in chloralkali and wood preserving plants	Nervous system (weakness, tremors, convulsions), skin sores, porphyria, decrease in thyroid hormones, carcinogen
Mirex	Dechlorane	Insecticides	Nervous system (trembling, tiredness, weakness), harmful effects on stomach, liver, intestines, kidneys, eyes, and thyroid
Toxaphene	Agricide Maggot Killer, Alltox, Camphofene Huilex, Geniphene, Hercules 3956, Hercules Toxaphene, Motox; Penphene, Phenicide, Phenatox, Strobane-T, Synthetic 3956, Toxakil	Pesticides	Nervous system (convulsions), liver and kidney damage, effect on immune system, liver and thyroid cancer
Polychlorinated Biphenyls (PCBs)	Aroclor	Industrial chemicals as coolants and lubricants in transformers and capacitors; Plastics; By-products of inks, dyes and paints	Chloracne, liver damage, anemia, stomach and thyroid gland injuries, effect on immune system, behavior and reproductive system, carcinogen
Polychlorinated dibenzo-p-dioxins	N/A	By-products from pesticides and incomplete combustion of waste; automobile emissions; peat; coal; wood	Chloracne, liver damage, increased risk of diabetes, reproductive, immune, enzyme, and developmental defects, carcinogen
Polychlorinated dibenzofurans	N/A	By -products from incomplete combustion of waste, production of PCBs; automobile emissions	Chloracne, reproductive, immune, enzyme, and developmental defects, carcinogen

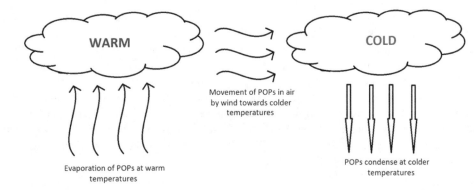

Figure 1. Grasshopper effect of POPs.

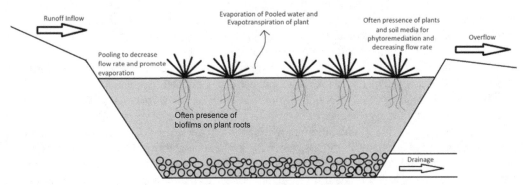

Figure 2. Example of stormwater containment system.

Stormwater containment systems are engineered facilities that allow stormwater runoff to pool within a confined system. The systems have traditionally been used to decrease the flow rate of stormwater to prevent flooding and erosion in urban streams (Davis 2008). However, as runoff is temporarily retained and slowly passes through these stormwater facilities, microorganisms and biota have the potential to treat runoff and sediment. An example of a stormwater containment system is a green infrastructure facility, which uses engineered soil and specialized plants to decrease the flow rate and treat pollutants in stormwater runoff (Figure 2). Green infrastructure facilities are effective for the treatment and removal of many of pollutants that include some POPs (Davis 2008). In this chapter, treatment options of POPs including microbial bioremediation and phytoremediation within stormwater containment systems are discussed.

2. Bioremediation

Bioremediation is a process where microorganisms and/or plants transform contaminated sites to reduce or eliminate the pollutant's harmful effects to the environment. The main objective of bioremediation processes is to remove and/or modify the pollutant to a form that is less or non-toxic to humans and the environment (Glazer and Nikaido 2007). Microorganisms and plants can treat pollutants in a variety of ways including extraction, degradation, filtration, volatilization, and stabilization. Generally, bioremediation is defined as treating contamination on site (*in situ*) or treating contamination after removal from the original contamination site (*ex situ*). Examples of *in situ* bioremediation include addition of nutrients, reducing agents, microorganisms, and/or plants to contaminated sites to optimize conditions for degradation of POPs (Glazer and Nikaido 2007). An example of *ex situ* bioremediation is the use of above ground reactors that contain microbes that degrade POPs and other pollutants in soil or water that was excavated from the contaminated site (Flickinger and Drew 1999). There are several advantages to each of these techniques. *In situ* bioremediation techniques usually cause lesser disturbance to the environment and are cheaper to apply since the contaminated area is not excavated using heavy machinery. In contrast, *ex situ* bioremediation treatments are often easier to manage and the treatment is more predictable as the contaminated soil can be moved to a contained and controlled environment for the treatment (Vidali 2001). Bioremediation within a stormwater containment system is an *in situ* bioremediation technique, where microorganisms, biota, and/or nutrients can enhance the capabilities of extraction, degradation, filtration, volatilization, and stabilization of POPs in the soil inside the storm water system or the runoff that slowly passes through the system.

2.1 Effect of bioavailability on bioremediation of POPs

An important consideration when performing bioremediation is the bioavailability of the POP, meaning to what extent do microorganisms and plants have access to the POP and in which chemical form (i.e., attached to particles) does the POP exist in the environment (Antizar-Ladislao 2010). Low bioavailability of the pollutant will give microorganisms and plant roots less ability to transport the pollutant to their cells and roots, thus limiting transformation. The treatment of Polychlorinated Biphenyls (PCBs), listed in the Stockholm Convention, is affected by bioavailability. There are 10 different PCB homologs, each containing from 1–10 chlorine atoms attached to two benzene rings. PCB homologs with a higher degree of chlorination have a higher log octanol water partitioning coefficients (log K_{ow}, Table 2) and are more hydrophobic (Hawker and Connell 1988). Due to the increasing hydrophobic structure of PCBs, generally, the bioavailability of PCB contaminants within an aqueous solution decreases as log K_{ow} increases (Hawker and Connell 1988). Another example of the importance of bioavailability is if the pollutant is highly volatile with a high vapor pressure. In this situation, the pollutant will tend to move into the air more readily than into water. Though this may decrease the bioavailability to microorganisms for treatment in soil media, both water-mobile and air-mobile organic pollutants can readily diffuse through plants thus increasing transformation (Pilon-Smits 2005). The vapor pressure of PCBs decreases as the number of chlorine atoms increases (Table 2). However, PCBs generally have low vapor pressures, resulting in low accounts of volatilization (Westcott and Bidleman 1981). The bioavailability of POPs varies based on the chemical properties and their interaction with the aqueous and/or soil media.

Through recent advances in technology, it is possible to increase the bioavailability of the pollutant to microorganisms without changing its or the soil media's properties. For example, biofilms used in bioremediation processes can increase the bioavailability of PCBs because they can increase the cell density and activity of the microorganisms, therefore increasing the chance of the microorganisms interacting with PCBs (Kjellerup and Edwards 2013).

2.2 Advantages and disadvantages of bioremediation

There are many advantages of using bioremediation to treat POPs compared to other types of treatment options such as physical excavation or chemical treatments. Bioremediation containment systems have the ability to transform contaminants from hazardous to less toxic compounds (Vidali 2001). Most bioremediation techniques do not require removal or transport of contaminated media and are therefore less expensive and more sustainable compared to other cleanup options for POPs (Megharaj et al. 2011, Vidali 2000). Lastly, bioremediation techniques cause little disturbance to the environment and the public (Glazer and Nikaido 2007). Many bioremediation facilities have plants and are aesthetically pleasing.

There are also disadvantages to the use of bioremediation, since bioremediation is largely affected by the bioavailability of the pollutant (Boopathy 2000, Megharaj et al. 2011). In addition, not all

Table 2. **Example molecular structure, log octanol water partitioning coefficients (log K$_{ow}$), and vapor pressure of PCB homologs** (Agency for Toxic Substances and Disease Registry 2000b, Robertson and Hansen 2015).

PCB homolog	Example molecular structure	Log K$_{ow}$	Vapor pressure (PA)
Monochlorobiphenyl	 Congener 2	4.7	1.1
Dichlorobiphenyl	 Congener 11	5.1	0.24
Trichlorobiphenyl	 Congener 28	5.5	0.054
Tetrachlorobiphenyl	 Congener 81	5.9	0.012
Pentachlorobiphenyl	 Congener 99	6.3	2.6×10^{-3}
Hexachlorobiphenyl	 Congener 138	6.7	5.8×10^{-4}

Table 2 contd. ...

...Table 2 contd.

PCB homolog	Example molecular structure	Log K_{ow}	Vapor pressure (PA)
Heptachlorobiphenyl	Congener 170	7.1	1.3×10^{-4}
Octachlorobiphenyl	Congener 203	7.5	2.8×10^{-5}
Nonachlorobiphenyl	Congener 206	7.9	6.3×10^{-6}
Decachlorobiphenyl	Congener 209	8.3	1.4×10^{-6}

compounds are transformed via bioremediation. Moreover, some degradation products from organic pollutant biodegradation can be even more toxic than their original chemical form (Boopathy 2000, Vidali 2001). The use of bioremediation is also highly specific; thus certain types of microorganisms and plant species will only be able to successfully treat certain POPs (Vidali 2001). Successful bioremediation also requires that the species can not only survive but thrive in the environment. Bioremediation also takes a long time to produce a clean or safe site (Vidali 2001). The biodegradation half-life time of PCBs varies from 3 to 37 years, increasing with higher degrees of chlorination and lower bioavailability (Sinkkonen and Paasivirta 2000). Lastly, it is difficult to show the positive effect of *in situ* bioremediation of POPs because of the various uncontrollable factors such as present ecosystem interaction in the remediation site (Madsen 1991). The advantages and disadvantages of bioremediation, excavation, and chemical treatments are summarized in Table 3.

Table 3. Advantages and disadvantages of bioremediation, excavation, and chemical treatments (Boopathy 2000, Megharaj et al. 2011, Vidali 2001).

	Description	Advantages	Disadvantages
Bioremediation (*In Situ*)	Use of natural biological activity from biota and microorganisms to remove from or degrade POPs in the media onsite	• Don't need to move contaminated media for treatment • Saves money on removal and transport costs • Aesthetics • Minimal disturbance to public and environment • Provide habitats for wildlife	• Largely affected by bioavailability of POP • Biodegradation may cause product that is more toxic than original POP • Highly specific; Not all compounds can be treated with bioremediation • Plant/Microbe species used must be able to survive in the toxic environment • Takes long • Uncontrollable factors on site
Excavation	Removing, usually by digging with heavy machinery, contaminated media from site using	• Usually results in full removal of contaminant from site • Quick and simple solution	• Disturbance to public and environment • Expensive to remove media and transport • Use of heavy machinery • Excavated media still contaminated. Further treatment necessary
Chemical Treatment (*In Situ*)	Use of specialized chemicals as reduction or oxidation agents to remediate POPs	• Don't need to move contaminated media for treatment • Save money on transport costs of contaminated soil	• Not all compounds can be treated with chemicals • Some chemical used for treatment may be harmful to humans and the environment • Uncontrollable factors on site • May need to restrict access to area during treatment • Risk of chemical byproducts

3. Bioremediation of POPs using microorganisms

The goal of biodegradation of POPs is to break down POPs to a form that is either completely harmless or less harmful. Bioremediation techniques using microorganisms can initiate or enhance the biodegradation process within stormwater contaminant systems (Abramowicz 1995, Erickson 1997). Various anaerobic bacteria have the ability to remove chlorine atoms from PCB homologues (organohalide respiration), thus decreasing the toxicity level of the pollutant. An example of anaerobic POP biodegradation is performed by *Dehalobium chlorocoercia* DF-1, which is an anaerobic bacterium that dechlorinates PCBs (Kjellerup et al. 2012), resulting in less toxic and less hydrophobic forms (Erickson 1997). The effectiveness of bioremediation depends on the presence of indigenous microorganisms in addition to the physical and chemical properties of the pollutant and soil matrix, such as the bioavailability and solubility and whether the product of biodegradation is even more toxic (Vogel 1996). Microbial activity is also effected by environmental conditions such as temperature, humidity, and ionic strength, as a high ionic strength increases the adherence of pollutants to a cell and therefore influences the rate of transformation (Shonnard et al. 1994, Vogel 1996). Various microbial bioremediation approaches applied for removing POPs are detailed below.

3.1 Bioaugmentation

Bioaugmentation is the process of adding microorganisms that have proven capabilities of transforming POPs to the soil media (Boopathy 2000, Megharaj et al. 2011). However, though bioaugmentation is usually successful when tested in a laboratory setting, the approach might experience disadvantages when introduced to a contaminated site. A common delivery method for bioaugmentation to, for instance contaminated groundwater, is the introduction of microorganisms via injection wells, where the microorganisms can be washed away or clog the well heads over time (Maxwell and Baqai 1995). In addition, competition with and predation from the naturally occurring microbial communities have shown to be a problem in many sites. Therefore, the combination of bioaugmentation and the use of biofilms might be a solution for some of these issues. When dechlorinating, bacteria used to transform PCBs were grown as a biofilm on activated carbon surfaces; they were not easily washed away and were not consumed by other microorganisms thus increasing their chances of survival (Kjellerup and Edwards 2013). Therefore, more research into the use of biofilms for bioaugmentation is required to ensure the biodegrading microorganisms' survival when added to natural soil media.

3.2 Biostimulation

Biostimulation is the process of stimulating naturally occurring microbial populations in the soil by providing nutrients that will enhance the microbial activity, thus cause an increase in POP transformation (Boopathy 2000). Anaerobic bacteria are often present and can degrade POPs in soil due to the existing environmental conditions. As anaerobic bacteria need electron acceptors other than oxygen and carbon sources (other than the contaminant) that can be used for co-metabolism with POPs, stimulation of the microbial activity can be performed. Addition of alternative election acceptors such as nitrate, sulfate or iron, resulting in anaerobic conditions and/or electron donors such as easily degradable organic substances like alcohols or fatty acids present in molasses, and compost to the soil media is practiced (Perelo 2010). In groundwater, tetrachloroethylene, a volatile chlorinated organic compound, can be readily dechlorinated after the injection of methanol and acetate as they act as electron donors (Major et al. 2002). A disadvantage of biostimulation is that the addition of these chemicals may affect the surrounding environment. A major advantage of biostimulation is that it increases the chance of the indigenous microorganisms' survival because they are naturally present in that environment and do not need to adapt, as would be the situation for bioaugmentation approach.

3.3 Bioventing

Bioventing is the process of treating contaminated sites by adding oxygen through the soil within a stormwater bioretention site to stimulate aerobic microbial activity (Boopathy 2000). Some POPs can be biodegraded more effectively using aerobic bacteria such as lower chlorinated PCBs (Abramowicz 1995). The aerobic bacteria can degrade the PCBs by breaking the biphenyl ring structure and subsequently mineralize the compounds to carbon dioxide and water. Bioventing is done by establishing a vacuum on the site or by vapor extraction via wells in the vadose zone, thus creating negative pressure and accelerating volatilization of hydrocarbons in the soil in addition to making oxygen available as electron acceptor. The increased rate of biodegradation results from lowering the water table and creating a larger vapor phase in the soil layer (Hoeppel et al. 1991). Advantages of bioventing are the cheaper process and increased efficiency for treating pollutants in the deep soil (Semple et al. 2001). A drawback can be that obligate anaerobic bacteria that can be involved in the degradation process will become inactive due to the presence of oxygen. Bioventing is commonly applied for the treatment of light petroleum hydrocarbon spills (Caliman et al. 2011).

4. Bioremediation of POPs using plants

POPs can also be treated using biological processes in plants and the rhizosphere. Phytoremediation is the treatment of pollutants using plants; a series of phytoremediation approaches are discussed in the following sections.

Phytoextraction is the process of removing pollutants from the soil and accumulating them within plant tissue (Pilon-Smits 2005). As these plants absorb toxic pollutants into their tissue, they must have the ability to tolerate high concentrations of pollutants and have a rapid growth rate to produce biomass (Salt et al. 1995). Unfortunately, this process can take a long time and can cause toxicity health concerns to wildlife that might ingest these plants, if they are not monitored and their access restricted (Pilon-Smits 2005, Salt et al. 1995). There is also concern over plant biomass disposal after phytoextraction due to the amount of toxic pollutants now within the plant itself. Though phytoextraction is often applied for extraction of heavy metals, it can be effective in removing POPs from the soil as well. For example, pumpkin (*Cucurbita pepo* ssp *pepo* cv. Howden) has the ability to take up and translocate high amounts of PCBs into the roots and shoots (Whitfield Åslund et al. 2007).

Rhizofiltration is the process of removing pollutants in water through adsorption in roots of plants, often grown in hydroponic setup (Aken et al. 2010). This approach can effectively transform POPs including PCBs, when combined with other microbial remediation techniques (Aken et al. 2010). This phytoremediation technique can be largely effected by biological processes in the plants such as intracellular uptake, vacuolar deposition and translocation to shoots, which makes the rhizofiltration less efficient as it creates more contaminated plant residue (Salt et al. 1995). Thus, plants that are effective translocators are often ineffective rhizofiltrators for pollutants.

Phytostabilization is the process of stabilizing the pollutant in soil by either preventing erosion, leaching, runoff or making the pollutant less bioavailable and therefore reducing exposure to the food chain (Campos et al. 2008, Cunningham 1996, Pilon-Smits 2005). At highly polluted sites, it is not uncommon to see a lack of vegetative cover due to the toxicity, before the remediation efforts are initiated. Lack of vegetation leads to increased spreading of the contaminant through wind erosion. Plants used for phytostabilization have the ability to populate soil sites in conditions that have high concentrations of the pollutant and therefore have the ability to prevent wind erosion (Salt et al. 1995). Grasses are efficient to prevent wind erosion, due to their adaptability to harsh conditions, whereas trees can function as a hydraulic barrier to prevent leaching or runoff of POPs. Poplar trees are particularly effective hydraulic barriers as they are deep rooted and transpire at high rates, therefore creating a powerful upward flow (Pilon-Smits 2005).

Phytostimulation: Certain plant species have the ability to facilitate microbial biodegradation of POPs, which is called phytostimulation. The plants can help promote microbial growth by providing larger root surface areas for microorganisms to bind to and produce exudates that together enhance biofilm formation (Fletcher and Hegde 1995). Therefore, plants with larger root structures are often more effective for phytostimulation. An example of a plant that uses phytostimulation to treat POPs is the mulberry tree (*Morus* sp.). Mulberry trees can produce phenolic compounds that stimulate microbial gene expression thus influencing degradation of PCBs and PAHs in soil (Fletcher and Hegde 1995, Leigh et al. 2002, McCutcheon and Schnoor 2004).

Phytovolatilization is the use of plants to volatilize the pollutants through transpiration (Campos et al. 2008, Salt et al. 1995). This process is effective because phytovolatilization of the contaminant usually results in dilution or photochemical decay in the atmosphere. However, the risk of degrading air quality in urban areas must also be considered (Limmer and Burken 2016, Pilon-Smits 2005). There are two different types of phytovolatilization, direct and indirect. Direct phytovolatilization is when the plant takes up and translocates the contaminants until it volatilizes from the stem or leaves (Limmer and Burken 2016). Indirect phytovolatilization is when the plant roots result in an increase in volatile contaminant flux. For example, a large tree can lower the water table, expanding the vadose zone and causing volatile contaminants to diffuse through the air more quickly than water thus increasing the flux (Limmer and Burken 2016).

Phytodegradation is defined as the process of degrading POPs to a harmless or less toxic form directly through enzymatic activities in the root or shoot of the plant (Pilon-Smits 2005). Plant enzymes have the ability to catabolize organic pollutants by converting them into inorganic compounds or degrading them to a stable, less toxic form to be eliminated or stored within the plant (Campos et al. 2008, McCutcheon and Schnoor 2004). For example, Elodea (*Elodea canadensis*) and kudzu (*Pueraria thunbergiana*) have the ability to dechlorinate p,p'- and o,p'-DDT, creating the less toxic products o,p'- and p,p'-DDD (Garrison et al. 2000).

5. Microbial remediation assisted phytoremediation

A variety of bioremediation techniques including bioventing, biostimulation, phytoextraction, phytovolatilization, and phytostabilization can advantageously be applied in combination (Figure 3). Figure 3 also shows biodegradation using biofilms that form on plant roots, a bioremediation technique that uses both microbial and phytoremediation approaches. Optimization of the remediation process is possible via utilization of plants and microorganisms, since plants have the ability to assist bacteria in microbial remediation. Microorganisms that have the ability to biodegrade POPs can be combined with plants to achieve optimal remediation results. When the soil is too toxic for plants and bacteria to survive, the combination of the two approaches may increase their individual chances of survival because many biodegrading bacteria have the ability to bind to plant roots and colonize in the plant rhizosphere (Glick 2010). This not only increases the chance of survival, but together, they can also increase the bioavailability of the pollutant by forming biofilms on the plant roots (Glick 2010). In addition, microorganisms can assist in the phytoremediation of POPs by promoting plant growth, as many microorganisms have the ability to fight against phyto-pathogens by forming biofilms, fixing nitrogen supply to plants, synthesizing siderophores to supply iron, synthesizing phytohormones, solubilizing minerals and/or synthesizing enzymes to alter plant ethylene levels (Glick 1995, 2003, 2004, 2010).

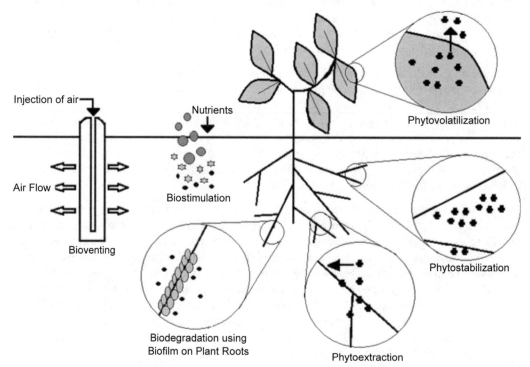

Figure 3. Bioremediation techniques including biovent, biostimulation, phytoextraction, phytovolatalization, phytostabilization and biofilm assisted biodegradation on plant roots.

6. Conclusions

In conclusion, the treatment of POPs can be done successfully using microbial remediation and/or phytoremediation in stormwater bioretention sites. Biofilms play a large role in the bioavailability of the pollutant while performing bioremediation of POPs, as they have the ability to expose POP degraders to a higher amount of POP, thus enhancing the rate of degradation. Microbial remediation techniques, including biodegradation, bioaugmentation, biostimulation, and bioventing and phytoremediation techniques including phytoextraction, rhizofiltration, phytostabilization, phytostimulation, phytovolatilization and phytodegradation for removing POPs are effective methods of decreasing POPs in the environment. However, POPs can still remain present in the environment due to limitations of each technique. Therefore, combinations of treatment techniques have been studied to overcome these challenges. Further research regarding the potential benefits and risks is required. A combination of these treatment techniques can be successfully implemented in stormwater containment systems such as green infrastructure to treat POPs in runoff.

References

Abramowicz, D.A. 1995. Aerobic and anaerobic PCB biodegradation in the environment. Environ Health Perspect 103: 97–99.

Agency for Toxic Substances and Disease Registry. 1994. ATSDR—Public Health Statement: Chlordane.

Agency for Toxic Substances and Disease Registry. 1995. ATSDR—Public Health Statement: Mirex and Chlordecone.

Agency for Toxic Substances and Disease Registry. 1996. ATSDR—Public Health Statement: Endrin.

Agency for Toxic Substances and Disease Registry. 1998. ATSDR—Public Health Statement: Chlorinated Dibenzo-P-Dioxins.

Agency for Toxic Substances and Disease Registry. 2000a. ATSDR—Public Health Statement: Polychlorinated Biphenyls (PCBs).

Agency for Toxic Substances and Disease Registry. 2000b. Toxicological Profile for Polychlorinated Biphenyls (PCBs).

Agency for Toxic Substances and Disease Registry. 2002a. ATSDR—Public Health Statement: Aldrin and Dieldrin.

Agency for Toxic Substances and Disease Registry. 2002b. ATSDR—Public Health Statement: DDT, DDE, and DDD.

Agency for Toxic Substances and Disease Registry. 2007. ATSDR—Public Health Statement: Heptachlor and Heptachlor Epoxide.

Agency for Toxic Substances and Disease Registry. 2014. ATSDR—Public Health Statement: Toxaphene.

Agency for Toxic Substances and Disease Registry. 2015. ATSDR—Public Health Statement: Hexachlorobenzene.

Aken, B.V., P.A. Correa and J.L. Schnoor. 2010. Phytoremediation of Polychlorinated Biphenyls: New Trends and Promises. Environ. Sci. Technol. 44: 2767–2776. https://doi.org/10.1021/es902514d.

Antizar-Ladislao, B. 2010. Bioremediation: Working with Bacteria. Elements 6: 389–394. https://doi.org/10.2113/gselements.6.6.389.

Boopathy, R. 2000. Factors limiting bioremediation technologies. Bioresource Technology 74: 63–67. https://doi.org/10.1016/S0960-8524(99)00144-3.

Caliman, F.A., B.M. Robu, C. Smaranda, V.L. Pavel and M. Gavrilescu. 2011. Soil and groundwater cleanup: benefits and limits of emerging technologies. Clean Techn Environ Policy 13: 241–268. https://doi.org/10.1007/s10098-010-0319-z.

Campos, V.M., I. Merino, R. Casado and L. Gómez. 2008. Phytoremediation of organic pollutants: a review. Spanish Journal of Agricultural Research 6: 38–47. https://doi.org/10.5424/sjar/200806S1-372.

Cunningham, S.D. 1996. Phytoremediation of soils contaminated with organic pollutants. pp. 55–114. *In*: Advances in Agronomy.

Davis, A. 2008. Field Performance of Bioretention: Hydrology Impacts. Journal of Hydrologic Engineering 13: 90–95. https://doi.org/10.1061/(ASCE)1084-0699(2008)13: 2(90).

Erickson, M.D. 1997. Analytical Chemistry of PCBs, Second Edition. CRC Press.

Fiedler, H., 2002. Persistent Organic Pollutants. Springer Science & Business Media.

Fletcher, J.S. and R.S. Hegde. 1995. Release of phenols by perennial plant roots and their potential importance in bioremediation. Chemosphere 31: 3009–2016.

Flickinger, M.C. and S.W. Drew. 1999. Encyclopedia of Bioprocess Technology. John Wiley.

Garrison, A.W., V.A. Nzengung, J.K. Avants, J.J. Ellington, W.J. Jones, D. Rennels and N.L. Wolfe. 2000. Phytodegradation of p,p'-DDT and the Enantiomers of o,p'-DDT. Environ. Sci. Technol. 34: 1663–1670. https://doi.org/10.1021/es990265h.

Glazer, A.N. and H. Nikaido. 2007. Microbial Biotechnology: Fundamentals of Applied Microbiology. Cambridge University Press.

Glick, B.R. 2010. Using soil bacteria to facilitate phytoremediation. Biotechnology Advances 28: 367–374. https://doi.org/10.1016/j.biotechadv.2010.02.001.

Glick, B.R. 2004. Bacterial ACC Deaminase and the Alleviation of Plant Stress. pp. 291–312. *In*: Microbiology, B.-A. in A. (Ed.). Academic Press.

Glick, B.R. 2003. Phytoremediation: synergistic use of plants and bacteria to clean up the environment. Biotechnology Advances, Proceedings of the Moo-Young Symposium on Biotechnology and Bioengineering 21: 383–393. https://doi.org/10.1016/S0734-9750(03)00055-7.

Glick, B.R. 1995. The enhancement of plant growth by free-living bacteria. Can. J. Microbiol. 41: 109–117. https://doi.org/10.1139/m95-015.

Hawker, D.W. and D.W. Connell. 1988. Octanol-water partition coefficients of polychlorinated biphenyl congeners. Environ. Sci. Technol. 22: 382–387. https://doi.org/10.1021/es00169a004.

Hoeppel, R.E., R.E. Hinchee and M.F. Arthur. 1991. Bioventing soils contaminated with petroleum hydrocarbons. Journal of Industrial Microbiology 8: 141–146. https://doi.org/10.1007/BF01575846.

Jones, K.C. and P. de Voogt. 1999. Persistent organic pollutants (POPs): state of the science. Environmental Pollution 100: 209–221. https://doi.org/10.1016/S0269-7491(99)00098-6.

Kamrin, M.A. 1997. Pesticide Profiles: Toxicity, Environmental Impact, and Fate. CRC Press.

Kjellerup, B. and S. Edwards. 2013. Application of Biofilm Covered Activated Carbon Particles as a Microbial Inoculum Delivery System for Enhanced Bioaugmentation of PCBs in Contaminated Sediment. Goucher College.

Kjellerup, B.V., P. Paul, U. Ghosh, H.D. May and K.R. Sowers. 2012a. Spatial Distribution of PCB Dechlorinating Bacteria and Activities in Contaminated Soil. Applied and Environmental Soil Science 2012. https://doi.org/10.1155/2012/584970.

Leigh, M.B., J.S. Fletcher, X. Fu and F.J. Schmitz. 2002. Root Turnover: An Important Source of Microbial Substrates in Rhizosphere Remediation of Recalcitrant Contaminants. Environ. Sci. Technol. 36: 1579–1583. https://doi.org/10.1021/es015702i.

Limmer, M. and J. Burken. 2016. Phytovolatilization of Organic Contaminants. Environ. Sci. Technol. 50: 6632–6643. https://doi.org/10.1021/acs.est.5b04113.

Madsen, E.L. 1991. Determining *in Situ* Biodegradation: Facts and Challenges. Environmental Science & Technology American Chemical Society 25: 1662–1673.

Major, D.W., M.L. McMaster, E.E. Cox, E.A. Edwards, S.M. Dworatzek, E.R. Hendrickson, M.G. Starr, J.A. Payne and L.W. Buonamici. 2002. Field Demonstration of Successful Bioaugmentation To Achieve Dechlorination of Tetrachloroethene To Ethene. Environ. Sci. Technol. 36: 5106–5116. https://doi.org/10.1021/es0255711.

Maxwell, C.R. and H.A. Baqai. 1995. Remediation of Petroleum Hydrocarbons by Inoculation with Laboratory-Cultured Microorganisms.

McCutcheon, S.C. and J.L. Schnoor. 2004. Phytoremediation: Transformation and Control of Contaminants. John Wiley & Sons.

Megharaj, M., B. Ramakrishnan, K. Venkateswarlu, N. Sethunathan and R. Naidu. 2011. Bioremediation approaches for organic pollutants: A critical perspective. Environment International 37: 1362–1375. https://doi.org/10.1016/j.envint.2011.06.003.

National Center for Biotechnology Information. 2010. ENDRIN [WWW Document]. PubChem Compound Database. URL https://pubchem.ncbi.nlm.nih.gov/compound/46174049 (accessed 1.26.17).

Perelo, L.W. 2010. Review: *In situ* and bioremediation of organic pollutants in aquatic sediments. Journal of Hazardous Materials 177: 81–89. https://doi.org/10.1016/j.jhazmat.2009.12.090.

Pilon-Smits, E. 2005. Phytoremediation. Annual Review of Plant Biology 56: 15–39. https://doi.org/10.1146/annurev.arplant.56.032604.144214.

Robertson, L.W. and L.G. Hansen. 2015. PCBs: Recent Advances in Environmental Toxicology and Health Effects. University Press of Kentucky.

Salt, D.E., M. Blaylock, N.P.B.A. Kumar, V. Dushenkov, B.D. Ensley, I. Chet and I. Raskin. 1995. Phytoremediation: A Novel Strategy for the Removal of Toxic Metals from the Environment Using Plants. Nat. Biotech. 13: 468–474. https://doi.org/10.1038/nbt0595-468.

Semple, K.T., B.J. Reid and T.R. Fermor. 2001. Impact of composting strategies on the treatment of soils contaminated with organic pollutants. Environmental Pollution 112: 269–283. https://doi.org/10.1016/S0269-7491(00)00099-3.

Shonnard, D.R., R.T. Taylor, M.L. Hanna, C.O. Boro and A.G. Duba. 1994. Injection-attachment of Methylosinus trichosporium OB3b in a two-dimensional miniature sand-filled aquifer simulator. Water Resour. Res. 30: 25–35. https://doi.org/10.1029/93WR02402.

Sinkkonen, S. and J. Paasivirta. 2000. Degradation half-life times of PCDDs, PCDFs and PCBs for environmental fate modeling. Chemosphere 40: 943–949. https://doi.org/10.1016/S0045-6535(99)00337-9.

Stockholm Convention. 2008. The 12 Initial POPs under the Stockholm Convention [WWW Document]. Stockholm Convention. URL http://chm.pops.int/TheConvention/ThePOPs/The12InitialPOPs/tabid/296/Default.aspx (accessed 1.26.17).

Vidali, M. 2001. Bioremediation. An overview. Pure and Applied Chemistry 73: 1163–1172.

Vogel, T.M. 1996. Bioaugmentation as a soil bioremediation approach. Current Opinion in Biotechnology 7: 311–316. https://doi.org/10.1016/S0958-1669(96)80036-X.

Westcott, J.W. and T.F. Bidleman. 1981. Determination of polychlorinated biphenyl vapor pressures by capillary gas chromatography. Journal of Chromatography A 210: 331–336. https://doi.org/10.1016/S0021-9673(00)97844-0.

Whitfield Åslund, M.L., B.A. Zeeb, A. Rutter and K.J. Reimer. 2007. *In situ* phytoextraction of polychlorinated biphenyl—(PCB)contaminated soil. Science of The Total Environment 374: 1–12. https://doi.org/10.1016/j.scitotenv.2006.11.052.

CHAPTER 12

Emerging Trends in Bioremediation of Explosive Industry Wastewater

S. Mary Celin,* Anchita Kalsi, Pallavi Bhanot,
Ila Chauhan and *Pritam Sangwan*

1. Introduction

'Pollution control'—one spoke in the wheel of environmental quality improvement—is essential to sustain the earth's life supporting system. Treatment of wastewater or control of waste should be considered as the inevitable last step in any industry. Among the various industries responsible for disturbing the sustainability of our earth, explosive industry, which comes under the defence sector, is one that contributes to air, water and soil pollution. Ammunition industries are producers of large quantity of toxic and harmful wastes. Many of these wastes are similar to those produced by large civilian industrial organizations, however, some are peculiar to the army mission. For example, the defence sector has major responsibility for energetic materials—propellants and explosives and for nuclear, biological and chemical (NBC) warfare defence. These tasks and activities have strong influence on the environment, generating toxic and harmful wastes at various stages of production. Critical equilibrium of various components of the ecosystem is being disturbed by the generation of hazardous waste generated from explosives manufacturing industries.

2. Explosives and environmental contamination by explosive industry waste water

By definition, explosives are solid or liquid substances, alone or mixed with one another, are in a metastablestate and are capable of undergoing a rapid chemical reaction without the participation of external reactants such as atmospheric oxygen (Meyer 1987). Nitro aromatic explosives consist of trinitrotoluene (TNT) in various degrees of purity and 2,4- and 2,6-isomers of dinitrotoluene (DNT). TNT has wide applications in shells, bombs, grenades, demolition explosives and propellant

Center For Fire, Explosive and Environment Safety (CFEES), Brig.S.K. Mazumdar Road, Timarpur, Delhi-110054.
* Corresponding author: celinkumaran@cfees.drdo.in

compositions. DNT is mainly used in the production of polyurethane foams and polymers in the manufacture of explosives and as a modifier of smoke less powders. Aromatic nitramines viz., Tetryl Aliphaticnitramines viz., Hexahydro 1,3,5-triniro 1,3,5-triazine (RDX) and 1,3,5,7-Tetranitro-1,3,5,7-tetraazacyclo octane (HMX) are used extensively as a booster charge in many munition formulations, especially in artillery shells or as a component in solid fuel rocket propellants. Nitrocellulose, Nitroglycerine (NG, glycerol trinitrate), Pentaerythritol tetranitrate (PETN) and Ethylene glycol dinitrate (EGDN) are the main nitric acid esters used as detonating agents, an important component in dynamites and multibase propellants (Urbanski 1990).

Pollutants from explosive industry are primarily produced by wastes from the explosive manufacture, such as acids used in nitration. Pollutants may also be produced during incorporation of the explosives in munitions, and in clean-up and disposal operations. The actual nature of the wastewater is dependent on the type of manufacturing process. The types of pollutants found in the explosive industry waste-water and their effect on the environment (Urbanski 1990) are listed in Table 1.

Aqueous waste containing dissolved high explosives originate from numerous munitions processes. Pink water is a generic term for munition wastewater produced at load, assemble, and pack (LAP) facilities, projectile washout facilities, cutting and hogging out operations, as well as demilitarization activities. Pink water has varying concentrations of TNT, RDX, HMX and DNT. These nitro organics make up approximately one third of the polar organic fraction. TNT wastewater is classified by the USEPA as hazardous waste (USEPA 1989). TNT is quite mobile in soil. Thus, residues of TNT in soil can be a source of ground water contamination both in the army installations and beyond installation boundaries.

The coloration arises from photolysis of α-TNT or hydrolysis in the presence of alkali (Sanders and Slagg 1978, Walsh et al. 1973). Depending on the formulation used, levels of RDX and HMX vary in the pink water. Pink water may also contain 2,4- DNT, which is a by product TNT manufacture. Other impurities normally present are acetylated decomposition derivatives of RDX and HMX. Depending on the explosives being handled at a particular manufacturing facility, pink water may contain all four nitro organics of interest, or TNT and 2,4-DNT or only RDX and HMX. The actual nature of the wastewater is dependent on the type of process in which the effluent is originating. These dissolved materials make the water highly toxic and therefore untreatable via classical biological water treatment systems. The nitro explosives present in munition wastewater are classified as priority pollutants by the USEPA.

Many of the explosive constituents are carcinogenic and cause several ailments to human beings. TNT is classified as a priority pollutant having a drinking water discharge limit of 1.0 μg/L (EPA 1989). Lifetime health advisory for explosive chemicals in drinking water is given in Table 2. TNT is acutely toxic to humans and animals and is mutagenic in the Ames test (Won et al. 1976). The

Table 1. Pollutants present in explosive industry wastewater and its effect on environment.

Pollutant	Effect on environment
Propellants and explosive	Dangerous, toxic, cannot be land filled or open burned
Nitrates	Toxic, increase solids content, eutrophication
Sulfates	Toxic, increase solids content, odorific at anaerobic conditions
Phosphates	Eutrophication
Acetates and organic esters	Toxic, increase dissolved oxygen demand, increased acidity
Nitro bodies (pink water)	Toxic
Red water	Toxic
Process sludge (solid waste)	Hazardous
Inert contaminants	Toxic, unsightly

Table 2. Lifetime health advisory for explosive chemicals in drinking water (http://www.epa.gov/ost/drinking/standards/dwstandards.pdf.).

Sl. No.	Compound	Limits (µg/L)
1.	2,4,6-trinitrotoluene (TNT)	`1.0
2.	Royal demolition explosive (RDX)	2.0
3.	High Melt Explosive (HMX)	400
4.	2,4-dinitrotoluene (2,4-DNT)	0.17
5.	2.6-dintrotoluene (2,6-DNT)	0.0068

major toxicological effects of exposure to RDX are nausea, irritability, convulsions, unconsciousness and amnesia.

Owing to the hazardous nature of nitro-organic contaminants present in explosive industry wastewater, it is essential that wastewater, before being discharged into the main water streams, is treated to remove these toxic nitro compounds.

3. Bioremediation and its prospects for treating explosives contaminated wastewater

The treatment of explosives contaminated wastewater is a major environmental challenge. To address this, several strategies have been explored and implemented with varying degree of success (Saez et al. 2012, Zhang et al. 2013, Oh et al. 2014, Roh et al. 2015, Reddy et al. 2016). Environmentally compatible disposal methods available for different energetic materials are highly recommended. The disposal and recycling processes for explosives must factor in the following:

- Environmental Safety
- Technical feasibility
- Economical operation.

Traditionally, the wastewater from explosive industry is pre-treated by filtration to remove suspended solids. Technologies for further treatment of wastewater include physical treatment, chemical destruction and biological transformation. Among various methods of remediation of explosives contamination, activated carbon adsorption has been found to be effective. But the technology comes with drawbacks like, incomplete regeneration of activated carbon and high temperatures employed in the process. Treatment by advanced oxidation processes results in generation of intermediates which require further follow-up treatment.

The goal of any remediation strategy is the conversion of targeted pollutants to non-toxic environmentally benign products. In all the above mentioned conventional treatment processes, the contaminant is only transferred from one phase to another, requiring an additional follow-on process for destruction, also leaving the possibility of future human contact. Therefore, the need exists for development of an environment friendly and sustainable technology that converts the contaminants to environmentally benign non-toxic compounds. These eco-friendly waste treatment technologies have established themselves as a frontline area of research. In nature, diverse types of microorganisms and energy sources are available. These microbes have the capability to break down various compounds to harvest energy. In this process, the microbes' take-up even the hazardous organic compounds for their growth in turn leading to its degradation. They are empowered to inhabit various ecological niches and pursue unusual metabolic and physiological activities. This incredible versatility harbored by microbes renders these recalcitrant compounds, a part of biogeochemical cycles.

The present chapter attempts to explore the recent trends and advances made in utilizing the abilities of biological agents in treating explosive industry wastewater.

4. Bacterial cultures in explosives bioremediation

Microorganisms break contaminants through reactions as a part of metabolic process for growth and development. Microorganisms utilize nitrogen and carbon contents of explosive chemicals for its metabolic needs. Microorganisms used in remediation process may be indigenous to a contaminated site or may be isolated from elsewhere. Bioremediation of explosive chemicals by microorganisms is reported to occur by aerobic and anaerobic processes. Both bacteria and fungi have proven to be potential bioremediators of explosives.

Cyplik et al., in 2012 applied bioremediation for treatment of effluents, contaminated with nitroglycerine (4.8 mg l^{-1}) and nitroglycol (1.9 mg l^{-1}) from an explosive manufacturing plant. A microbial consortia isolated from soil under anaerobic conditions was used for the same. Nitroglycerine and nitroglycol were removed with complete removal of nitrate. Activity of dehydrogenases in activated sludge was measured to assess the toxicity of the effluent sample. Ames test was also conducted to determine the mutagencity which was found to not exceed 0.5.

When TNT contaminated wastewater was subjected to bioremediation by *Raoultella terrigena*, it was postulated that TNT uptake by bacterial cells was due to diffusion and it was reduced by nitroreductases. The products of this reduction were ADNT (10–20%) in the solution. It was also observed that, 80–90% of the initial TNT was converted to intra- or intermolecular coupling products which remain in the cell in form of insoluble tetranitrozoxytoluenes or bound to proteins. *R. terrigena* HB, during TNT transformation process, formed brownish cells that could be removed by centrifugation or filtration (Claus et al. 2007). RDX in aquifer microcosms was removed using Fe(0) in flow through column packed with steel wool. The hydrogen produced at the cathode due to corrosion of Fe(0) by water, was used by anaerobic bacteria to further reduce RDX. The anaerobic consortium was found to further degrade the hetrocyclic intermediates produced during reaction of RDX with Fe(0). The initial Fe(0) treatment helped in reduction of the toxicity of the wastewater, thereby enhancing biodegradation (Wildman and Alvarez 2001).

Achromobacter spanius, a novel strain of bacteria isolated from TNT contaminated soil, was found to be effective in degrading TNT from the aqueous solutions (Gumuscu and Teckiney 2013). Complete removal of 100 mg/ L TNT within 20 h under aerobic conditions was reported. Degradation of TNT was observed at varying environmental conditions between pH 4.8–8.0 and 4–43°C. However, the most efficient degradation was at pH 6.0–7.0 and 30°C. This study suggested possible TNT degradation via DNT conversion route.

2,4-Dinitroanisole (DNAN), a key component of a new class of melt cast formulations designed for use in insensitive munitions, is a low sensitivity replacement for TNT, Nocardioides bacterial strain. JS1661 was evaluated for its ability to mineralize DNAN in non-sterile soil, aqueous media and in a fluidized bed bioreactor. Various concentration of DNAN studied under this treatment. DNAN was completely degraded under all tested conditions with little or no accumulation of DNP (Dinitro phenol) and concomitant release of nitrite. The results of the study revealed that the strain could survive well in different explosive concentrations as well as environmental conditions. Hence, it could be used for bioaugmentation (Karthikeyana and Spain 2016).

5. Fungi as an agent of bioremediation of explosive compounds

Fungi has evolved as the decomposers in the global cycle of life and death and provide nourishment to the other biota that is present in soil, and are thus considered as vigorous agent present in the soil food web. They have innate capability to breakdown molecules (living or non-living object), disassembling long-chained toxins into simpler and less toxic constituents (Mishra and Malik 2014, Sharma and Malviya 2016). They function very often in correlation with bacteria and they have an astonishing potential to clean up the contaminated environments. Fungi have been used to clean waterways, soil, and even radioactive waste (Mokeeva et al. 1998). Breakdown of environmental

Table 3. Summary of potential bacterial species employed for treating explosive contaminants in wastewater.

Sl. No.	Explosive chemical	Bacterial species employed	Reference
1.	TNT	*Achromobacter spanius*	Gumuscu and Teckiney 2013
		Pseudomonas fluorescens	Pak et al. 2000
		Pseudomonas sp. JLR11	Nunez et al. 2000
		Mycobacterium vaccae	Vanderberg et al. 1995
		Enterobacter	Bae et al. 1995
		Desulfovibrio sp.	Boopathy et al. 1993
		Desulfovibrio sp.	Preuss et al. 1993
		Pseudomonas fluorescens	Naumova et al. 1982
		Pseudomonas putida	Kearney et al. 1983
		Veillonella alkalescens	McCormick et al. 1976, 1981
		Pseudomonas sp.	Won et al. 1974
2.	DNT	*Pseudomonas* sp.	Spanggord et al. 1991
		Achromobacter xylosoxidans	Keenan et al. 2004
		Burkholderia cepacia R34	Johnson et al. 2000
		Desulfovibrio sp. strain B	Boopathy et al. 1993
		Escherichia coli	Petersen et al. 1979
3.	RDX	*Clostridium* sp. EDB2*Geobacter metallireducens*	Bhushan et al. 2004
		Acetobacterium malicum strain HAAP-1	Kwon and Finneran 2008
		Rhodococcus sp.	Adrian and Arnett 2004
		Enterobacter cloacae	Nejidat et al. 2008
		Gordonia and *Williamsia* spp.	Pudge et al. 2003.
		Clostridium acetobutylicum	Thompson et al. 2005
		Klebsiella pneumoniae strain SCZ-1	Zhang and Hughes 2003
		Rhodococcus sp. Strain DN22	Zhao et al. 2002
		Desulfovibrio sp.	Fournier et al. 2002
			Boopathy et al. 1998,
			Mc Cormick et al. 1981
4.	HMX	*Pseudomonas* sp. (HPB1), *Bacillus* sp. (HPB2, HPB3)	Singh et al. 2009
		Caldicellulosiruptor owensensis	Huang 1998
		Methylobacterium sp.	Aken et al. 2004

contaminants by Fungi is termed as mycoremediation. The ability of these fungi to form extended mycelia networks, the low specificity of their catabolic enzymes and their independence from using pollutants as a growth substrate make these fungi well suited for bioremediation process.

Bioremediation of explosives using fungi has gained an increased interest in last few decades due to the chemical toxicity and environmental persistence of nitro explosives. Fungi help in the remediation of explosives by using its lignin degrading enzyme. There are some other species of fungi which have been evaluated as the engines of the decomposition process than any other fungal species like the white rot fungus *Phanerochaete chrysosporium* which carries out the mineralization of nitro explosive compounds like TNT. It is a lignin degrading fungus which produces extracellular peroxidases, small organic molecules and hydrogen peroxide (Spain 1995). The lignolytic system is nonspecific and can biodegrade compounds including nitro-organics such as TNT.

Tsai (1991), suggested that the white rot fungal system is effective in causing some bioconversion changes in the red water waste stream. The red water was first treated with UV radiation to make it sensitive to bioremediation. It was then incubated with whole fungal culture or extracellular enzyme preparations of ligninase for a week to reduce its color intensity as well as cytotoxicity.

Fournier et al. (2006) studied the ability of the white rot fungi, *Phanerochaete chrysosporium and Irpex lacteus*, for the bioremediation of nitramine CL-20. For this purpose, they used ^{15}N and ^{14}C labeled CL-20. They found that both the fungus could efficiently degrade the nitramine under aerobic conditions with the release of nitrous oxide (N_2O) and carbon dioxide (CO_2). When pure manganese peroxidase was used, CL-20 broke down to form an intermediate, glyoxal which was easily decomposed by *P. chrysosporium.*

Lee et al. (2009) monitored the ability of white rot fungus—*Irpex lacteus* to transform TNT to other compounds. They observed a significant decrease in the TNT concentration over the time period of experiment. At the end of 48 hours study, about 95% TNT had been degraded. Authors also observed formation of two other intermediates such as 4-amino-2,6-dinitrotoluene (4-ADNT) and 2-amino-4,6-dinitrotoluene (2-ADNT) using GC-MS. Finally, this study concluded that the white rot fungus could degrade both the TNT and its metabolites formed in the process.

6. Immobilized cells in bioremediation

In bioremediation, maintaining a high population of the biomass is of utmost importance. For this purpose, cell immobilization is employed (Bayat et al. 2015). Cell immobilization can be defined as "the physical confinement or localization of intact cells to a certain region of space with preservation of some desired catalytic activity" (Karel et al. 1985). Immobilization also improves the survival and stability of the organisms in the polluted environment. This has also led to the use of immobilized cells in several other biotechnological processes like production of useful chemicals, treatment of wastewaters and bioremediation of polluted sites. Numerous biotechnological processes utilize immobilized technology such as industrial processes, medicine, biosensors and many more.

The immobilized cell systems have the following several advantages (Beyat et al. 2015):

1) Providing cell reuse and reducing the costly processes of cell recovery and cell recycle.

2) Elimination of cell washout problems at high dilution rates.

3) High flow rates allow high volumetric productivities.

4) Providing suitable micro environmental conditions.

5) Improving genetic stability.

6) Protection against shear damage.

7) High resistance to toxic chemicals, pH, temperature, solvents and heavy metals.

8) Decline of maturation time for some products.

6.1 Carrier selection

Immobilization involves confinement of the cells on to a solid support. This support is the biocarrier. Carrier selection is a very important aspect of immobilization. Following criteria are to be considered while choosing an appropriate carrier (Zacheus et al. 2000):

1) Non toxic and biodegradable

2) High surface area

3) Long shelf life

4) Low cost

5) High mechanical, biological and chemical stability.

The biocarriers can either be organic (celluloses, dextran andagarose) or inorganic (zeolite, clay and ceramics). Moreover, organic carriers can be natural (alginate, agar and agarose) or synthetic (acrylamide, poluyurethane and resins) (Lu et al. 2009).

6.2 Types of immobilization techniques

Immobilization techniques can be divided into four major categories based on the physical mechanism employed: (i) attachment or adsorption on solid carrier surfaces, (ii) entrapment within a porous matrix, (iii) self aggregation by flocculation (natural) or with crosslinking agents (artificially induced), and (iv) cell containment behind barriers (Pilkington et al. 1998).

6.3 Immobilized cell bioreactors

Immobilization of microbial cells on a solid support and applying them for wastewater treatment has gained special attention in the last few decades. In cell immobilization biological and physical stability of microorganisms increases as it restricts cellular mobility and retains catalytic activity (Saez et al. 2012). The immobilized cell bioreactor have been suggested to be much better than free culture bioreactor owing to many advantages like continuous reactor operation without risk of cell washout, protection of cells from toxic substrates, high growth rate of cell and enhanced gas–liquid mass transfer rate (Sokol et al. 2004, Gonzalez et al. 2001). The immobilized cell systems are reported to be more tolerant to fluctuations in pH, temperature, and presence of inhibitor compounds (Venkatasubramanian and Veith 1979, Tallur et al. 2009, Zheng et al. 2009).

Tope et al. (1999) compared the efficiency of resting cells and immobilized cells of *Arthrobacter* sp. in the transformation of TNT to its byproducts 4-amino-2,6-dinitrotoluene (4-ADNT), 2-amino-4,6-dinitrotoluene (2-ADNT) and also 2,4-diamino-6-nitrotoluene (2,4-DANT). They carried out the experiments at 30°C in phosphate buffer with pH 7.2 and TNT concentration of 60 mg L^{-1}. The resting cells led to complete transformation of TNT within 36 hours. They then immobilized the cells of *Arthrobacter* sp. in Barium Alginate. They found that the immobilization improved the transformation. TNT was completely transformed within 24 h.

Ullah et al. (2010) performed a study on the biodegradation of the explosive TNT using a *Bacillus* sp. YER1. They isolated the *Bacillus* sp. from the red water effluent of the TNT manufacturing site and performed a comparative study between free and immobilized culture. They used charcoal and polystyrene for immobilization *Bacillus* sp. The study was conducted for 168 hours. They also evaluated the performance of both the biocarriers at different temperatures, pH and surfactant. The optimum temperature for both charcoal and polystyrene was found to be 37°C. Charcoal showed best performance at pH 7 whereas polystyrene showed optimum at pH 5. Also, they found that presence of the surfactant tween 20 enhanced the biotransformation of TNT with both charcoal and polystyrene.

Wang et al. in 2010 used a combined process of immobilized microorganism and biofilter for the degradation of TNT in aqueous solution. Two upflow and submerged biological filters were set up. Each reactor was made of polymethyl methacrylate. A self-made patented FPUFS carrier was used for immobilization. The synthesized polymer carrier was cubical with dimensions as 20 mm × 20 mm × 20 mm. The pores in the carrier enhanced mixing of air, wastewater and carrier and acted as sites for microbial immobilization. The microbial consortium named B925 was used in the study. Later, this microbial consortium was found to be (based on 16S rRNA gene sequencing) comprising of *Pseudomonas* sp., *Flavobacterium* sp., *Chryseobacterium* sp. and *Rimerella* sp. Their TNT degrading ability has already been studied before. The degradation rate of TNT varied greatly at the early stage, which indicated that the microorganisms in the reactor were not well accommodated to TNT. After an operation for 20 days, the system entered the stable reaction operation stage. Their results show that this technique was successful in degrading TNT. They also studied the intermediated produced during the anaerobic reaction using GC/MS. GC/MS analysis identified 2-amino-4,6-dinitrotoluene (2-A-4,6-DNT), 4-amino-2,6-dinitrotoluene (4-ADNT), 2,4-diamino-6-nitrotoluene (2,4-DA-6-NT) and 2,4-diamino-6-nitrotoluene (2,6-DA-4-NT) as the main anaerobic degradation products. Ethanol was added to the water sample in the beginning of the study as an electron donor. It was shown to play a major role in TNT degradation.

Table 4. Summary of Immobilization technologies used with their advantages and disadvantages.

Sl. no.	Immobilization technique	Principle	Advantages	Disadvantages
1.	Adsorption	Physical interaction between the cell and the insoluble solid matrix via weak forces, viz., Vander Waals forces, ionic bonds, hydrophobic interactions and hydrogen bonds	Mild, easy, cheap, reversible, No chemical treatment required	Leakage of the cells, unstable interactions, low reproducibility
2.	Covalent binding	Covalent bond formation between the inorganic carrier and the cell via a cross linking agent	Stronger association useful in enzyme immobilization, reversible	Rarely used for whole cell immobilization as the cross linking agents lead to reduction in cell viability. It can also lead to loss of enzyme activity
3.	Entrapment	Capturing of cells within a support matrix	Fast, cheap, mild reactions involved. No cell leakage	Irreversible, high cost, deactivation of support during immobilization
4.	Encapsulation	Enveloping the organism within various forms of spherical, semi permeable membrane	Protects the biocatalyst from extreme conditions due to microencapsulation. Prevents leakage, hence higher efficiency to perform specific function	Irreversible

(Ferenandes et al. 1998, Soysal et al. 2004, Groboillot et al. 1994)

6.4 Biogranulation

Biogranulation is another form of natural immobilization. The immobilized systems can be artificial or natural. In artificial systems, the cells are entrapped in a matrix, whereas in natural systems, the cells attach themselves on to the carrier surfaces spontaneously and colonize to form a biofilm. In biogranulation, the cells are attached to each other and also entrapped in the extracellular polymeric substances (EPS) matrix made by the microorganisms. This demonstrates the ability of the microorganism to attach to one another and form self immobilized granules. These granules are compact and dense. By co-aggregation, the different bacteria interact to form an aggregate of consortium held within a polymeric matrix (Venugopalan et al. 2005). However, it is different from the formation of biofilm because no carrier or supporting material is needed to develop granules (Liu et al. 2002).

Advantages of microbial granules include:

1) Compact and dense microbial structure
2) Good settling ability
3) High biomass retention in the bioreactor
4) Ability to withstand fluctuating organic loads.

6.4.1 Mechanism of biogranulation

There are four stages involved in biogranulation, viz., formation of seed sludge, followed by compact aggregates, granular sludge and finally, mature spherical granules (Sondhi et al. 2010).

Granulation involves two major processes, the auto-aggregation and co-aggregation of microorganisms. Auto-aggregation is the cell-cell interaction between genetically identical cells, whereas, co-aggregation is the cell-cell interaction between genetically distinct cells (Liu and Tay 2002).

The microbial granulation largely depends on:

1) Aggregating abilities of the individual bacterial strains
2) Intercellular communication between the cells
3) Cell surface hydrophobicity
4) EPS released by the cells.

6.4.2 Types of biogranulation

There are two types of biogranulation, viz., anaerobic granulation and aerobic granulation.

Anaerobic granulation is commonly observed in upflow anaerobic sludge blanket (UASB) type reactors. UASB reactors are commonly used in waste water treatment plants worldwide (Boonsawang et al. 2008, Rajagopal et al. 2013, Torres et al. 2009). The disadvantages of this technology include: requirement of long start up period of 2–4 months, and high operating temperature of 30–35°C. It is unsuitable for low strength organic wastewater and for ammonium and phosphorus removal.

Aerobic granulation is commonly observed in sequencing batch reactors (SBR) (Nancharaiah and Kiran Kumar Reddy 2018). It requires cyclic feeding and starvation, high shear stress and short settling time. Aerobic granules have been developed in SBRs on degradable carbon sources such as glucose, acetate and ethanol (Keller et al. 2001). SBRs are used for nutrient removal (Cassidy and Belia 2005) and also to treat industrial and other hazardous wastewaters.

6.4.3 Microbial granules in bioremediation of explosives

Kiran Kumar Reddy et al. (2017) cultivated aerobic granules in sequencing batch reactors (SBR) for the removal of 2,4-dinitrotoluene (DNT). They cultivated the granules by feeding the SBR with acetate and DNT. The granules were formed in 30 days since the startup. Alternately, they also cultivated granules using acetate as carbon source and then acclimatized them with DNT. Aerobic granules developed by both the approaches were evaluated for their potential to remove DNT. Both the approaches showed a rapid removal of 2,4-DNT with more than 90% of 10 mg/L of DNT removed in 24 hours. These granules also led to the removal of ammonium-nitrogen, phosphorus and organic carbon. HPLC analysis indicated that the biotransformation of 2,4-DNT led to the formation of 2-amino-4-nitrotoluene. They also compared the efficiency of granules under anaerobic conditions. The results indicated that the aerobic conditions gave much better and higher transformation rates than anaerobic conditions.

Zhang et al. (2013) evaluated aerobic granules for the treatment of HMX production wastewaters. They used sequencing batch reactor for this purpose. They observed a high performance of the reactor after 40 days. Organic carbon removal was found to be up to 97.57% and nitrogen removal was found to be upto 80%. UV spectra results indicated the degradation of HMX from the waste water. Chun et al. in 2010 studied the degradation of RDX using anaerobic granules from a UASB reactor. They obtained anaerobic granules from mesophilic UASB reactor treating food wastewater. The granules were observed to be black and spherical with a diameter of 0.45–2 mm these granules were then acclimatized to RDX by gradually increasing its concentration. They performed the RDX biodegradation batch experiments in different conditions, viz., RDX as sole substrate, under nitrogen rich state, sulphate enriched condition, nitrate enriched condition and glucose enriched condition. The extra nitrogen had no significant effect on RDX degradation, whereas sulphate enriched condition had a slight negative effect. Also, presence of nitrate led to the inhibition of RDX degradation. Glucose on the other hand enhanced the biodegradation of RDX by anaerobic granules.

7. Biochar as an ecofriendly adsorbent

Biochar has gained considerable attention in recent years due to the impressive results of this material for pollution remediation, global warming mitigation, improving soil fertility and recycling of agricultural waste. It is an ancient practice over 2000 years old, to convert agricultural waste into soil enhancer. Biochar is a form of charcoal, also known as Agichar or Black Carbon. It's a carbonaceous residue generated by the pyrolysis of the organic materials. It involves the thermal decomposition of organic materials in oxygen limited environment that results in a stable solid material which is rich in carbon content that can be sequestered in the soil to sustainably enhance its environmental value. Biochar has been effectively used as an adsorbent, have potential to inhibit the leaching of nitrate and fecal bacteria into waterways, reduces N_2O and CH_4 emissions from the soil and immobilizes pesticides, herbicides, heavy metals and explosives in the soil. Thus, biochar is defined as a solid by-product containing carbon rich material derived from incomplete combustion and different thermal processes.

It is a highly heterogeneous material with chemical composition and physical properties that vary widely depending upon the pyrolysis conditions and biomass feedstock compositions. For example, manure based biochar will have greater nutrient content than that formed from woodchips and on the other hand, wood based biochar will exhibit more stability. When pyrolysis is done at very high temperature, the biodegradability of biochar decreases. In contrast to this, high temperature increases micro porosity, C/O and C/H ratio and adsorptive capacity of the biochar.

This subset of black carbon has two very specific and unique properties that makes it stand out among other organic amendments and influences the fate and transport of the organic pollutants. These two properties are adsorption and stability. These two pillars can be effectively used to address the most urgent environmental problems of today. Biochar has the potential for mitigating anthropogenic green house effect. It can be effectively used as an appropriate approach for improving water quality, to ensure soil fertility and to enhance agricultural productivity.

Biochar has been investigated as an electron transfer mediator for the reduction of organic contaminants. It has been observed that reduction of nitroaromatic compounds, such as DNT, and heterocyclic nitramines, such as RDX was greatly enhanced by biochar (oh et at. 2013).

The feasibility of using biochar as a sorbent to remove nitro explosives from contaminated water has been investigated through batch processes. Some biochars have been found to effectively adsorb nitro explosives and their sorption capacity was found to be proportional to its surface area and carbon content (Oh and Seo 2014).

Oh et al. (2013) reported the reduction of RDX and DNT present in waste water using poultry litter biochar. Poultry litter biochar had small surface area and consisted of redox active compounds. Using both, biochar and dithiothreitol enhanced the removal of DNT and RDX from the solution. In addition to this, reduction products 2A4NT, 4A2NT, DAT and Formaldehyde of DNT and RDX, respectively, were also detected. This clearly demonstrated the reduction of nitro explosives using biochar.

Adsorption of TNT and RDX by biochar alginate beads was studied using batch system and continuous fixed bed columns at pH 6.0. The batch study indicated the adsorption of TNT and RDX on alginate beads. These beads were prepared by dropping alginate solution containing biochar in $CaCl_2$ solution. The maximum adsorption capacities of alginate beads from langmuir equation were found to be 90.09 mg/g and 28.09 mg/g for TNT and RDX, respectively. Ion exchange and electrostatic interactions between explosives and sorbent were found to be the reason for sorption. Fixed bed column studies for RDX and TNT were done to check the breakthrough time. Column studies of TNT and RDX have been found to be in good agreement with obtained results using Thomas model, used for designing adsorption column. Overall results suggested that biochar alginate beads could be applied as ecofriendly adsorbent for removal of nitro explosives (Roh et al. 2015).

Nehrenheim et al. (2008) investigated contaminated wastewater from Swedish disposal industry which was treated with organic by product–pink bark (*Pinus silvestris*) as a adsorbent of an explosive, TNT (2,4,6 trinitrotoluene). Pine bark is a by-product from a global industry, namely manufacturing pulp and paper, cellulose and other woods products. The basic idea of this study was to establish that pine bark could be a potential adsorbent for treatment of TNT contaminated wastewater. In this study, the shaking procedure was carried out in accordance to the activated carbon (AC) filtration step at the disposal factory. All the samples from the experiment were analyzed by means of HPLC. Approximately 80% of the TNT was retained by the pine bark. This basically means that the waste water was poured into the pine bark and filtered out immediately. The results indicated that the adsorption process was relatively rapid and that a major part can be retained by the pine bark within just a few seconds.

Oh and Seo (2015), used biochar derived from raw rice for removing DNT, TNT and RDX from wastewater through batch experiments and found that the sorption capacity of explosives depends upon the surface area and carbon content of the biochar. They observed that the maximum sorption capacity of the nitro explosives increased when temperature of pyrolysis elevated from 250°C to 900°C. It was also found that surface treatment of biochar with acid or oxidant also increased the sorption capacity of biochar for TNT and DNT.

8. Hybrid systems a sustainable solution for treating explosives wastewater

Hybrid systems utilizing physical, chemical and biological processes in combination offer a sustainable strategy for coping with the refractory and complex nature of contaminants present in wastewater. Physical treatment processes such as adsorption on to activated carbon and chemical processes such as advanced oxidation process using UV irradiation, ozone treatment and zero valent iron oxidation have been proven successful for treating explosives contaminated waste water. But, they do not give a sustainable treatment solution as the contaminant is merely transferred from one phase to the other, whereas biological methods in combination with these physico-chemical treatment processes have demonstrated complete destruction of the explosive contaminants with the formation of environmentally benign end products.

Pink water consisting of dissolved trinitrotoluene (TNT) and cyclotrimethylenetrinitramine (RDX), and its by-products was successfully treated by a combined anaerobic treatment in a fluidized bed reactor (FBR) and physisorption using granular activated carbon (GAC). It was observed that, bacteria developed into a biofilm on GAC during the treatment. Bench scale studies using an anerobic bacterial consortium with ethanol as sole source of carbon were performed for transformation of TNT. It was found that, TNT transformed to TAT, which further degraded to undetectable end products. Similarly, RDX was fund to be transformed to mononitroso, dinitroso and trinitroso derivatives (MNX, DNX. TNX), before being mineralized to methanol and formaldehyde. Gas produced as a by-product of anaerobic degradation was collected at the top of the column, and sent to a flare to burn off the methane (Maloney et al. 2002).

Fluidized bed biofilm or fixed bed biofilm reactors in combination with physical adsorption processes used GAC and resin for treating perchlorate in waste water. This hybrid process was found to be efficient in reducing perchlorate to acceptable levels (Venkatesan et al. 2010, Xiao et al. 2010).

As reported by Oh et al. (2005), an integrated iron reduction—activated sludge process is a viable option for treating RDX contaminated wastewater. Findings of batch and column experiments showed rapid and complete removal of RDX by pre-treatment with zero valent iron, which readily made RDX more amenable to biodegradation by a mixed culture. Removal of RDX from waste water was evaluated by yet another combined process wherein TiO_2-nanotubes (NTs)/SnO_2-Sb anode was employed as pretreatment followed by anaerobic/aerobic activated sludge process. Findings of the study revealed that the electrocatalytic process significantly enhanced the biodegradability of RDX. The toxic and refractory organic pollutants were effectively removed in the combined treatment process (Chen et al. 2011).

Hexanitrostilbene (HNS) is an important heat resistant explosive, used in military and aerospace fields. Waste water that emanates during the synthesis and purification of HNS has high chemical oxygen demand (COD) and chromaticity. Composition of the wastewater includes pyridine and its derivatives, as well as some nitroaromatic compounds (NACs) that are acutely toxic and mutagenic. This waste water on treatment by an integrated anaerobic–aerobic bioreactor system with immobilized bacteria resulted in complete removal of nitroaromatic compounds like nitrobenzene and dinitrobenzene and up to 80% mineralization of pyridine. *Pseudomonas* species from anaerobic filter was found to play a major role in degrading aromatic compounds, while pyridine was mainly removed by the order *Rhizobiales* sp. in biological aerated filter (Liu et al. 2012).

Hybrid treatment process consisting of electrocatalytic reaction in combination with anoxic–oxic process was demonstrated to treat hexahydro-1,3,5-trinitro-1,3,5-triazine (RDX) wastewater. Electrolytic system having optimized process parameters of $5 \, g \, L^{-1} \, Na_2SO_4$ of electrolyte concentration and $20 \, mA \, cm^{-2}$ of current density was proven as efficient pretreatment system which was followed by the anoxic-oxic activated sludge process (Chen et al. 2011).

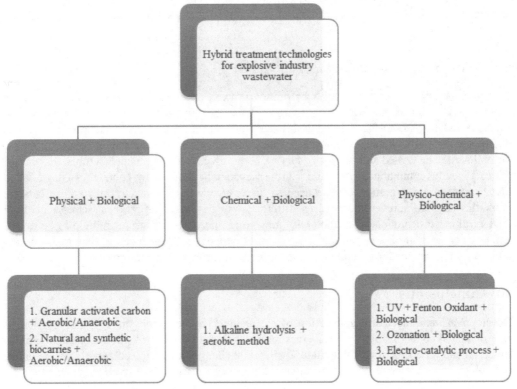

Figure 1. Summary of hybrid systems employed in bioremediation of explosive contaminated aqueous waste.

9. Future prospects in bioremediation of explosive industry waste water

9.1 Role of bioinformatics and biotechnology

Advances in the field of bioinformatics are revolutionizing the field of bioremediation. Bioinformatics is the combination of biology and information technology. It focuses on the cellular and molecular level of biosystems and finds application in modern biotechnology. It incorporates statistical tools and algorithms to establish and analyze relationship between biological data sets, structures, profiles and biochemical pathways involved in the biodegradation pathway of the xenobiotic compound. Proteomics deals with the study of role of proteins. In bioremediation, proteins play a major role in degrading the pollutant. The expressed proteins change with the different environmental conditions. They may be down or up regulated in presence of the explosive contaminant. Proteomics can help in studying the physiological changes in an organism during bioremediation and hence help in finding the concerned gene and its regulation. Hence, the regulation of the concerned gene can help in up regulation of bioremediation (Fulekar and Sharma 2008). Genomics gives an insight into the metagenomic population which can help in bioremediation. Studies on microbial adaptation of toxic environments may give rise to trace new metagenomic communities useful for efficient bioremediation. Specific functions and interactions of microbial communities with respect to contamination-degrading capabilities can be a result of environmental-based gene switching in the metagenomes.

The genetic potential of uncultivable microorganisms to remediate pollutants are now explored by metagenomic analyses (Martin et al. 2006). Recent technological breakthroughs in *de novo* sequencing of microbial metagenomes based on pyrosequencing have provided rapid and relatively inexpensive methods to generate microbial community profiles simultaneously from different environmental samples.

9.2 Role of indigenous microbial community

The inherent degradation capabilities of proficient explosive degrading microorganisms in the contaminated environment can prove successful for the implementation of bioremediation of explosive industry wastewater. The natural sludge, soil and wastewater from contaminated site have high population of well-adapted microbes. These microbes can withstand high concentrations of the pollutant to be treated and can easily survive in the adverse environment. Microbial enzymes of these indigenous communities play a crucial role as metabolic catalysts for contaminant degradation. Metagenomics and genomics can be used to discover new microbial enzymes whose catalytic properties can be improved/modified by different strategies based on rational, semi-rational and random directed evolution. Thus, naturally inhabiting microbial consortia can help in better and efficient degradation on a large scale. Also, control parameters are reduced as the natural flora is utilized for bioremediation, leading to a cost- effective prospect for the purpose.

10. Conclusions

Development and successful application of a suitable bioremediation strategy for treatment of environmental contamination by explosives require a thorough and in-depth understanding of the structural and functional diversity of natural microbial communities. Molecular tools and genomics technology are an integral part of the monitoring process. Various biological methods involving both bacterial and fungal cultures have been proposed for the bioremediation of explosives. These methods have varying degree of adaptability and success in such adverse environmental conditions. Application of free bacterial cultures gives promising results but remains questionable in large scale field applications. Utilization of immobilized cell reactors has been found to give an edge to the free cultures, increasing their efficiency and applicability. Fungi, with their ability to survive in harsh environmental conditions, and effective Lignin Degrading Peroxidases (LDP) are a promising tool for the removal of explosive compounds. Adopting an integrated treatment approach for bioremediation of explosives seems to be a very promising tool for the transformation and mineralization of explosives as it brings the best of physical, chemical and biological aspects to one place. Bioremediation promises a cleaner, safer, environmentally sound and economically efficient removal of recalcitrant chemicals from the environment.

References

Adrian, N.R. and C.M. Arnett. 2004. Anaerobic biodegradation of hexahydro-1,3,5-trinitro-1,3, 5-triazine (RDX) by *Acetobacterium malicum* strain HAAP-1 isolated from a methanogenic mixed culture. Current Microbiology 48: 332–340.

Bae, B., R.L. Autenrieth and J.S. Bonner. 1995. Aerobic biotransformation and mineralization of 2,4,6-trinitrotoluene. pp. 231–238. *In*: R.E. Hinchee, R.E. Hoeppel and D.B. Anderson (eds.). Bioremediation of Recalcitrant Organics. Columbus, Ohio: Battelle Press.

Bayat, Z., M. Hassanshahian and S. Cappello. 2015. Immobilization of microbes for bioremediation of crude oil polluted environments: a mini review. The Open Microbiology Journal 9: 48–54. https://dx.doi.org/10.2174% 2F1874285801509010048.

Bhushan, B., A. Halasz, S. Thiboutot, G. Ampleman and J. Hawari. 2004. Chemotaxis-mediated biodegradation of cyclic nitramine explosives RDX, HMX, and CL-20 by *Clostridium* sp. EDB2. Biochemical and Biophysical Research Communications 316: 816–821.

Boonsawang, P., S. Laeh and N. Intrasungkha. 2008. Enhancement of sludge granulation in anaerobic treatment of concentrated latex wastewater. Songklanakarin Journal of Science and Technology 30(Suppl. 1): 111–119.

Boopathy, R., C.F. Kulpa and M. Wilson. 1993. Metabolism of 2,4,6-trinitrotoluene (TNT) by *Desulfovibrio* sp. (B strain). Applied Microbiology and Biotechnology 39: 270–275.

Boopathy, R., M. Gurgas, J. Ullian and J. Manning. 1998. Metabolism of explosive compounds by sulfate-reducing bacteria. Current Microbiology 37: 127–131.

Cassidya, D.P. and E. Belia. 2005. Nitrogen and phosphorus removal from an abattoir wastewater in a SBR with aerobic granular sludge. Water Research 39: 4817–4823.

Chen, Y., L. Hong, W. Han, L. Wang, X. Sun and J. Li. 2011. Treatment of high explosive production wastewater containing RDX by combined electrocatalytic reaction and anoxic–oxic biodegradation. Chemical Engineering Journal 168: 1256–1262.

Claus, H., T. Bausinger, I. Lehmler, N. Perret, G. Fels, U. Dehner, J. Preuss and H. König. 2007. Transformation of 2,4,6-trinitrotoluene (TNT) by *Raoultella terrigena*. Biodegradation 18: 27–35.

Cyplik, P., R. Marecik, A. Piotrowska-Cyplik, A. Olejnik, A. Drożdżyńska and L. Chrzanowski. 2012. Biological Denitrification of high nitrate processing wastewaters from explosives production plant. Water, Air and Soil Pollution 223(4): 1791–1800.

Fournier, D., A. Halasz, J. Spain, P. Fiurasek and J. Hawari. 2002. Determination of key metabolites during biodegradation of hexahydro-1,3,5-trinitro-1,3,5-triazine with *Rhodococcus* sp. strain DN22. Applied and Environmental Microbiology 68: 166–172.

Fournier, D., F. Monteil-Rivera, A. Halasz, M. Bhatt and J. Hawari. 2006. Degradation of CL-20 by white-rot fungi, Chemosphere 62: 175–181.

Gonzalez, G., M.G. Herrera, M.T. Garcia and M.M. Pena. 2001. Biodegradation of phenol in a continuous process: comparative study of stirred tank and fluidized-bed bioreactors. Bioresource Technology 76: 245–251.

Groboillot, A., D.K. Boadi, D. Poncelet and R.J. Neufeld. 1994. Immobilization of cells for application in the food industry. Critical Reviews in Biotechnology 14: 75–107.

Gumuscu, B. and T. Tekinay. 2013. Effective biodegradation of 2,4,6-trinitrotoluene using a novel bacterial strain isolated from TNT-contaminated soil. International Biodeterioration and Biodegradation 85: 35–41.

Huang, C.Y. 1998. The anaerobic biodegradation of the High Explosive Octahydro-1,3,5,7-tetranitro-1,3,5, 7-tetrazocine (HMX) by an extremely thermophilic anaerobe *Caldicellulosiruptor owensensis*, Sp. Nov. Ph.D thesis, University of Claifornia Los Angeles (UCLA), USA.

Johnson, G.R., R.K. Jain and J.C. Spain. 2000. Properties of the trihydroxytoluene oxygenase from *Burkholderia cepacia* R34: an extradiol dioxygenase from the 2,4-dinitrotoluene pathway. Archives of Microbiology 173: 86–90.

Karel, S.F., S.B. Libicki and C.R. Robertson. 1985. The immobilization of whole cells: engineering principles. Chemical Engineering Science 40: 1321–1354.

Karthikeyana, S. and J.C. Spain. 2016. Biodegradation of 2,4-dinitroanisole (DNAN) by *Nocardioides* sp. JS1661 in water, soil and bioreactors. Journal of Hazardous Materials 312: 37–44.

Kearney, P.C., Q. Zeng and J.M. Ruth. 1983. Oxidative pretreatment accelerates TNT metabolism in soils. Chemosphere 12: 1583–159.

Keenan, B.G., T. Leungsakul, B.F. Smets and T.K. Wood. 2004. Saturation mutagenesis of *Burkholderia cepacia* R34 2,4-Dinitrotoluene Dioxygenase at DntAc valine 350 for synthesizing nitrohydroquinone, methylhydroquinone, and methoxyhydroquinone. Applied and Environmental Microbiology 70: 3222–3231.

Keller, J., S. Watts, W. Battye-Smith and R. Chong. 2001. Full-scale demonstration of biological nutrient removal in a single tank SBR process. Water Science and Technology 43: 355–362.

Kiran Kumar Reddy, G., M. Sarvajith, Y.V. Nancharaiah and V.P. Venugopalan. 2016. 2,4-Dinitrotoluene removal in aerobic granular biomass sequencing batch reactors. International Biodeterioration and Biodegradation 119: 56–65.

Kwon, M.J. and K.T. Finneran. 2008. Biotransformation products and mineralization potential for hexahydro-1,3,5-trinitro-1,3,5-triazine (RDX) in abiotic versus biological degradation pathways with anthraquinone-2,6-disulfonate (AQDS) and *Geobacter metallireducens*. Biodegradation 19: 705–715.

Lee, S., S.Y. Lee and K.S. Shin. 2009. Biodegradation of 2,4,6-trinitrotoluene by white rot fungus *Irpex lacteus*. Mycobiology 37(1): 17–20.

Liu, G., Z. Ye, H. Li, R. Che and L. Cui. 2012. Biological treatment of hexanitrostilbene (HNS) produced wastewater using an anaerobic–aerobic immobilized microbial system. Chemical Engineering Journal 213: 118–124.

Liu, Y. and J.H. Tay. 2002. The essential role of hydrodynamic shear force in the formation of biofilm and granular sludge. Water Research 36(7): 1653–1665.

Maloney, S.W., N.R. Adrian, R.F. Hickey and R.L. Heine. 2002. Anaerobic treatment of pinkwater in a fluidized bed reactor containing GAC. Journal of Hazardous Materials 92: 77–88.

Martin, H.G., N. Ivanova, V. Kunin, F. Warnecke, K.W. Barry, A.C. McHardy, C. Yeates, S. He, A.A. Salamov, E. Szeto, E. Dalin, N.H. Putnam, H.J. Shapiro, J.L. Pangilinan, I. Rigoutsos, N.C. Kyrpides, L.L. Blackall, K.D. McMahon and P. Hugenholtz. 2006. Metagenomic analysis of two enhanced biological phosphorus removal (EBPR) sludge communities. Nature Biotechnology 24: 1263–1269.

McCormick, N.G., F.E. Feeherry and H.S. Levinson. 1976. Microbial transformation of 2,4,6-TNT and other nitroaromatic compounds. Applied and Environmental Microbiology 31: 949–958.

McCormick, N.G., J.H. Cornell and A.M. Kaplan. 1981. Biodegradation of hexahydro-1,3,5-trinitro-1,3,5-triazine. Applied and Environmental Microbiology 42: 817–823.

Meyer, R. 1987. Explosives. 3rd Ed. 452 Pages, Publisher: VCH Publ. Inc., New York, N.Y., USA.

Mishra, A. and A. Malik. 2014. Novel fungal consortium for bioremediation of metals and dyes from mixed waste stream. Bioresource Technology 171: 217–226.

Mokeev, A.N., V.A. Iljin and N.B. Gradova. 1998. Biotechnological degradation of the radioactive cellulose containing waste. Journal of Molecular Catalysis B: Enzymatic 5: 441–445.

Nancharaiah, Y.V. and G. Kiran Kumar Reddy. 2018. Aerobic granular sludge technology: mechanisms of granulation and biotechnological applications. Bioresource Technology 247: 1128–1143.

Naumova, R.P., T.O. Belousova and R.M. Gilyazova. 1982. Microbial transformation of 2,4,6-trinitrotoluene. Prikl Biokhim Mikrobiol 18(1): 85–90.

Nejidat, A., L. Kafka, Y. Tekoah and Z. Ronen. 2008. Effect of organic and inorganic nitrogenous compounds on RDX degradation and cytochrome P-450 expression in *Rhodococcus* strain YH1. Biodegradation 19: 313–320.

Oh, S.Y., J.G. Son and P.C. Chiu. 2013. Biochar-mediated reductive transformation of nitro herbicides and explosives. Environmental Toxicology and Chemistry 32(3): 501–508.

Oh, S.Y. and Y.D. Seo. 2014. Sorptive removal of nitro explosives and metals using biochar. Journal of Environmental Quality 43(5): 1663–1671.

Oh, S.Y. and Y.D. Seo. 2015. Factors affecting sorption of nitro explosives to biochar: pyrolysis temperature, surface treatment, competition, and dissolved metals. Journal of Environmental Quality 44(3): 833–840.

Pak, J.W., K.L. Knoke, D.R. Noguera, B.G. Fox and G.H. Chambliss. 2000. Transformation of 2, 4, 6-trinitrotoluene by purified xenobiotic reductase B from *Pseudomonas fluorescens* I-C. Applied and Environmental Microbiology 66: 4742–4750.

Pilkington, P.H., A. Margaritis, N.A. Mensour and I. Russell. 1998. Fundamentals of immobilized yeast cells for continuous beer fermentation: a review. Journal of the Institute of Brewing 104: 19–31.

Preuss, A., J. Fimpel and G. Dickert. 1993. Anaerobic transformation of 2,4,6-trinitrotoluene (TNT). Archives of Microbiology 159: 345–353.

Pudge, I.B., A.J. Daugulis and C. Dubois. 2003. The use of Enterobacter cloacae ATCC 43560 in the development of a two-phase partitioning bioreactor for the destruction of hexahydro-1,3,5-trinitro-1,3,5-triazine (RDX). Journal of Biotechnology 100: 65–75.

Rajagopal, R., N.M. Cata Saady, M. Torrijos, J.V. Thanikal and T.S. Hung. 2013. Sustainable agro-food industrial wastewater treatment using high rate anaerobic process. Water 5(1): 292–311.

Roh, H., M.R. Yu, K. Yakkala, J.R. Koduru, J.K. Yang and Y.Y. Chang. 2015. Removal studies of Cd(II) and explosive compounds using buffalo weed biochar-alginate beads. Journal of Industrial and Engineering Chemistry 26: 226–233.

Saez, J.M., C.S. Benimeli and M.J. Amoroso. 2012. Lindane removal by pure and mixed cultures of immobilized actinobacteria. Chemosphere 89: 982–987.

Sanders, O. and N. Slagg. 1978. U.S. Army Research and Development Command (ARRADCOM) Technical report. Arl CD-TR-T8025.

Sharma, S. and P. Malaviya. 2016. Bioremediation of tannery wastewater by chromium resistant novel fungal consortium. Ecological Engineering 91: 419–425.

Singh, R., P. Soni, P. Kumar, S. Purohit and A. Singh. 2009. Biodegradation of high explosive production effluent containing RDX and HMX by denitrifying bacteria. World Journal of Microbiology and Biotechnology 25(2): 269–275.

Sokół, W. and W. Korpal. 2004. Determination of the optimal operational parameters for a three-phase fluidized bed bioreactor with a light biomass support when used in treatment of phenolic wastewaters. Biochemical Engineering Journal 20: 49–56.

Sondhi, A., S. Guha, C.S. Harendranath and A. Singh. 2010. Effect of Aluminum (Al3+) on granulation in upflow anaerobic sludge blanket reactor treating low-strength synthetic wastewater. Water Environment Research 82: 715–724.

Spain, J.C. 1995. Biodegradation of nitroaromatic compounds. Annual Review of Microbiology 49: 523–555.

Spanggord, R.J., J.C. Spain, S.F. Nishino and K.E. Mortelmans. 1991. Biodegradation of 2,4-dinitrotoluene by a *Pseudomonas* sp. Applied and Environmental Microbiology 57: 3200–3205.

Tallur, P.N., V.B. Megadi and H.Z. Ninnekar. 2009. Biodegradation of *p*-cresol by immobilized cells of *Bacillus* sp. strain PHN 1. Biodegradation 20: 79–83.

Thompson, K.T., F.H. Crocker and H.L. Fredrickson. 2005. Mineralization of the cyclic nitramine explosive hexahydro-1,3,5-trinitro-1,3,5-triazine by *Gordonia* and *Williamsia* spp. Applied and Environmental Microbiology 71: 8265–8272.

Tope, A.M., K. Jamil and T.R. Baggi. 1999. Transformation of 2,4,6-trinitrotoluene (TNT) by immobilized and resting cells of *Arthrobacter* sp. Journal of Hazardous Substance Research 2(3): 1–8. https://doi.org/10.4148/1090-7025.1013.

Torres, P., J.A. Rodríguez, L.E. Barba, L.F. Marmolejo and C.A. Pizarro. 2009. Combined treatment of leachate from sanitary landfill and municipal wastewater by UASB reactors. Water Science and Technology 60(2): 491–495.

Tsai, T.S. 1991. Biotreatment of red water—a hazardous waste stream from explosive manufacture with fungal systems. Hazardous Waste and Hazardous Materials 8(3): https://doi.org/10.1089/hwm.1991.8.231.

Ullah, H., A. Ali Shah, F. Hasan and A. Hameed. 2010. Biodegradation of trinitrotoluene by immobilized *Bacillus* sp. Pakistan Journal of Botany 42(5): 3357–3367.

Urbanski, T. 1990. Chemistry and Technology of Explosives Vol. 1, Pergamon Press, U.K.

USEPA. U.S. Environmental Protection Agency. 1989. Trinitrotoluene Health Advisory. Environmental Protection Agency, Criteria and Standards Division, Office of Drinking Water, Washington, D.C.

Vanderberg, L.A., J.J. Perry and P.J. Unkefer. 1995. Catabolism of 2,4,6-trinitrotoluene by *Mycobacterium vaccae*. Applied Microbiology and Biotechnology 43: 937–945.

Venkatasubramanian, K. and W.R. Veith. 1979. Introduction. Page 1 *In*: K. Venkatasubramanian (ed.). Immobilized Microbial Cells. Am. Chem. Soc. Symp. Ser. 106. The Society, Washington, DC. Tamilnadu. M.Sc. (Agri.) Thesis, Tamil Nadu Agricultural University, Coimbatore, India.

Venkatesan, A.K., M. Sharbatmaleki and J.R. Batista. 2010. Bioregeneration of perchlorate laden gel-type anion-exchange resin in a fluidized bed reactor. Journal of Hazardous Materials 177: 730–737.

Walsh, J.T., R.C. Chalk and C.Jr. Merritt. 1973. Application of liquid chromatography to pollution abatement studies of munition wastes. Analytical Chemistry 46(7): 1215–20.

Wang, Z., Z. Ye, M. Zhang and X. Bai. 2010. Degradation of 2,4,6-trinitrotoluene (TNT) by immobilized microorganism-biological filter. Process Biochemistry 45: 993–1001.

Won, W.D., R.J. Heckly, D.J. Glover and J.C. Hoffsommer. 1974. Metabolic disposition of 2,4,6-trinitrotoluene. Applied Microbiology 27(3): 513–516.

Won, W.D., L.H. Disalvo and J. Ng. 1976. Toxicity and mutagenicity of 2,4,-6-trinitrotoluene and its microbial metabolites. Applied and Environmental Microbiology 31(4): 576–80.

Xiao, Y., D.J. Roberts, G. Zuo, M. Badruzzaman and G.S. Lehman. 2010. Characterization of microbial populations in pilot-scale fluidized-bed reactors treating perchlorate and nitrate-laden brine. Water Research 44: 4029–4036.

Zacheus, O.M., E.K. Livanainen, T.K. Nissinen, M.J. Lehtola and P.J. Martikainen. 2000. Bacterial biofilm formation on polyvinyl chloride, polyethylene and stainless steel exposed to ozonated water. Water Research 34(1): 63–70.

Zhang, J.H., M.H. Wang and X.M. Zhu. 2013. Treatment of HMX-production wastewater in an aerobic granular reactor. Water Environmental Research 85: 301–307.

Zheng, C., J. Zhou, J. Wang, B. Qu, J. Wang, H. Lu and H. Zhao. 2009. Aerobic degradation of nitrobenzene by immobilized of *Rhodotorula mucilaginosa* in polyurethane foam. Journal of Hazardous Materials 168: 298–303.

CHAPTER 13

Electrocatalytic Biofilm (ECB)
Functional Role in Energy and Product Valorization

J. Shanthi Sravan,[1,2] *Sulogna Chatterjee,*[1,2]
Manupati Hemalatha[1,2] *and S. Venkata Mohan*[1,2,*]

1. Introduction

Biofilm is a complex, three-dimensional matrix structure attached to the submerged surface through propagation in the form of clumps (Mclean et al. 2010, Rozendal et al. 2008, Venkata Mohan et al. 2019). The formation of biofilm involves a complex series of events involving interactions between physical and biological processes. Initially, the free-swimming bacterial cells attach themselves on the electrode surface and form cell clusters called microcolonie (Figures 1 and 2). The cells produce an extracellular polymeric substances matrix to adhere and initiate the formation of the biofilm (Krishna et al. 2019, Li et al. 2017a, Barraud et al. 2015). The microcolonies promote the coexistence of diverse bacterial species and their metabolic states (Malvankar and Lovely 2012, Orellana et al. 2013). Physiological mutuality among the bacterial cells is a major factor in shaping up the structure and function of biofilm, and ensures that they become very efficient microbial communities. Conventional biofilm may form on a wide variety of surfaces, including living tissues, industrial or potable water system piping, or natural aquatic systems. The solid-liquid interface between a surface and an aqueous medium (e.g., water and blood) provides an ideal environment for the attachment and growth of microorganisms.

During the complex process of adhesion, the microbial cells alter their phenotype in response to the proximity of a conductive surface (Istanbullu et al. 2012, Babauta et al. 2012). The composition of the microbial community is a critical parameter for the formation of electrocatalytic biofilm (ECB), as it influences the electron transport mechanisms on the electrode surface (Arunasri et al. 2016). The electrode oxidation and the ECB formation are dependent on the microbial community and the conductive materials used. The charge, current holding capacity, conductivity and physico-

[1] Bioengineering and Environmental Science lab, EEFF Department, CSIR-Indian Institute of Chemical Technology (CSIR-IICT), Hyderabad, 500 007, India.
[2] Academy of Scientific and Innovative Research (AcSIR), India.
* Corresponding author: vmohan_s@yahoo.com; svmohan@iict.res.in

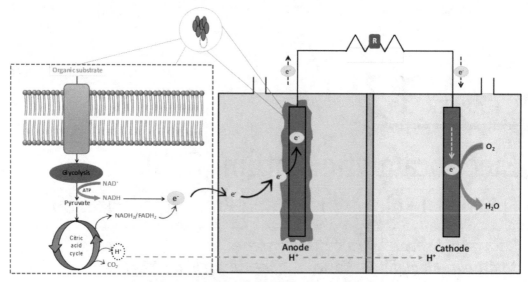

Figure 1. Representation of biofilm formation in bioelectrochemical systems.

Color version at the end of the book

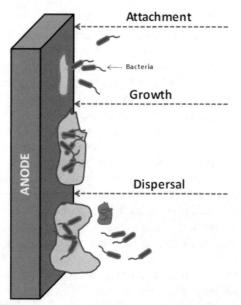

Figure 2. Representation of stages of biofilm formation on electrode surface.

Color version at the end of the book

chemical stability of the conductive materials also influence the ECB formation (Zhang et al. 2010). ECB developed on the electrode surfaces offer intrinsic electrocatalytic properties (Yang et al. 2012). This electrocatalytic property of the ECB has led to the development of novel microbial electrochemical technologies (MET) such as microbial fuel cells (MFC) and microbial electrolysis cells (MEC) (Venkata Mohan et al. 2016, Butti et al. 2016, Bajracharya et al. 2016, Venkata Mohan et al. 2014). The capability of some of these microorganisms to connect their metabolisms directly to external potential supply (electrofermentation) is a very exciting area of research that is in progress in exploring for the possibilities of EABs applications (Harnisch and Rabaey 2012, Chandrasekhar et al. 2015).

2. Biofilm and electrocatalytic behavior

ECB-associated organisms are very different in both phenotypic and electrogenic characteristics from their freely suspended counterparts (Li et al. 2017a). Electrocatalytic biofilm (ECB) is a fast expanding area of research involving several disciplines like electrochemistry, microbiology, biotechnology and chemical engineering (Du et al. 2007, Chandra et al. 2017). ECBs coated onto the electrodes have wide applications, in bioelectricity production, bioremediation, biosynthesis processes, biosensor design, and biogas production (Kim et al. 2015). ECB formation also reduces heavy metal concentration in wastewaters (e.g., selenium, tellurium, chromium and uranium) (Li et al. 2008, Nancharaiah et al. 2015). EABs are being studied in the METs towards the recovery of biobased product. The formation of EABs is beneficial as biofilm and no longer requires addition of external mediators because of the self conductive nature of the EABs through electrode surface. The efficiency of any MET directly depends on the electrogenic activity of the biocatalyst. The biocatalyst with high electron discharge capability is considered to be electrochemically active and is important for the functioning of MFCs and MECs. The electrogenic activity of the bacteria is influenced by the organic matter present in the wastewater, electron transfer rate from bacteria to the anode and the membrane efficiency in the transfer of hydrogen ions from anode compartment to the cathode compartment. The microorganisms that can deliver electrons from their oxidative metabolic pathways to their external environment are termed as exoelectrogens. *Geobacter* and *Shewanella* are known for their exoelectrogenic ability (Ringeisen et al. 2006, Gorby et al. 2006). The EAB form ECB on conductive materials, allowing a direct electron transfer via the electrode surface using its conductive property, without involvement of external mediators (Harnisch et al. 2011). This electrocatalytic property of biofilm has been related to the presence of some specific EAB strains involved in exchange of electrons with electrode surfaces, viz., *Geobacter sulfurreducens*, *Shewanella oneidensis* and *Rhodoferax ferrireducens* (Voordeckers et al. 2010, Marsili et al. 2008b). EABs can be obtained and isolated principally from natural sites such as soils or seawater and freshwater sediments or from samples collected from a wide range of different microbially rich environments (e.g., sewage sludge, activated sludge and industrial/domestic effluents).

2.1 Electron transfer

The electron transfer mechanism to the intermediate electron acceptor (anode) is majorly by two mechanisms, viz., direct electron transfer (DET) and mediated electron transfer (MET) (Figure 3). DET mechanism is by the physical contact of the microbial cells with the anode, without the involvement of any redox mediators. The outer membrane cytochromes (C-type) are implicated in the direct transfer of electrons from NADH to the anode (Patil et al. 2012). MET is associated with mediators for the exocellular electron transfer from biocatalyst to the anode. The mediators may be artificially added or naturally excreted soluble shuttlers or primary and secondary metabolites from bacterial metabolism (Krishna et al. 2019, Kuchi et al. 2018). Bacteria such as *Shewanella* and *Pseudomonas* secrete some chemical species such as flavins called shuttle molecules, to transfer electrons from the outer membrane of bacteria to the electrodes. Natural and artificial mediators such as phenazines, phenoxazines and quinines are found to be efficient mediators in the exoelectron transfer mechanism in bioelectrochemical systems. In ECBs, electron transfer through biofilm to the anode is associated with the presence of electrically conductive pili (Reguera et al. 2006) or pilus-like appendages (Gorby et al. 2006). This nanowire like structures may facilitate development of thicker ECBs which allow direct inter-species electron transfer and avoids the need for soluble mediators resulting in higher anode performances (Reguera et al. 2006).

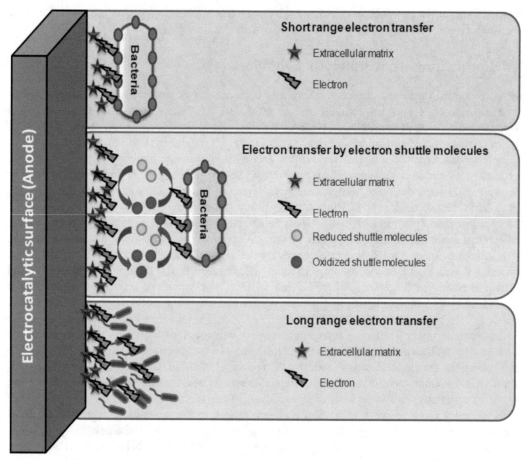

Figure 3. Electron transfer mechanisms and role of biofilm in bioelectrochemical systems.

Color version at the end of the book

Geobacter, Rhodoferax and *Shewanella* are reported to use conductive appendages (nanowires-20 nm in length), which are the networks of cellular outgrowth for their exocellular electron transfer through membrane bound organelles (Logan 2009). The conductivity of these bacterial nanowires was found to be much higher than synthetic metallic nanostructures. The membrane bound cell organelles, viz., cytochromes and nanowires will help in electron transfer from outer membrane of bacterial cell to the external electron acceptor (anode) (Kumar et al. 2017). The specific electron transfer proteins could be altered towards enhanced electroactivity of the biocatalysts. The efficient electron transfer from biocatalyst to electrode is an important criterion as electron loss is a major concern in bioelectrochemical systems. It could be achieved by the external supply of vitamins like riboflavin, but addition of redox mediators is not cost effective and sustainable (Kim et al. 2015). Therefore, facilitation of improved electron donation by producing electron shuttles and permeabilizing the cell surface by integrating membrane porins results in increased electron transfer and bioelectrogenic activity.

2.2 Electrocatalytic bacteria (ECB)

ECB are a phylogenetically diverse group of microbes. The most prevalent bacteria in ECB of METs majorly include *Geobacter* sp., *Shewanella* sp., *Rhodoferax ferrireducens, Aeromonas hydrophila, Pseudomonas aeruginosa, Clostridium butyricum* and *Enterococcus gallinarum*. Specifically, *Geobacter* and *Shewanella* sp. are discussed for their electroactive nature (Kumar et al. 2017, Voordeckers et al. 2010). *Geobacter sulfurreducens*, a gram negative bacterium, has the ability to form thin and thick biofilms on the electrode. It oxidizes acetate completely into protons and electrons and also has the ability to reduce minerals such as Fe(III). In thin monolayer biofilm (> 50 μm), the electron transfers to the electrode through the outer-membrane c-Cyts or through the secreted riboflavin which interacts with the c-Cyts to shuttle the electrons out of the cell (Inoue et al. 2010, Malvankar and Lovely 2012). The thick layer biofilm produces conductive pili which are proteinaceous and made of monomer units of PilA encoded by PilA gene (Inoue et al. 2010). Pili plays a major role in electron transfer mechanism and in biofilm formation. Outer-membrane multi-copper proteins such as OmpB or OmpC are majorly required for Fe(III) oxide reduction and their absence results in inhibition of iron reduction (Orellana et al. 2013, Inoue et al. 2010). *G. sulfurreducens* is able to transfer the electrons via direct electron transfer rather than using shuttles. The electrons generated in the cytoplasm are transferred by direct contact to the extracellular electron acceptors. It anaerobically oxidizes the organic carbon completely to carbon dioxide and water and transfer electrons to the terminal electrons acceptors like metal ions, elemental sulfur or fumarate (Dan Sun et al. 2016, Pozo et al. 2016, Kiely et al. 2011). The highly conductive pili of *G. sulfurreducens* are involved in the long-range electron transfer responsible for a 10-fold increase in electricity production (Reguera et al. 2006).

Shewanella species, a γ-proteobacteria, possess the ability to reduce iron and manganese, and use these metals as electron acceptors. Thus, exoelectrogens can reduce the various substrates and transfer electrons exogenously by outer membrane c-Cyts containing the MtrCAB complex (Kotloski et al. 2013, Gorby et al. 2006). These exoelectrogens secretes flavins, i.e., riboflavin and flavin mononucleotide. Both these flavin molecules interact with c-Cyts and form complexes (flavin-c-Cyts) and help the electrons to hop across the membrane (from periplasm to outer surface of the cell) (Voordeckers et al. 2010). Fluorescence emission spectra showed the presence of riboflavin and quinone derivatives in the cell-free supernatant of exoelectrogens. It is reported that removal of riboflavin from biofilm decreases the electron transfer rate by > 70% (Marsili et al. 2008a). The exoelectrogens such as *Geobacter* sp. and *Shewanella* sp. produces pili that are conductive accounting to the production of electricity (Gorby et al. 2006, Pham et al. 2008).

2.3 Electrode-microbe interaction in the context of biofilm

The electrode-microbe interaction has several characteristics that are important in the attachment process of microorganisms and biofilm formation. The roughness of the solid surface is usually an important parameter in extending the microbial biofilm thickness (Venkata Mohan et al. 2008b). The higher surface area of the electrode due to the rough surface weakens the shear forces paving way to easy formation of the biofilm (Artyushkova et al. 2015). The physicochemical properties of the surface may also exert a strong influence on the rate and extent of bacterial attachment. Most studies suggest that microorganisms attach more rapidly to hydrophobic, non-polar surfaces like Teflon, plastics, etc., than to hydrophilic materials such as glass or metals (Aracic et al. 2014). The hydrophobic interactions occurring between the microbe and the electrode are hypothesized to enable the microorganisms to overcome the active repulsive forces within a certain distance from the electrode surface, resulting in an irreversible attachment. The larger the surface area of the electrode, the more is the possibility for ECBs to develop. This results in more possibility for electron generation and producing higher amount of bioelectricity, biohydrogen or other platform chemicals (Harnisch et al. 2011).

3. Factors influencing ECB formation

The bacterial biofilm formed at the anode acts as an efficient electron donor in the microbial environment. The biofilm formation from a microbiologist perspective depends on various operating parameters involved in the microbial environment.

3.1 pH

The anodic reactions produce protons that acidify the ECB which can affect the MET performance. The performance response of well developed biofilm on the variation of the pH-environment was assessed and reported in various studies, in order to mimic the influence of a changing pH in the wastewater (Patil et al. 2011). Longer exposure times and especially highly alkaline conditions lead to an irreversible biofilm detachment that cannot be re-established when the biofilm is exposed to pH 7 (Rozendal et al. 2008). The operational window is limited to pH-values between pH 6 and 9, which is well in accordance with the pH window for the formation of ECB (Rabaey and Rozendal 2010). At pH 7, biofilm growth was two-fold more active on electrodes than at pH 6 and 8 (Torres et al. 2008).

3.2 Hydrodynamics

The electrode surface exposed to a medium will affect the electrogenic rate kinetics. The hydrodynamic interactions of the microbial cells with electrode surfaces will also alter the biofilm characteristics (Harnisch et al. 2011). The flow velocity in the microbe-electrode interaction also has an inverse effect on the thickness of the biofilm formed (Glaven and Tender 2012). The low linear flow velocities are the optimum conditions for multi-layered attachment of the biofilm bacteria onto the electrode surface (Harnisch and Rabaey 2012). The rate of settling and layered association with the submerged surface of the biofilm attachment will depend largely on the flow velocity characteristics of the liquid. Cell size and cell motility are also key parameters which help in the association of microbes with the electrodes (Kumar et al. 2015).

3.3 Aqueous medium

Characteristics of the aqueous medium, such as pH, nutrient levels, ionic strength, and temperature may play a role in the rate of biofilm formation on a substratum (Patil et al. 2012). Bacterial attachment and biofilm formation in different aqueous systems are also temperature dependent (Kumar et al. 2015). This effect may be due to water temperature or other unmeasured, seasonally affected parameters. The increase in the concentration of several cations (sodium, calcium, lanthanum, ferric iron, etc.) affects the attachment onto the electrode surfaces due to the repulsive forces between the negatively charged bacterial cells and the electrode surfaces (Harnisch and Rabaey 2012). The number of attached bacterial cells on the electrode surface also increases with an increase in the nutrient concentration.

3.4 Cell surface

The hydrophobicity of the cell surface is an important factor in the adhesion of microbes to the electrode surface. The hydrophobic interactions tend to increase with an increase in non-polar nature of one or both the surfaces involved (i.e., the microbial cell surface and the electrode surface) (Babauta et al. 2012). Microbes are majorly negatively charged due to the presence of non-flagellar appendages and EPS, which contribute to cell surface hydrophobicity influencing the rate and extent of biofilm formation (Erable et al. 2010). Most fimbriae have high proportion of hydrophobic amino acid residues and play a role in cell surface hydrophobicity and attachment, probably by overcoming the initial electrostatic repulsion barrier that exists between the microbe and the electrode surface (Harnisch and Rabaey 2012). The treatment of adsorbed cells with proteolytic enzymes also facilitates microbe-electrode interaction, providing evidence for the role of proteins in attachment (Istanbullu et al. 2012). Attachment of biofilm on the electrode surfaces will occur most readily on rougher, more hydrophobic, and adhesive coated surfaces. An increase in flow velocity, water temperature, or nutrient concentration may also equate to increased attachment, if these factors do not exceed critical levels (Ritcher et al. 2008). Properties of the cell surface, specifically the presence of fimbriae, flagella, and surface-associated polysaccharides or proteins are important and provide a competitive advantage for one organism over another in a mixed community and are important in cell attachment and biofilm formation.

4. Applications of ECB

Electrochemically active biofilm (ECB) has potential applications in production of bioelectricity, biogas and platform chemicals along with bioremediation and biosensor design (Figure 4).

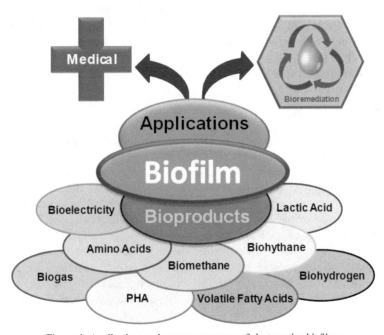

Figure 4. Applications and resource recovery of electroactive biofilms.

Color version at the end of the book

4.1 Bioelectricity

The best known application is microbial fuel cell (MFC) that has been well known as a means that generates electricity by oxidizing biodegradable organic matter in the presence of microorganisms as a biocatalyst in fuel cell type setup. The biocatalyst present in the anode chamber of MFC generates electrons and protons through anaerobic respiration of organic substrate. Electrons transfer through the electrodes connected across an external circuit from anode to the cathode. Proton exchange membrane separates the anode and cathode chambers and allows proton diffusion to the cathode chamber, where they combine with the electron acceptor (Li et al. 2017b). The potential difference between the respiratory system and electron acceptor generates the current and voltage needed to generate electricity (Venkata Mohan et al. 2007, 2008a,b, 2014). Most MFCs have been operated using anaerobic or facultative aerobic bacteria which oxidize various substrates. Power production in MFC also depends on the ability of biofilm formation which varies with bacterial species, substrate concentration, cathode catalysts and MFC configuration (Kumar et al. 2017, Nikhil et al. 2018). ECBs which are conductive in nature are capable of producing good bioelectricity (Ritcher et al. 2008). *Geobacter* and *Shewanella* spp. are reported to perform long-range electron transport through microbial nanowires for bioelectricity production (Kumar et al. 2015, Ringeisen et al. 2006). ECB based MFCs can potentially produce EABs of much lesser quantity than their suspended counterparts in the activated sludge and thereby decreases sludge production by 50–70% and also reduce operational cost by 20–30% (Wang and Ren 2013, Fan et al. 2012). The effect of anodic biofilm growth and extent of its coverage on the anodic surface of a single chambered mediator-less MFC was evaluated for bioelectricity generation (Venkata Mohan et al. 2008b). Three MFCs (plain graphite electrodes, air cathode, Nafion membrane) were operated separately with variable biofilm coverage [control, anode surface coverage (ASC), 0%], partially developed biofilm [PDB, ASC ~ 44%, 90 days] and fully developed biofilm [FDB, ASC ~ 96%, 180 days] under acidophilic conditions (pH 6) at room temperature. Higher specific power production and yield and substrate removal efficiency were observed, especially with FDB operation. pH affects the electron transfer kinetics of anodic biofilms (Yuan et al. 2011). The apparent electron transfer rate constant (k_{app}) and exchange current density (i_0) are greater whereas the charge transfer resistance (R_{ct}) is smaller at pH 9.0 than at other conditions. However, alkaline conditions benefit biofilm formation in MFCs and demonstrate that electrochemical interactions between bacteria and electrodes in MFCs are greatly enhanced under alkaline conditions (Yuan et al. 2011).

4.2 Bioremediation

High biomass density promotes the mineralization process by maintaining favorable pH conditions, solute concentration and redox potential in the proximity of the cells (Singh et al. 2006, Barkay et al. 2001). MET attracted more attention in wastewater treatment because of its relatively higher treatment efficiency (Erable et al. 2010). EAB on anode oxidizes substrate to release electrons by direct or indirect extracellular electron transfer (EET). EABs as cathodic catalysts enhances the cathode reduction rate resulting in enhanced performance (Kumar et al. 2017, Pozo et al. 2016). Biocathode has wide range of applications. The different electrode conditions will influence the structure of EAB community that finally decides the performance of MET. Azo dye as a common pollutant in dyeing wastewater can be decolorized by bioelectrochemical systems (BES), and electrons from the anode can be utilized for breaking down the chromogenic azo bond of azo dye (Nagendranatha Reddy and Venkata Mohan 2016). The degradation of methyl orange, an azo dye, by EABs in the presence of nanoparticles has enhanced degradation rate (Kalathil and Chaudhuri 2016). The EABs acted as the electron generator while nanoparticles functioned as the electron carrier agents to enhance degradation rate of the dye (Kalathil and Chaudhuri 2016). BES has also emerged as a new technology platform for removing and recovering metals from metallurgical wastes, process streams and wastewaters. Biodegradation of organic matter by EAB at the anode has been successfully coupled to cathodic reduction of various metal ions. Leaching of Co(II) from $LiCoO_2$ particles, and removal of metal ions, i.e., Co(III/II), Cr(VI), Cu(II), Hg(II), Ag(I), Se(IV/VI), Te(VI) and Cd(II) from aqueous solutions have been demonstrated in recent years (Nancharaiah et al. 2015).

4.3 Biofuel

Currently, hydrogen is produced mostly from non-renewable fossil fuels, which is unsustainable in the long run. Bio-hydrogen production from non-fossil fuels will provide a greener chemical approach in hydrogen economy. Fermentative hydrogen production appears to be a promising technology but the low hydrogen yield hinders its practical applications. In this context, MEC is emerging as a potential system for biohydrogen production on the similar principle of electrolysis of water but with a decreased energy requirement (Chiranjeevi et al. 2014, Liu et al. 2010). The anode part of a MEC is similar to a MFC, where the EAB growing on the anode surfaces breakdown the organic substrates to CO_2, electrons and protons (Arunasri et al. 2016, Venkata Mohan and Lenin Babu 2011, Logan et al. 2008, Liu et al. 2005). The electrons and protons travel through the external circuit and membrane, respectively, and combine at the cathode to generate biohydrogen. The cathode is similar to a water electrolyzer and also the external electricity required for MECs to produce biohydrogen is much lower than required for water electrolysis (Liu et al. 2010). Methane is also produced in MECs, primarily after hydrogen gas is evolved from the cathode through acetoclastic methanogenesis (Clauwaert et al. 2008, Logan and Rabaey 2012). But comparatively, a kilogram of biomethane has lower energy and economic value than the same mass of biohydrogen. In MECs, wide range of organic substrates, such as complex mixtures of biomass including municipal, animal wastewater, etc., can be utilized by electroactive microorganisms for production of biohydrogen and biomethane.

4.4 Platform chemicals

During the oxidation of organic matter in wastewater, EAB transfer electrons to the anode which subsequently flows to the cathode, where they can be used for the production of value added products. The EAB formed on the anode can produce bioplastics (polyhydroxybutyrate (PHB), polyhydroxyalkanoates (PHA)), short chain carboxylic acids (VFA-C_2–C_6), biosolvents (alcohols), amino acids and lactic acids from organic substrates. Production of polyhydroxyalkanoates (PHA) was reported at the cathode under microaerophilic conditions (Logan and Rabaey 2012, Venkata Mohan et al. 2012, Rabaey and Rozendal 2010). Microbial electrochemical synthesis (MES) is the specific term used for the production of platform chemicals from organic matter, wastewater and other renewable resources (Liu et al. 2010, 2005, Logan et al. 2008). Other compounds having physico-chemical properties such as phenazine, hydrogen peroxide (H_2O_2), caustic soda, etc., can also be produced but are strain dependent (Rabaey et al. 2005, Rabaey and Rozendal 2010, Pham et al. 2009, Marsili et al. 2008a, b). CO_2 produced during various processes can be utilized as renewable carbon source for the production of biofuels and platform chemicals via microbial bioconversion which can lead to the reduction of the global CO_2 emissions. It can also be applicable for other BES for recycling and reusing the emitted CO_2 from wastewater for the generation of bioenergy- bioCH_4 (Villano et al. 2010) and biodiesel (Powell and Hill 2010) or the production of chemical intermediates like formic acid (Zhao et al. 2012).

4.5 Biosensor

Biosensor is a miniaturized, biological analytical recognition device with a bioreceptor used to detect an analyte. The ECB coated on the electrode surface can react with the analyte for specific detection. These miniaturized prototypes are electrochemical biosensors specifically applicable for diagnostics and are successful to an extent (Patil et al. 2010). They have applications in various fields including diagnostics, testing of food, soil, water, environment, etc., and have potential for research and development. Biosensor for *in situ* monitoring of nitrate/nitrite concentrations has been designed which is based on diffusion of nitrate/nitrite inside the concentrated mass of bacteria through a membrane. Bacteria convert these ions into nitrous oxide which can be electrochemically detected (Larsen et al. 2000). This kind of biosensor has great significance during wastewater treatment. Novel MFC design based on a single chamber reactor in order to use it like a BOD sensor has also been developed (Kumlanghan et al. 2007). The specific reactivity and ability of biofilms to form stable polymer-like hydrogel aggregates of microorganisms have been studied to form systems analogous to electrocatalytic redox-polymer modified electrodes (Lotowska et al. 2016). Growth of biofilms has been demonstrated with use of *Yersinia enterocolitica*, a robust Gram-negative rod-shaped bacteria known to be resistant to pH changes (4–10) and temperature variations (0–40°C). Charge distribution and propagation within the biofilm have been enhanced by introduction of multi-walled carbon nanotubes. The biofilm-based hybrid matrices have exhibited electrocatalytic activity during electro-reductions of oxygen and hydrogen peroxide. These electrode designs are relevant to biosensing and for the development of alternate cathode materials for biofuel cells or bio-batteries (Lotowska et al. 2016). Enzymatic electrocatalysis is also an area of research to receive extensive attention for the developments of amperometric biosensors and enzymatic fuel cells (Kumar et al. 2017).

5. Future perspectives

The ability of ECBs to exchange the electrons directly with the conductive materials has opened up new perspectives in MFC application. ECBs are being explored as a new area of interest involving interdisciplinary aspects like electrochemistry, microbiology and chemical engineering domains. Their applications are being expanded in the field of renewable energy towards achieving sustainability. The use of ECBs in MFCs can be established as one of the platform technology that can efficiently influence the bacterial electro-metabolism towards product synthesis. ECBs can also be useful as electrochemical biosensors as they are extremely specific and sensitive with capability to adapt to the exposed environment. The specificity of the ECBs in product synthesis, biosensor development and waste remediation is best suitable for their application in remote areas.

The presence of ECBs on the electrodes has a key role in the electron transfer influencing the overall performance of MFCs. It involves the direct transfer of electrons between the bacteria and the conductive materials which expanded the interest related to MFCs. Presence of ECBs no longer require the addition of external mediators and thus make the system less expensive, less polluting and more sustainable over time. It increases the transitional potential of MFC as a technology for bioelectricity generation and while achieving sustainability. Extending the interference of EABs and biofilm formation to the biocathodes can also lead to fuel cell sustainability while overcoming many of the obstacles in MFCs. Overcoming the challenges involved in MFC operation has a wider range of applications in waste remediation with generation of bioenergy and biobased products. The capabilities of ECBs in MFCs need further establishment through continuous research to develop it as a transitional technology for commercialization. The commercialization aspect can improve the sanitation, a major issue among all the developing countries and has a role to play in addressing the global environmental problems.

6. Conclusions

The chapter summarizes the specific characteristics of ECBs along with their mechanisms, operational parameters and the wider range of applications. ECBs generated by EABs have a potential role in driving the microbial electrochemical technologies towards societal benefit in bioremediation and bioenergy sector. The interaction of microbe and electrode on the anode results in the development of EABs eliminating non-EABs for achieving specificity. ECB in MFCs has a positive impact on the energy generation with higher significant conversion efficiency. Though there is a good impact of ECB in MFC operation, it also inter-depends on the electrometabolism, electrode materials, substrate and the reactor design considerations among others. These multiple challenges need to be addressed prior to implementation at a larger scale of operation. ECBs coated on electrodes have recently become popular in various fields like bioremediation, platform chemical synthesis, biofuel production and biosensor design. ECB formation in the anode of an MFC has the capability of generating excess reducing equivalents by oxidation of organic substrates. The higher generation of reducing equivalents helps in producing considerable amount of bioelectricity. The presence of ECBs in METs does not require an input of additional/external energy while possessing a wide range of applications in waste/wastewater treatment domain for bioenergy and platform chemical production in the form of DST INSPIRE Fellowship. JSS and MH duly acknowledges CSIR for providing fellowship.

Acknowledgements

The authors would like to acknowledge the Director, CSIR-IICT for encouragement and support. Financial support from Department of Science and Technology (New Indigo Project; DST/IMRCD/ New Indigo/Bio-e-MAT/2014/(G)/(ii)). SC acknowledges Department of Science and Technology for providing research fellowship.

References

Aracic, S., L. Semenec and A.E. Franks. 2014. Investigating microbial activities of electrode-associated microorganisms in real-time. Frontiers in Microbiology 663(5): 1–7.

Artyushkova, K., J.A. Cornejo, L.K. Ista, S. Babanova, C. Santoro, P. Atanassov and A.J. Schuler. 2015. Relationship between surface chemistry, biofilm structure, and electron transfer in *Shewanella* anodes. Biointerphases 10(1): 019013. doi: 10.1116/1.4913783.

Arunasri, K., J.A. Modestra, D.K. Yeruva, K.V. Krishna and S. Venkata Mohan. 2016. Polarized potential and electrode materials implication on electro-fermentative di-hydrogen production: Microbial assemblages and hydrogenase gene copy variation. Bioresource Technology 200: 691–698.

Babauta, J., R. Renslow, Z. Lewandowski and H. Beyenal. 2012. Electrochemically active biofilms: facts and fiction. A review. Biofouling 28(8): 789–812.

Bajracharya, S., M. Sharma, G. Mohanakrishna, X.D. Benneton, D.P.B.T.B. Strike, P.M. Sarma and D. Pant. 2016. An overview on emerging bioelectrochemical systems (BESs): Technology for sustainable electricity, waste remediation, resource recovery, chemical production and beyond. Renewable Energy 98: 153–170.

Barraud, N., S. Kjelleberg and S.A. Rice. 2015. Dispersal from microbial biofilms. Microbiology Spectrum 3(6): MB-0015-2014. doi:10.1128/microbiolspec.MB-0015-2014.

Chandra, R., J.S. Sravan, M. Hemalatha, B. Sai Kishore and S.Venkata Mohan. 2017. Photosynthetic synergism for sustained power production with microalgae and photobacteria in a biophotovoltaic cell. Energy and fuels 31(7): 7635–7644.

Chandrasekhar, K., K. Amulya and S. Venkata Mohan. 2015. Solid phase bio-electrofermentation of food waste to harvest value-added products associated with waste remediation. Waste Management 45: 57–65.

Chiranjeevi, P., A. Naresh Kumar and S. Venkata Mohan. 2014. Critical assessment of biofilm and suspended growth reactor configurations for acidogenic biohydrogen production using wastewater as a function of redox microenvironment. International Journal of Hydrogen Energy 39(14): 7561–7571.

Clauwaert, P., R. Toledo, H.D. Vander, R. Crab, W. Verstraete, H. Hu, K.M. Udert and K. Rabaey. 2008. Combining biocatalyzed electrolysis with anaerobic digestion. Water Science and Technology 57(4): 575–9.

Dan Sun, D., J. Chen, H. Huang, W. Liu, Y. Ye and S. Cheng. 2016. The effect of biofilm thickness on electrochemical activity of *Geobacter sulfurreducens*. Journal of Hydrogen Energy, 1–6.

Erable, B., N.M. Duteanu, M.M. Ghangrekar, C. Dumas and K. Scott. 2010. Application of electro-active biofilms. Biofouling 26: 57–71.

Fan, Y., S.K. Han and H. Liu. 2012. Improved performance of CEA microbial fuel cells with increased reactor size. Energy and Environmental Science 5: 8273–8280.

Glaven, S.M.S. and L.M. Tender. 2012. Study of the Mechanism of catalytic activity of *G. Sulfurreducens* biofilm anodes during biofilm growth. ChemSusChem 5: 1106–1118.

Gorby, Y.A., S. Yanina, J.S. McLean, K.M. Rosso, D. Moyles, A. Dohnalkova, T.J. Beveridge, I.S. Chang, B.H. Kim, K.S. Kim, D.E. Culley, S.B. Reed, M.F. Romine, D.A. Saffarini, E.A. Hill, L. Shi, D.A. Elias, D.W. Kennedy,

K.G. Pinchu, K. Watanabe, S. Ishii, B. Logan, K.H. Nealson and J.K. Fredrickson. 2006. Electrically conductive bacterial nanowires produced by *Shewanella oneidensis* strain MR-1 and other exoelectrogens. Proceedings of the National Academy of Sciences USA 103: 11358–11363.

Harnisch, F., F. Aulenta and U. Schröder. 2011. Microbial fuel cells and bioelectrochemical systems: industrial and environmental biotechnologies based on extracellular electron transfer. Comprehensive Biotechnology 2: 643–659.

Harnisch, F. and K. Rabaey. 2012. The diversity of techniques to study electrochemically active biofilms highlights the need for standardization. ChemSusChem 5(6): 1027–1038.

Inoue, K., C. Leang, A.E. Franks, T.L. Woodard, K.P. Nevin and D.R. Lovley. 2010. Specific localization of the c-type cytochrome OmcZ at the anode surface in current producing biofilms of *Geobacter sulfurreducens*. Environmental Microbiology Reports 3: 211–217.

Istanbullu, O., J. Babauta, H.D. Nguyen and H. Beyenal. 2012. Electrochemical biofilm control: mechanism of action. Biofouling 28(8): 769–778.

Kalathil, S. and R.G. Chaudhuri. 2016. Hollow palladium nanoparticles facilitated biodegradation of an azo dye by electrically active biofilms. Materials (Basel) 9(8): 653. https://dx.doi.org/10.3390%2Fma9080653.

Kiely, P.D., J.M. Regan and B.E. Logan. 2011. The electric picnic: synergistic requirements for exoelectrogenic microbial communities. Current Opinion in Biotechnology 22: 378–385.

Kim, B., J. An, D. Fapyane and I.S. Chang. 2015. Bioelectronic platforms for optimal bio-anode of bio-electrochemical systems: from nano- to macro scopes. Bioresource Technology 195: 2–13.

Kotloski, N.J. and J.A. Gralnick. 2013. Flavin electron shuttles dominate extracellular electron transfer by *Shewanella oneidensis*. mBio. 4: 00553–12. DOI: 10.1128/mBio.00553-12.

Krishna, K.V., K. Swathi, M. Hemalatha and S. Venkata Mohan. 2019. Bioelectrocatalyst in Microbial Electrochemical Systems and Extracellular Electron Transport. In Microbial Electrochemical Technology 117–141.

Kuchi, S., O. Sarkar, B. Sai Kishore, G. Velvizhi and S. Venkata Mohan. 2018. Stacking of microbial fuel cells with continuous mode operation for higher bioelectrogenic activity. Bioresource Technology 257: 210–216.

Kumar, A., L.H.H. Hsu, K. Paul, F. Barriere, P.N.L. Lens, L. Lapinsonniere, J.H.V. Lienhard, U. Schroder, X. Jiang and D. Leech. 2017. The ins and outs of microorganism-electrode electron transfer reactions. Nature Reviews in Chemistry 1: 0024.

Kumar, R., L. Singh, Z.A. Wahid and M.F.M. Din. 2015. Exoelectrogens in microbial fuel cells toward bioelectricity generation. International Journal of Energy Research 39: 1048–1067.

Kumlanghan, A., J. Liu, P. Thavarungkul, P. Kanatharana and B. Mattiasson. 2007. Microbial fuel cell-based biosensor for fast analysis of biodegradable organic matter. Biosensors and Bioelectronics 22: 2939–2944.

Larsen, L.H., L.R. Damgaard, T. Kjær, T. Stenstrom, A. Lynggaard-Jensen and N.P. Revsbech. 2000. Fast responding biosensor for on-line determination of nitrate/nitrite in activated sludge. Water Research 34: 2463–2468.

Li, J., L. Hu, L. Zhang, D. Ye, X. Zhu and Q. Liao. 2017a. Uneven biofilm and current distribution in three-dimensional macroporous anodes of bio-electrochemical systems composed of graphite electrode arrays. Bioresource Technology 228: 25–30.

Li, J., H. Li, J. Zheng, L. Zhang, Q. Fu, X. Zhu and Q. Liao. 2017b. Response of anodic biofilm and the performance of microbial fuel cells to different discharging current densities. Bioresource Technology 233: 1–6.

Liu, H., S. Grot and B.E. Logan. 2005. Electrochemically assisted microbial production of hydrogen from acetate. Environmental Science and Technology 39: 4317–20.

Liu, H., H. Hu, J. Chingnell and Y. Fan. 2010. Microbial electrolysis: novel technology for hydrogen production from biomass. Biofuels 1: 129–42.

Logan, B.E., D. Call, S. Cheng, H.V. Hamelers, T.H. Sleutels and A.W. Jeremiase. 2008. Microbial electrolysis cells for high yield hydrogen gas from organic matter. Environmental Science and Technology 42: 8630–40.

Logan, B.E. 2009. Exoelectrogenic bacteria that power microbial fuel cells. Nature Reviews in Microbiology 7(5): 375–81.

Logan, B.E. and K. Rabaey. 2012. Conversion of wastes into bioelectricity and chemicals by using microbial electrochemical technologies. Science 337: 686–90.

Lotowska, W.A., J.A. Rutkowska, E. Seta, E. Szaniawska, A. Wadas, S. Sek, A. Raczkowska, K. Brzostek and P.J. Kulesza. 2016. Bacterial-biofilm enhanced design for improved electrocatalytic reduction of oxygen in neutral medium. Electrochimica Acta 213: 314–323.

Malvankar, N.S. and D.R. Lovley. 2012. Microbial nanowires: a new paradigm for biological electron transfer and bioelectronics. ChemSusChem 5: 1039–1046.

Marsili, E., J.B. Rollefson, D.B. Baron, R.M. Hozalski and D.R. Bond 2008a. Microbial biofilm voltammetry: direct electrochemical characterization of catalytic electrode-attached biofilms. Applied and Environmental Microbiology 74(23): 7329–7337.

Marsili, E., D.B. Baron, I.D. Shikhare, D. Coursolle, J.A. Gralnick and D.R. Bond. 2008b. Shewanella secretes flavins that mediate extracellular electron transfer. Proceedings of National Academy of Sciences USA 105(10): 3968–3973.

Mclean, J.S., G. Wanger, Y. Gorby, M. Wainstein, J. McQuaid, S.I. Ishii, O. Bretschger, H. Beyenal and K.H. Nealson. 2010. Quantification of electron transfer rates to a solid phase electron acceptor through the stages of biofilm formation from single cells to multicellular communities. Environmental Science and Technology 44(7): 2721–2727.

Nagendranatha Reddy, C. and S. Venkata Mohan. 2016. Integrated bio-electrogenic process for bioelectricity production and cathodic nutrient recovery from azo dye wastewater. Renewable Energy 98: 188–196.

Nancharaiah, Y.V., S. Venkata Mohan and P.N.L. Lens. 2015. Metals removal and recovery in bioelectrochemical systems: a review. Bioresource Technology 195: 102–114.

Nikhil, G.N., D.N.S. Krishna Chaitanya, S. Srikanth, Y.V. Swamy and S. Venkata Mohan. 2018. Applied resistance for power generation and energy distribution in microbial fuel cells with rationale for maximum power point. Chemical Engineering Journal 335: 267–274.

Orellana, R., J.J. Leavitt, L.R. Comolli, R. Csencsits, N. Janot, K.A. Flanagan, A.S. Gray, C. Leang, M. Izallalen and T. Mester. 2013. U (VI) reduction by a diversity of outer surface c-type cytochromes of *Geobacter sulfurreducens*. Applied and Environmental Microbiology 79: 6369–6374.

Patil, S.A., F. Harnisch, B. Kapadnis and U. Schröder. 2010. Electroactive mixed culture biofilms in microbial bioelectrochemical systems: The role of temperature for biofilm formation and performance. Biosensors and Bioelectronics 26(2): 803–808.

Patil, S.A., F. Harnisch, C. Koch, T. Hübschmann, I. Fetzer, A. Carmona-Martínez, S. Müller and U. Schröder. 2011. Electroactive mixed culture derived biofilms in microbial bioelectrochemical systems: The role of pH on biofilm formation, performance and composition. Bioresource Technology 102(20): 9683–9690.

Patil, S.A., C. Hägerhäll and L. Gorton. 2012. Electron transfer mechanisms between microorganisms and electrodes in bioelectrochemical systems. Bioanalytical Reviews 4(2-4): 159–192.

Pham, T.H., N. Boon, P. Aelterman, P. Clauwaert, L. De Schamphelaire, L. Vanhaecke, K. De Maeyer, M. Höfte, W. Verstraete and K. Rabaey. 2008. Metabolites produced by *Pseudomonas* sp. enable a Gram positive bacterium to achieve extracellular electron transfer. Applied Microbiology and Biotechnology 77(5): 1119–1129.

Pham, T.H., P. Aelterman and W. Verstraete. 2009. Bioanode performance in bioelectrochemical systems: recent improvements and prospects. Trends in Biotechnology 27(3): 168–178.

Powell, E.E. and G.A. Hill. 2010. Carbon dioxide neutral, integrated biofuel facility. Energy 35: 4582–4586.

Pozo, G., L. Jourdin, Y. Lu, J. Keller, P. Ledezma and S. Stefano Freguia. 2016. Cathodic biofilm activates electrode surface and achieves efficient autotrophic sulfate reduction. Electrochimica Acta 213: 66–74.

Rabaey, K., N. Boon, M. Höfte and W. Verstraete. 2005. Microbial phenazine production enhances electron transfer in biofuel cells. Environmental Science and Technology 39(9): 3401–3408.

Rabaey, K. and R.A. Rozendal. 2010. Microbial electrosynthesis-revisiting the electrical route for microbial production. Nature Reviews in Microbiology 8(10): 706–716.

Reguera, G., K.P. Nevin, J.S. Nicoll, S.F. Covalla, T.L. Woodard and D.R. Lovley. 2006. Biofilm and nanowire production leads to increased current in *Geobacter sulfurreducens* fuel cells. Applied and Environmental Microbiology 72(11): 7345–7348.

Ringeisen, B.R., E. Henderson, P.K. Wu, J. Pietron, R. Ray, B. Little and J.M. Jones-Meehan. 2006. High power density from a miniature microbial fuel cell using *Shewanella oneidensis* DSP10. Environmental Science and Technology 40(8): 2629–34.

Rozendal, R.A., H.V.M. Hamelers, K. Rabaey, J. Keller and C.J.N. Buisman. 2008. Towards practical implementation of bioelectrochemical wastewater treatment. Trends in Biotechnology 26(8): 450–459.

Singh, R., D. Paul and R.K. Jain. 2006. Biofilms: implications in bioremediation. Trends in Microbiology 14(9): 389–397.

Venkata Mohan, S., G. Velvizhi, K.V. Krishna and M.L. Babu. 2014. Microbial catalyzed electrochemical systems: A bio-factory with multi-facet applications. Bioresource Technology 165: 355–364.

Venkata Mohan, S., S. Raghavulu Veer, S. Srikanth and P.N. Sarma. 2007. Bioelectricity production by mediator less microbial fuel cell under acidophilic condition using wastewater as substrate: Influence of substrate loading rate. Current Science 92(12): 1720–1726.

Venkata Mohan, S., G. Mohanakrishna, B. Reddy Purushotham, R. Saravanan and P.N. Sarma. 2008a. Bioelectricity generation from chemical wastewater treatment in mediator less (anode) microbial fuel cell (MFC) using selectively enriched hydrogen producing mixed culture under acidophilic microenvironment. Biochemical Engineering Journal 39(1): 121–13.

Venkata Mohan, S., S.V. Raghuvulu and P.N. Sarma. 2008b. Influence of anodic biofilm growth on bioelectricity production in single chambered mediatorless microbial fuel cell using mixed anaerobic consortia. Biosensors and Bioelectronics 24(1): 41–47.

Venkata Mohan, S., S. Srikanth and M.V. Reddy. 2012. Microaerophilic microenvironment at biocathode enhances electrogenesis with simultaneous synthesis of polyhydroxyalkanoates (PHA) in a bioelectrochemical system (BES). Bioresource Technology 125: 291–299.

Venkata Mohan, S. and C. Reddy Nagendranatha. 2016. Integrated bio-electrogenic process for bioelectricity production and cathodic nutrient recovery from azo dye wastewater. Journal Renewable Energy 98: 188–196.

Venkata Mohan, S., J.S. Sravan, B. Sai Kishore, K.V. Krishna, J.A. Modestra, G. Velvizhi, A.N. Kumar, S. Varjani and A. Pandey. 2019. Microbial Electrochemical Technology: Emerging and Sustainable Platform. In Microbial Electrochemical Technology 3–18.

Villano, M., F. Aulenta, C. Ciucci, T. Ferri, A. Giuliano and M. Majone. 2010. Bioelectrochemical reduction of CO_2 to CH_4 via direct and indirect extracellular electron transfer by a hydrogenophilic methanogenic culture. Bioresource Technology 101: 3085–3090.

Voordeckers, J.W., B.C. Kim, M. Izallalen and D.R. Lovley. 2010. Role of *Geobacter sulfurreducens* outer surface c-type cytochromes in reduction of soil humic acid and anthraquinone-2, 6-disulfonate. Applied and Environmental Microbiology 76: 2371–2375.

Wang, H. and Z.J. Ren. 2013. A comprehensive review of microbial electrochemical systems as a platform technology. Biotechnology Advances 31: 1796–1807.

Yang, Y., M. Xu, J. Guo and G. Sun. 2012. Bacterial extracellular electron transfer in bioelectrochemical systems. Process Biochemistry 12: 1707–1714.

Yuan, Y., B. Zhao, S. Zhou, S. Zhong and L. Zhuang. 2011. Electrocatalytic activity of anodic biofilm responses to pH changes in microbial fuel cells. Bioresource Technology 102: 6887–689.

Zhang, Y., M.D. Merrill and B.E. Logan. 2010. The use and optimization of stainless steel mesh cathodes in microbial electrolysis cells. International Journal of Hydrogen Energy 35: 12020–12028.

Zhao, H.Z., Y. Zhang, Y.Y. Chang and Z.S. Li. 2012. Conversion of a substrate carbon source to formic acid for CO_2 emission reduction utilizing series stacked microbial fuel cells. Journal of Power Sources 217: 59–64.

CHAPTER 14

Titania Nanoparticles Biofilm Formation in Environment Bacteria

A Possible Defense Mechanism

Jyoti Kumari,[#] Ankita Mathur,[#] N. Chandrasekaran
and *Amitava Mukherjee**

1. Introduction

A nanomaterial, nanopowder, nanocrystal or nanocluster is an ultrafine particle with a minimum of one dimension less than 100 nm (Dalai et al. 2012). Due to various possible applications in biomedical, electronic, optical, and other fields, study of nanomaterials is at present an area of widespread scientific/technical research. Nanopowders are of enormous scientific value as they efficiently link bulk materials with minute or atomic structures. When the size reaches the nanoscale range, and the proportion of atoms at the surface of matter becomes considerable, the properties of materials modify. The interesting, and occasionally unforeseen, properties of nanomaterials are primarily owing to the characteristics of the surface of the material influencing the properties, instead of the bulk properties.

Titanium dioxide nanoparticles (TiO_2 NPs) have an extensive variety of commercial purpose, predominantly in consumer goods (Newman et al. 2009). By 2020, nano-based products or goods are projected to reach a \$3 trillion market size, employing six million human resources (Liu et al. 2013). Consumption of TiO_2 NPs in various industries has led to increased production of TiO_2 NPs, resulting in contamination of the ecosystem by industrial discharge and nanoparticles release into surroundings by domestic and various other means. Previous reports suggested that the entry of TiO_2 NPs into water body directly or indirectly from various sources (sunscreen lotions, paints, medical use, dismantling of batteries, nano-coatings, groundwater remediation and food additives) (O'Brien and Cummins 2010), subsequently causes toxic effects on organisms present in the aquatic ecosystem.

Centre for Nanobiotechnology, Vellore Institute of Technology, Vellore, India.
\# Equal contribution
* Corresponding author: amit.mookherjea@gmail.com, amitav@vit.ac.in

For assessing nanoparticle toxicity, bacteria are widely used as test organisms. Being essential receptors, bacterial assays are commonly used to evaluate ecological nanotoxicology (Holden et al. 2012). For fast hazard/toxicological screening of NPs, bacteria can play the role of test organisms (Jin et al. 2010).

Due to high demand for TiO_2 NPs, and the associated escalating usage and risk to environmental exposure, a detailed study was done which focused on the influence of TiO_2 NPs on the water ecosystem. The main aim of the study was to obtain details about the behavior of TiO_2 NPs on different freshwater sediment and wastewater microorganisms. Tests were carried out using individual bacterial strains and mixture of these strains in the form of consortium. Filtered and sterilized lake water and wastewater was used as matrix without additional nutrient to imitate chemical composition of water ecosystem.

2. Materials

From Sigma-Aldrich, Titanium dioxide nanoparticles were purchased (Specifications-CAS no.: 637254, Anatase 99.7%, dehydrated titanium (IV) dioxide nanoparticles). DCFH-DA (2′,7′-Dichloro fluorescein diacetate) was procured from Sigma-Aldrich. Each and every additional compounds/chemical used all through the experimentation was of reagent rank.

3. Methods

3.1 Stability examination of nanoparticles in lakewater

Mean hydrodynamic size analysis of TiO_2 NPs by dynamic light scattering was done using 90 plus Particle Size Analyzer, Brookhaven Instruments Corporations, USA at time periods of 0, 2, 4, 6, and 24 h to understand the colloidal stability on lake water. In deionized water, TiO_2 NPs (100 mg/mL) stock dispersal was prepared by sonication for 10 min using 350W with the aid of an ultrasonic processor (Sonics, USA). By dilution of stock suspension of TiO_2 NPs made up of filtered lake water, a functioning 1 mg/mL concentration was prepared. Particle hydrodynamic size was measured at time interval of 0 h, 2 h, 4 h, 6 h and 24 h using dynamic light scattering method (Dalai et al. 2012).

A solution of 100 µg/mL TiO_2 NPs was made as stock facilitated with Millipore water, further subjected to ultrasonication for a time period of 10 min at 130 W. A working concentration of 0.25, 0.5 and 1 µg/mL was made in treated waste water by accessing with ample amount from stock suspension. Particle stability was examined at time period of 0, 6 and 12 h under dark condition and thereby effective diameter was evaluated with particle size analyzer (Brookhaven Instruments Corporation, USA).

3.2 Consortium development

Bacterial strains (*Bacillus altitudinis*, *Bacillus subtilis*, and *Pseudomonas aeruginosa*) were chosen to perform antagonistic/synergistic analyses for ternary consortium development. Single bacterial strain (*B. alitudinis*) was cultured in nutrient broth medium for 4 h incubation period, over 100 µL of culture broth of which was dispensed, followed by streaking of *B.subtilis* on the plate. A similar trend was pursued for *P. aeruginosa*, *B. subtilis*, *B. alitudinis*. Afterwards, plates were incubated at 30–37°C for 24 h incubation. Then the plates were scored for antagonistic or synergistic growth. Absence of zone of inhibition between the bacterial strains shows no competitive inhibition (antagonistic growth) (Samuel et al. 2012). Toxicity study of TiO_2 NPs was further carried out using the developed ternary consortium.

For the development of the consortium, five wastewater bacterial isolates (*Exiguobacterium acetylicum, Exiguobacterium indicum, Pseudomonas nitroreducens, Bacillus flexus* and *Brevudimonas diminuta*) were assayed for synergistic and antagonistic effects. One isolate was taken and allowed to grow in nutrient broth for 8 h. After the incubation period, 100 μl of one of the culture broth was poured on the surface of nutrient agar and a loop full of culture from second isolate was streaked on the nutrient agar plate. The plates were incubated at 37°C for 24 h and were further examined. All the five strains were checked accordingly, and the zone of inhibition was found to be missing, thereby depicting the absence of competitive inhibition (Samuel et al. 2012).

3.3 Cytotoxic effects of TiO₂ NPs on sediment, wastewater bacteria and consortium

Cell viability assessment was done with single strains and consortium to find out the toxic cause of TiO₂ NPs. Cell number found in control samples was considered as 100%. Cell loss in experimental samples was compared with control. After the interaction time (2, 4, 6 and 24 h) for different samples, the decrease in cellular viability was considered in percentage. The drop off in the viability of cells was compared with the control. On nutrient broth agar medium, standard plate count assay was performed to find out cell viability.

Examination of the effect of cell viability in five wastewater individual bacteria and their consortia after interaction with TiO₂ NPs was selected. All the individual bacteria and their consortia were grown till exponential phase in nutrient broth media. The culture was taken and centrifuged at 7000 g, 10 min; further, the cells were washed with treated waste water to eliminate growth media. Cell count of 5×10^8 cells were asserted throughout the experiments. Individual bacteria and their consortia were examined with 0.25, 0.5 and 1 μg/mL of TiO₂ NPs concentration. The experimental beakers were maintained under dark condition and kept at incubating shaker maintaining 37°C, 24 h at 120 rpm in a shaker incubator (Orbitech).

3.4 Reactive oxygen species (ROS) determination

Dichloro-dihydro-fluorescein diacetate (DCFH-DA) was used as the fluorescence probe for the measurement of Reactive Oxygen Species (ROS). Regarding ROS generation evaluation, a non-polar dye is converted to a hydrophilic derivative that is DCFH, which is not fluorescent; however, it later gets converted to fluorescent DCF as soon as it gets oxidized by intracellular ROS and peroxidases. ROS production was observed at 24 h in the NP-interacted as well as control bacterial cells following the procedure, the same as described by Wang and Joseph (Wang and Joseph 1999), with minor changes. Along with DCFH-DA (100 μM), incubation of bacterial cell pellet (5 mL) at 37°C for 30 min was carried out. Excitation and emission wavelengths of spectrofluorometer were 485 nm and 530 nm, correspondingly. Spectrofluorometer (SL174, ELICO) was used for measurement of fluorescence. To evaluate auto fluorescence action of TiO₂ NPs, which possibly gets hindered using DCFDA dye, NPs without bacterial cells were too analyzed as a negative control.

Similarly, in case of individual waste water bacteria and their consortia, 5 ml of suspension was incubated with dye DCFH-DA maintaining a final concentration of 100 μM for a temperature of 37°C, 30 min. Generation of ROS was measured with similar protocol as mentioned for sediment bacteria.

3.5 SOD assay

Superoxide dismutase, the action of test bacterial cultures, was assessed by the technique explained by Wintherbourn et al. (Winterbourn et al. 1975) employing riboflavin like O_2 producer. This process was established by a potential of SOD to hamper the decline of NBT (nitroblue tetrazolium) using superoxide, which is produced by the reaction of riboflavin which is photo-reduced with O_2 (Hossain et al. 2007). Centrifugation for 10 min at 7000 g was done for test cells, and the pellet that was collected was washed using PBS. Supernatant was collected from centrifuged (7000 x g for 10 min) samples. 0.1 M EDTA and 1.5 mM NBT were supplemented and set aside in dark condition. Then, the addition of riboflavin was performed followed by incubation for 15 min at natural room temperature. UV–vis spectroscopy (Systronics, India Ltd.) was used to find out the activity of SOD at 530 nm absorbance value.

To assess the effect of individual waste water bacteria, their consortium were interacted with highest concentration (1 μg/mL) of TiO_2 NPs and similar procedure was followed as mentioned above.

3.6 Assessment of membrane permeability

A general indicator of membrane permeability and cytotoxicity in the cells is LDH release. Interaction of bacterial cells with TiO_2 NPs was carried out for 24 h and centrifuged for 10 min at 7000 g. The LDH amount in the solution supernatant was calculated pursuing the typical procedure by Brown et al. (2001). 30 mM sodium pyruvate and Tris–HCl 0.2 M were supplemented to the supernatant (100 μL). NADH (100 mL) of 6.6 mM was supplemented before use. UV–vis spectroscopy was used to examine LDH activity at 340 nm.

3.7 Exopolysaccharides measurement

Bacterial strains and consortium capability of EPS production were evaluated. In 100 mL nutrient broth medium, a loop of the strains and consortium were inoculated separately. At 30°C and oscillation at 180 rpm, all bacterial strains and their consortium were allowed to nurture for 24 h. Centrifugation for 10 min at 10,000 g was done with cultures, and the collection of the supernatant was done for each one. Ethanol of equivalent quantity was supplemented to the harvested supernatant and put to one side to the EPS at 4°C for settling down overnight. Afterward, the combination acquired was centrifuged at 10,000 g for 30 min. Furthermore, the pellets were harvested. By the phenol–sulphuric acid technique, the EPS collected was anticipated for all samples (Dubois et al. 1956).

3.8 Static biofilm formation measurement

All three bacterial strains and their consortium were grown overnight, and in new media, it was diluted to 1:100. In 96-well plate, 100 μL of diluted culture and 1 mg/mL TiO_2 NPs was added and incubated for 24 h at 30–37°C the stable environment. Control wells were kept by adding only NPs and NB medium (culture medium) with no inoculums. With distilled water subsequent incubation, the 96 microwell plates were washed to eliminate cells which are unattached and then set aside for 1 h at 37°C for aeration. Afterward, using 1% crystal violet at room temperature for 30 min, staining was completed. With disinfected distilled water, stained wells were washed for 5 times, and 95% of ethanol was added for 15 min to remove leftover crystal violet. Quantification of biofilm biomass was done at 590 nm. For single bacteria plus bacterial consortium, static biofilm test was done thrice (Ammendolia et al. 2014).

4. Results

4.1 Stability of TiO_2 nanoparticles in lake water and wastewater

Mean hydrodynamic size analysis of TiO_2 NPs by dynamic light scattering was done using 90 plus Particle Size Analyzer, Brookhaven Instruments Corporations, USA at time periods of 0, 2,

4, 6, and 24 h to understand the colloidal stability on lake water. In deionized water, TiO_2 NPs (100 mg/mL) stock dispersal was prepared by sonication for 10 min using 350 W with the aid of an ultrasonic processor (Sonics, USA). At different time gap, i.e., 0 h, 2 h, 4 h, 6 h, and 24 h, the hydrodynamic size investigation of TiO_2 NPs (1 µg/mL) was done to find out the aggregation performance in lake water medium. By rising exposure, the z-average ranges found was: 574 ± 0.292 nm, 576 ± 0.288 nm, 590 ± 0.255 nm, 601 ± 0.252 nm, and 753 ± 0.316 nm at 2nd, 4th, 6th and 24th h. Beyond 24 h, the particles appeared to reach micron range. The current analysis of TiO_2 NPs (1 µg/mL) showed that it was not aggregated in lake water.

The measurement for the size of nanoparticles in wastewater matrix under aerobic condition was analyzed under dark condition. The size of the nanoparticle in wastewater medium for a concentration of 0.25 µg/mL, ranged between 500 ± 0.02, 509 ± 0.01 nm and 620 ± 0.02 nm for a period of 0 to 12 h, respectively. The range of effective diameter for a concentration of 0.5 µg/mL was found to be 545 ± 0.01 nm, 635 ± 0.02 nm and 710 ± 0.01 nm and an increment in the effective diameter was observed to be 793 ± 0.01 nm, 838 ± 0.01 nm and 1250 ± 0.02 nm for a concentration of 1 µg/mL of TiO_2 NPs. The stability of NPs in wastewater medium depends on nanoparticle aggregation. The aggregation of nanoparticles is influenced by the presence of surfactants, cations and the organic matter in the medium. The reason behind toxicity is the reduction in specific surface area of nanoparticles as a result of the aggregation process (Gurr et al. 2005).

4.2 Bacterial isolation and classification from lake sediment and wastewater

Freshwater section of sediment was collected from VIT Green Lake, VIT University, Vellore, Tamil Nadu, India, and was kept at 4°C inside tubes of polypropylene and further microbiological investigation was done. A gram of collected sample was supplemented with sterile water (100 mL) and mixed actively followed by allowing the soil to gravitate. Using sterile distilled water, samples were serially diluted. To a plate containing nutrient agar, the suspension was added, followed by keeping it at 30°C inside the incubator for 24 h. The bacterial colonies with different morphologies were sub-cultured and tagged for further use. The isolated bacterial strains were further plated to obtain pure bacterial colonies. The bacterial strains were identified as *Bacillus altitudinis*, *Bacillus subtilis*, and *Pseudomonas aeruginosa* using 16S rRNA gene sequencing method (GenBank Accession number: KF929417, KF929418, and KF929419, respectively).

The collected wastewater was subjected to standard dilution in aseptic conditions. Primary isolation was carried out by streaking the sample on the surface of the nutrient agar plate. The plates were then further incubated at 37°C for 24 h. Single isolated colonies were selected and sub cultured on the nutrient agar plate. This process was repeated until pure cultures were obtained. The dominant bacterial isolates were selected for the study. According to the standard methods, morphology and Gram tests were accomplished with the assistance of microbial taxonomy (Sharifi-Yazdi et al. 2001). The genomic DNA was extracted employing phenol-chloroform method. Sequencing was performed with fluorescent dye terminator method (Big dye terminator cycle sequencing kit, ABI Prism 3.1) (Sambrook and Russel 2001). The sequences were examined with BLAST and aligned with help of CLUSTAL W facilitating neighbor joining method and phylogenetic tree was constructed. The sequences which showed 99% similarity in BLAST were identified. The five bacterial isolates identified were *Exiguobacterium acetylicum* (VITWW1), *Pseudomonas nitroreducens* (VITWW2), *Bacillus flexus* (VITWW3), *Brevundimonas diminuta* (VITWW4) and *Exiguobacterium indicum* (VITWW5). The 16SrRNA sequences were submitted to Genbank and the accession IDs obtained as KJ146070 for *Exiguobacterium acetylicum*, KT272871 for *Exiguobacterium indicum*, KT272873 for *Bacillus flexus*, KT272872 for *Brevundimonas diminuta* and KJ146071 for *Pseudomonas nitroreducens*.

4.3 Measurement of toxicity

4.4 Investigational arrangement of cytotoxic effects of TiO2 NPs sediment and wastewater bacteria and their consortium

With the help of standard plate count test, the toxicity of TiO_2 NPs towards individual bacterial strains and their consortium in lake water medium was measured. In *B. altitudinis*, *B. subtilis*, *P. aeruginosa*, and their consortium, an exposure-dependent and concentration reduction in cell viability was observed. In dark conditions for *B. altitudinis*, *B. subtilis*, *P. aeruginosa*, and the consortium, viability of TiO_2 NP-interacted (at 24 h and 1 µg/mL) cells was 74 ± 1.34%, 77 ± 1.17%, 74 ± 1.51%, 82.1 ± 0.73%, 73 ± 1.99%, 76 ± 0.59%, 72 ± 0.75%, and 80.1 ± 0.89%. With concentration and time, the viability of TiO_2 NP-treated cells was statistically noteworthy (p < 0.05) in all (*B. altitudinis*, *B. subtilis*, *P. aeruginosa* and the consortium) cases. Nonetheless, for all single isolates (*B. altitudinis*, *B. subtilis*, and *P. aeruginosa*), the cell mortality was more than consortium after TiO_2 NP interaction (1 µg/mL) for 24 h. However, it was not statistically considerable (p > 0.05). Previous study by Kumar et al. (2014) also reported cell viability decrease in the presence of TiO_2 NPs.

To estimate the toxicity of nanoparticles on individual isolates and their consortium, colony count assay was performed. With dark exposure, the viability exhibited a concentration dependent decline and at a concentration of 1 µg/mL TiO_2 NPs, it was constituted to be 76.10 ± 2.3, 71.9 ± 2.3, 66.5 ± 2.5, 63.0 ± 3.0 and 51.0 ± 2.0, 49 ± 1.9% for consortium and single isolates as *Exiguobacterium acetylicum, Pseudomonas nitroreducens, Exiguobacterium indicum* and *Brevundimonas diminuta, Bacillus flexus* respectively. A statistical significance (p > 0.05) was found between control and treated samples and also between individual isolates and their consortium. There are some reports observed after interaction of bacteria with NPs that a percent viability decline of 76% was observed for *E. coli* when interacted with 100 µg/mL of TiO_2 NPs under dark condition (Brunet et al. 2009). Interaction of NPs to the bacterial cell surface generates reactive oxygen species could be one of the mechanism for toxicity.

4.5 Evaluation of oxidative stress

4.5.1 Reactive oxygen species (ROS) determination

Dichloro-dihydro-fluorescein diacetate (DCFH-DA) was used as the fluorescence probe for the measurement of reactive oxygen species (ROS). ROS of the treated *B. altitudinis*, *B. subtilis*, and *P. aeruginosa* cells and consortium was 4.470.9, 4.470.1, 4.370.2, and 3.370.3 below dark conditions, correspondingly. Essentially, there was statistically irrelevant (p > 0.05) amount of ROS production for the single isolates when compared to the bacterial consortium.

The evaluation of toxicity was further evaluated with the measurement of oxidative stress especially the role of reactive oxygen species and release of superoxide ions. For the wastewater isolates consortium, minimum amount of ROS generation was found whereas for the individual isolates the quantified ROS was 3 X, 1.6 X, 1.6 X and 1.7 X, 1.5 X for *E. acetylicum, P. nitroreducens* and *E. indicum, B. diminuta* and *B. flexus*, respectively.

4.5.2 SOD assay

Superoxide dismutase production was measured in dark conditions for single strains and consortium. SOD concentration increase after TiO_2 NPs (1 µg/mL; 24 h) treatment was calculated for all strains and consortium. In dark conditions, SOD generation was 3.33 ± 0.3%, 3.24 ± 0.2%, 3.39 ± 0.2%, and 3.09 ± 03% for all strains and consortium. SOD level in test samples was less in consortium when compared to single strains, though the dissimilarity was not considerably statistical.

For the evaluation of antioxidant response with respect to the effect of action of NPs, SOD activity was done. The level of SOD estimated for single isolates was found to be 78 ± 0.9, 65 ± 0.8 and 67 ± 0.9, 37 ± 0.8 and 28 ± 0.1, 10 ± 0.9% with respect to control for consortium, *E. acetylicum, P. nitroreducens, E. indicum, B. diminuta* and *B. flexus*, respectively. This proves that the consortium possess higher ability to overcome the oxidative stress induced by the nanoparticles.

4.5.3 Assessment of membrane permeability

Lactate dehydrogenase assay was used to find out membrane permeability of single bacteria and the consortium after TiO_2 NP interaction (1 µg/mL, 24 h). *B. altitudinis, B. subtilis, P. aeruginosa* and consortium showed 4.3 ± 0.2%, 13.4 ± 0.1%, 12.5 ± 0.6% and 8.81 ± 0.1% of LDH in dark condition correspondingly when compared with control.

The attachment of nanoparticle to the bacterial surface often disrupts the permeability of the membrane which is detected by release of lactate dehydrogenase (LDH). The activity was found to be 10 ± 0.03, 13.50 ± 0.02 and 17.60 ± 0.01, 17.08 ± 0.02 and 20.08 ± 0.01, 20.45 ± 0.02% for consortium, *E. acetylicum, P. nitroreducens, E. indicum, B. diminuta* and *B. flexus*, respectively. A statistical significance ($p > 0.05$) was found between control and treated samples and also between individual isolates and their consortium for the ROS, SOD and LDH analysis.

The generation of ROS under dark condition may be due to the presence of carbon centered free radicals on the bacterial cell membrane (Fenoglio et al. 2009). In the present study, ROS activity was less for consortium as compared to single isolates and possesses higher SOD activity; this is due to the higher capacity of the enzyme to dismutate superoxide ions, thereby reducing the effect of superoxide ions. The change in the membrane permeability of the treated cells may be due to the attachment of nanoparticles to the bacterial cell membrane; this alters the redox potential of the bacterial cell member and further causes changes in the membrane permeability.

4.5.4 Exopolysaccharides withdrawal and measurement

The bacterial defiance in contrast to the toxicant was measured by formation of EPS under dark conditions by *B. altitudinis, B. subtilis, P. aeruginosa* cells, and the consortium after TiO_2 NP (1 µg/mL, 24 h) interaction. In control 0.02 ± 0.03, 0.04 ± 0.03, 0.06 ± 0.01, 0.09 ± 0.03 µg/mL, 0.12 ± 0.08, 0.12 ± 0.01, 0.20 ± 0.09, and 0.27 ± 0.03 µg/mL of EPS was produced under dark conditions for *B. altitudinis, B. subtilis, P. aeruginosa* cells, and the consortium (Table 1). Increased EPS formation was found in test bacterial consortium than in single bacterial strains, while the dissimilarity was not noteworthy statistically. On the other hand, the rise in EPS generation was lower for control when compared to all test samples which was interacted with TiO_2 NPs under dark conditions. Exopolysaccharides generation after TiO_2 NP exposure can support the endurance of the bacteria in unfavorable ecological conditions. Biofilm growth and EPS production

Table 1. EPS production by *B. altitudinis, B. subtilis, P. aeruginosa* cells and consortium, in dark conditions after 24 h incubation in TiO_2 NPs (1 µg/mL) (n = 3).

Freshwater sediment bacterial isolates	Control	1 µg/mL
Consortia	0.061 ± 0.07	0.85 ± 0.05
Bacillus altitudinis	0.040 ± 0.08	0.58 ± 0.08
Bacillus subtilis	0.041 ± 0.02	0.55 ± 0.02
Pseudomonas aeruginosa	0.048 ± 0.03	0.73 ± 0.04

Table 2. Illustration of EPS release for individual isolates and their consortium under dark condition showing increase in the EPS for control and varying concentration of TiO_2 NPs (0.25, 0.5 and 1 μg/mL).

Waste water bacterial isolates	Control	0.25 μg/mL	0.5 μg/mL	1 μg/mL
Consortium	1.23 ± 0.02	2.37 ± 0.008	2.38 ± 0.20	3.00 ± 0.01
Exiguobacterium acetylicum	0.128 ± 0.03	0.163 ± 0.02	2.10 ± 0.11	2.73 ± 0.130
Pseudomonas nitroreducens	0.33 ± 0.01	0.871 ± 0.01	1.52 ± 0.571	2.14 ± 0.126
Exiguobacterium indicum	0.02 ± 0.003	0.457 ± 0.02	0.281 ± 0.03	0.543 ± 0.08
Brevundimonas diminuta	0.015 ± 0.01	0.05 ± 0.03	0.236 ± 0.08	0.54 ± 0.09
Bacillus flexus	0.003 ± 0.01	0.038 ± 0.11	0.07 ± 0.003	0.45 ± 0.02

may protect the bacteria against host defenses. Few previous studies explained that the presence $FeCl_3$ enhances biofilm formation in *P. aeruginosa* (Patriquin et al. 2008, Borcherding et al. 2014).

These are the complex organic compounds released in response to combat the deleterious effects of NPs. The release of EPS quantitated was around 1.24 ± 0.02, 0.129 ± 0.01 and 0.34 ± 0.03, 0.23 ± 0.01, 0.016 ± 0.02 and 0.02 ± 0.01 A.U. for the control cells whereas after treatment with highest exposure dose of 1 μg/mL, the EPS release was quantitated to be 2.99 ± 0.01, 2.78 ± 0.02 and 2.15 ± 0.01, 0.55 ± 0.02, 0.48 ± 0.02 A.U. for consortium, *E. acetylicum, P. nitroreducens, E. indicum and B. diminuta, B. flexus,* respectively (Table 2). The release of EPS was found to be exposure dependent and a statistical significance (p > 0.05) was found between the release of control EPS significance was found between EPS of consortium and individual isolates after the treatment also. EPS plays an important role in the formation of biofilm and also helps in preventing dehydration of cells. EPS alleviates the cell adhesion.

4.6 Static biofilm formation measurement

The formation of biofilm after TiO_2 NPs (1 μg/mL) interaction with *B. altitudinis, B. subtilis* and *P. aeruginosa* cells, and consortium in stable condition was observed under dark conditions after 24 h incubation period. For *B. altitudinis, B. subtilis, P. aeruginosa,* and their consortium at 590 nm optical density, the formed biofilm was measured as 0.040 ± 0.08, 0.041 ± 0.02, 0.048 ± 0.03 and 0.061 ± 0.07 in control, and 0.058 ± 0.08, 0.055 ± 0.02, 0.073 ± 0.04 and 0.085 ± 0.05 under dark conditions (Table 3). Furthermore, the formation of biofilm in test consortium when compared to single test isolates was higher, while the dissimilarity was not considerably statistical.

Biofilms are usually complex of microorganisms residing in a highly hydrated structure known as extrapolymeric substances (EPS). EPS and biofilm act as a defense mechanism and protects the underlying cells within the free medium after interaction with NPs. The biofilm formation for the control cells was estimated to be 1.43 ± 0.02, 1.13 ± 0.01 and 0.95 ± 0.01, 0.902 ± 0.02, 0.45 ± 0.02 and 0.38 ± 0.02 A.U. whereas after treatment with highest dose of NPs, the biofilm formation was maximum and estimated to be around 2.99 ± 0.02, 2.17 ± 0.01 and 1.68 ± 0.02, 1.85 ± 0.03, 1.6 ± 0.02, 1.36 ± 0.01 A.U., respectively, For *E. acetylicum, P. nitoreducens, E. indicum,*

Table 3. Biofilm formation of *B. altitudinis, B. subtilis, P. aeruginosa* cells and consortium, in dark conditions after 24 h incubation in TiO_2 NPs (1 μg/mL) (n = 3).

Freshwater sediment bacterial isolates	Control	1 μg/mL
Consortia	0.09 ± 0.03	0.27 ± 0.03
Bacillus altitudinis	0.02 ± 0.03	0.12 ± 0.08
Bacillus subtilis	0.04 ± 0.03	0.12 ± 0.01
Pseudomonas aeruginosa	0.06 ± 0.01	0.20 ± 0.09

Table 4. Depiction of static biofilm formation for individual isolates and their consortium under dark condition showing increase in the formation of biofilm for control and varying concentration of TiO$_2$ NPs (0.25, 0.5 and 1 µg/mL).

Waste water bacterial isolates	Control	0.25 µg/mL	0.5 µg/mL	1 µg/mL
Consortium	1.43 ± 0.01	1.58 ± 0.02	2.06 ± 0.03	2.98 ± 0.01
Exiguobacterium acetylicum	1.30 ± 0.02	1.38 ± 0.03	1.60 ± 0.02	2.17 ± 0.02
Pseudomonas nitroreducens	1.00 ± 0.01	1.02 ± 0.02	1.53 ± 0.02	1.67 ± 0.03
Exiguobacterium indicum	0.90 ± 0.01	0.961 ± 0.02	1.50 ± 0.03	1.82 ± 0.03
Brevundimonas diminuta	0.50 ± 0.02	0.56 ± 0.01	1.11 ± 0.02	1.50 ± 0.38
Bacillus flexus	0.34 ± 0.01	0.453 ± 0.02	0.50 ± 0.02	1.35 ± 0.02

B. diminuta and *B. flexus* (Table 4). A statistical significance ($p > 0.05$) was obtained between single isolates and their consortium and also biofilm formed of the treated samples as compared to the control. According to a report, the interaction of bacteria with increase in dose of TiO$_2$ NPs leads to the increase in biofilm formation (Ammendolia et al. 2014).

4.7 Microscopic examination

SEM images were taken to further understand the initial structure of biofilm and its changes following exposure to TiO$_2$ NPs. The biofilm architecture of the consortium with and without TiO$_2$ NPs was examined under dark conditions after a 24 h incubation period. The representative SEM micrographs of the biofilm produced by consortium in the absence and presence of TiO$_2$ NPs at 1 µg/mL dose are shown in Figure 1. The untreated consortium biofilm appeared as smooth isolated bacterial cells without any damage and no structural abnormalities were noticed. An extracellular matrix, mainly constituted of the aggregate cells, was observed in dark-treated consortium. The biofilm mass exhibited a large three-dimensional structure composed of numerous bacterial cells adherent to one another and on the exopolysaccharides in case of the treated consortium. These images demonstrated that TiO$_2$ NPs can rapidly penetrate through the biofilm due to the adsorption by chemical components within EPS resulting in cellular damage (Figure 1).

The alteration in surface morphology of the bacterial cells when exposed to TiO$_2$ NPs was studied through scanning electron microscopy. The evenness of the membrane was decreased after the NP interaction, and the cells were found to form agglomerates.

The changes in the morphology of the bacterial cells embedded within biofilm was analyzed with SEM. Control biofilm depicted an intact bacterial cell within EPS matrix of the biofilm whereas

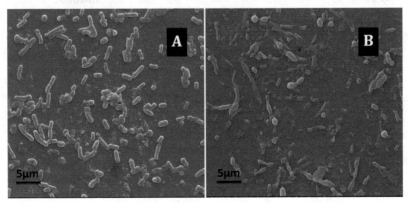

Figure 1. Scanning electron microscopic image: A—Before nanoparticles interaction (Control), B—After TiO$_2$ nanoparticle interaction.

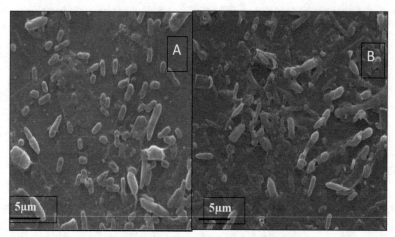

Figure 2. Portrays the bacterial cells within biofilm (A) Depicts the intact bacterial cells within biofilm (B) Represents the distortion of the morphology of the bacterial cells within biofilm when interacted with 1 μg/mL of TiO$_2$ NPs under dark condition.

after treatment with the highest exposure concentration of NPs, the bacterial cells within the biofilm exhibit a distortion in the cells embedded and injured within EPS (Figure 2).

5. Conclusion

In the current study, the different probable means of toxic effects of photo catalytic titania NPs towards the dominant freshwater and waste water isolate have been looked at in freshwater and waste water matrix, respectively. The results showed that the loss of viability of the bacterial cells was reliant on concentration of nanoparticles, the incubation time and also the type of condition. A likely cause for the nanoparticles toxicity can be the production of different types of ROS (hydroxyl and superoxide anions) caused by the nanoparticles interaction. LDH assay defined the membrane break and the consequent bio-uptake of TiO$_2$ NPs. The internalization and bio-uptake of the titania nanoparticles added extensively towards the toxic effects of TiO$_2$ nanoparticles. The role of Exopolysaccharides discharge and resultant biofilm formation as probable protection mechanism were evaluated between the individual bacterial isolates and their consortium. Therefore, the consortium of bacterial cells provides evidence to have improved abilities in resisting the toxic effects of titania nanoparticles. The future scenario could be the testing of bacterial cells that have a natural distinct potential of exopolysaccharides production and that illustrates no response against phototoxic effect of Titanium dioxide nanoparticles.

References

Ammendolia, M.G., F. Iosi, B. De Berardis, G. Guccione, F. Superti, M.P. Conte and C. Longhi. 2014. Listeria monocytogenes behaviour in presence of non-UV-irradiated titanium dioxide nanoparticles. PloS One 9(1): 9e84986. https://dx.doi.org/10.1371%2Fjournal.pone.0084986.

Borcherding, J., J. Baltrusaitis, H. Chen, L. Stebounova, C.M. Wu, G. Rubasinghege, I.A. Mudunkotuwa, J.C. Caraballo, J. Zabner, V.H. Grassian and A.P. Comellas. 2014. Iron oxide nanoparticles induce *Pseudomonas aeruginosa* growth, induce biofilm formation, and inhibit antimicrobial peptide function. Environmental Science Nano 1(2): 123–132.

Brunet, L., D.Y. Lyon, E.M. Hotze, P.J. Alvarez and M.R. Wiesner. 2009. Comparative photoactivity and antibacterial properties of C60 fullerenes and titanium dioxide nanoparticles. Environmental Science and Technology 43(12): 4355–4360.

Fenoglio, I., G. Greco, S. Livraghi and B. Fubini. 2009. Non-UV-induced radical reactions at the surface of TiO2 nanoparticles that may trigger toxic responses. Chemistry A European Journal 15(18): 4614–4621.

Gurr, J.R., A.S. Wang, C.H. Chen and K.Y. Jan. 2005. Ultrafine titanium dioxide particles in the absence of photoactivation can induce oxidative damage to human bronchial epithelial cells. Toxicology 213(1-2): 66–73.

Holden, P.A., R.M. Nisbet, H.S. Lenihan, R.J. Miller, G.N. Cherr, J.P. Schimel and J.L. Gardea-Torresdey. 2012. Ecological nanotoxicology: integrating nanomaterial hazard considerations across the subcellular, population, community, and ecosystems levels. Accounts of Chemical Research 46(3): 813–822.

Jin, X., Li, M., J. Wang, C. Marambio-Jones, F. Peng, X. Huang, R. Damoiseaux and E.M. Hoek. 2010. High-throughput screening of silver nanoparticle stability and bacterial inactivation in aquatic media: influence of specific ions. Environmental Science and Technology 44(19): 7321–7328.

Kumar, D., J. Kumari, S. Pakrashi, S. Dalai, A.M. Raichur, T.P. Sastry, A.B. Mandal, N. Chandrasekaran and A. Mukherjee. 2014. Qualitative toxicity assessment of silver nanoparticles on the fresh water bacterial isolates and consortium at low level of exposure concentration. Ecotoxicological and Environmental Safety 108: 152–160.

Liu, X., G. Chen, A.A. Keller and C. Su. 2013. Effects of dominant material properties on the stability and transport of TiO$_2$ nanoparticles and carbon nanotubes in aquatic environments: from synthesis to fate. Environmental Science: Processes and Impacts 15(1): 169–189.

Newman, M.D., M. Stotland and J.I. Ellis. 2009. The safety of nanosized particles in titanium dioxide–and zinc oxide–based sunscreens. Journal of the American Academy of Dermatology 61(4): 685–692.

O'Brien, N. and E. Cummins. 2010. Ranking initial environmental and human health risk resulting from environmentally relevant nanomaterials. Journal of Environmental Science and Health Part A 992–1007.

Patriquin, G.M., F. Banin, C. Gilmour, R. Tuchman, E.P. Greenberg and K. Poole. 2008. Influence of quorum sensing and iron on twitching motility and biofilm formation in *Pseudomonas aeruginosa*. Journal of Bacteriology 190(2): 662–671.

Sambrook, J. and D.W. Russell. 2001. Molecular cloning: A Laboratory Manual 17–3.32.

Samuel, J., M.L. Paul, M. Pulimi, M.J. Nirmala, N. Chandrasekaran and A. Mukherjee. 2012. Hexavalent chromium bioremoval through adaptation and consortia development from Sukinda chromite mine isolates. Industrial Engineering and Chemical Research 51(9): 3740–3749.

Sharifi-Yazdi, M.K., C. Azimi and M.B. Khalili. 2001. Isolation and identification of bacteria present in the activated sludge unit, in the treatment of industrial waste water. Iranian Journal of Public Health 30: 91–4.

Index

Color Plate Section

Chapter 2

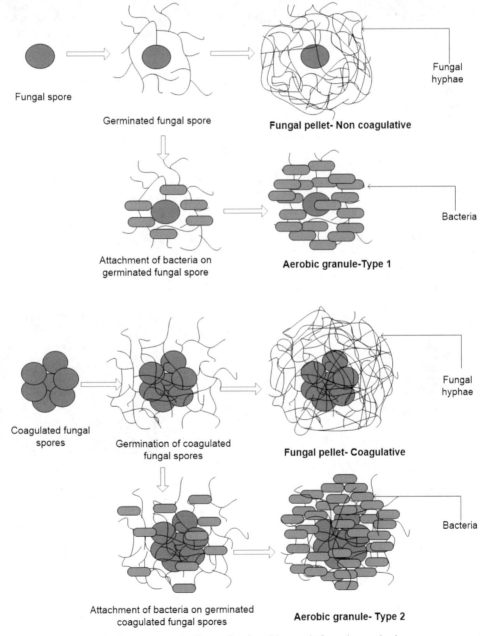

Figure 1. Proposed fungal pellet mediated aerobic granule formation mechanisms.

Chapter 3

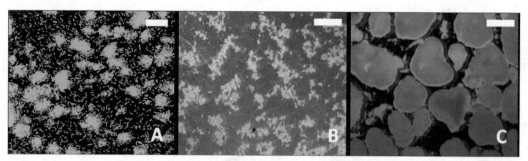

Figure 1. Examples of microbial communities. (A) *P. putida* biofilm on glass substratum. Bar = 20 µm. (B) Activated sludge flocs, bar = 2 mm. (C) Aerobic granular biomass, bar = 2 mm.

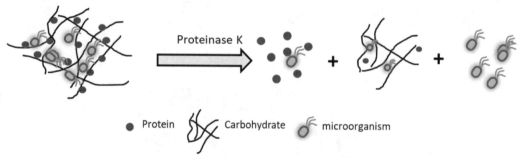

Figure 4. Diagram on mechanism of action of proteinase K on a single aerobic granule.

Chapter 4

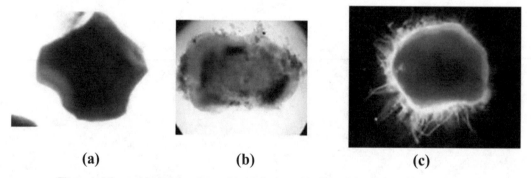

Figure 4. Microscopic pictures of anaerobic (a), fresh aerobic (b), and matured aerobic granules (c).

Chapter 6

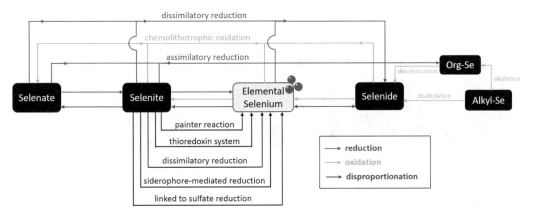

Figure 2. Selenium transformation involving different species and oxidation state (adopted and modified from Nancharaiah and Lens 2015). Transformation pathways are indicated through coloured arrows. Reduction pathway covers both assimilatory and dissimilatory reduction while oxidation is done through chemolithotroph pathway. Reduction pathways of selenite to elemental selenium pathway by microorganisms can occur through five different mechanisms.

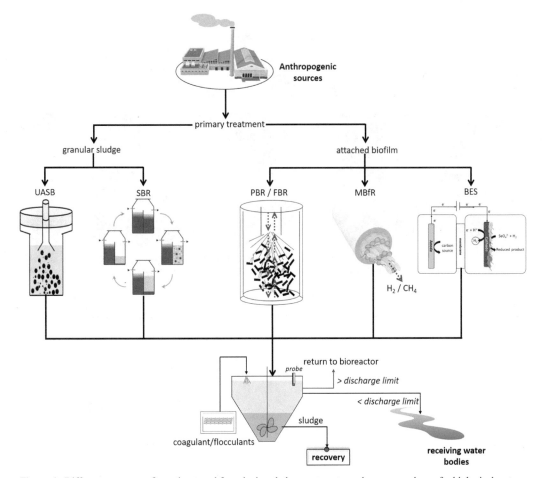

Figure 4. Different reactor configurations used for selenium-laden wastewater and recovery scheme for biological system.

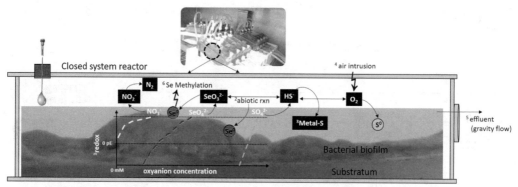

[1] Redox condition within the biofilm was not measured and was based on theoretical concept
[2] Formation of red color was observed when abiotic batch test was conducted between SeO_3^{2-} and HS^- (also observed by Hockin and Gadd 2003)
[3] Metal-sulfur precipitation can theoretical occur with iron (present in the growth medium at 7.5µM)
[4] Reactor was operated shut from open air but not in an anaerobic chamber (air intrusion is still possible)
[5] Reactor was at a 10° angle therefore no water accumulation occurred inside and effluent exits via gravity flow
[6] Volatilization of Se was considered negligible in this condition

Figure 5. Schematic representation of biofilm substrate consumption utilization exposed to nitrate, selenate and sulfate using DFR with possible interaction between oxyanions, by-products and biofilm detailed in the scheme.

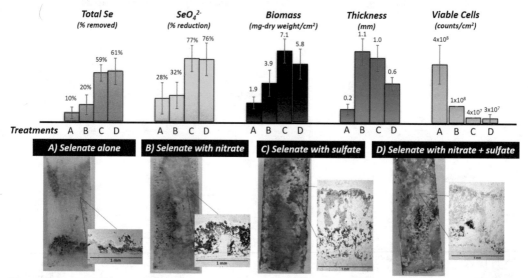

Figure 6. Impact of selenate reduction with co-contaminants to biofilm selenium removal performance, biofilm formation and growth. Data gathered and reconstructed from study conducted by Tan et al. (2018a) while biofilm images came from personal archive.

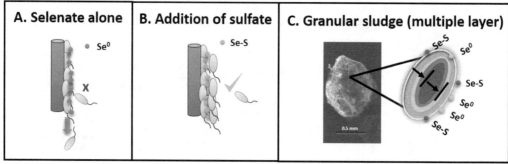

Figure 7. Proposed coping mechanism of (a-b) biofilm compared with (c) granular sludge system when treating wastewater containing either selenate or selenate with sulfate.

Chapter 7

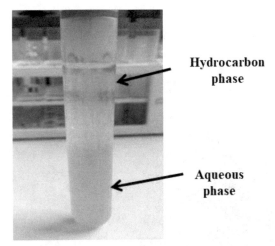

Figure 3. Phase separation of microbial cells by adding hydrocarbon.

Chapter 8

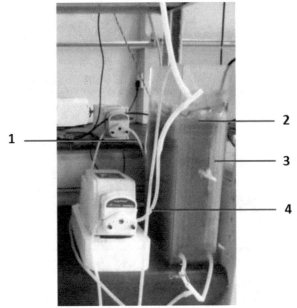

Figure 2. Experimental setup of the MBR system and the sample collection points: 1. Influent: synthetic wastewater; 2. Bioreactor: synthetic wastewater + activated sludge; 3. Foulant: biofilm sampled from the membrane; 4. Effluent (reclaimed water): achieved after MBR treatment.

Chapter 9

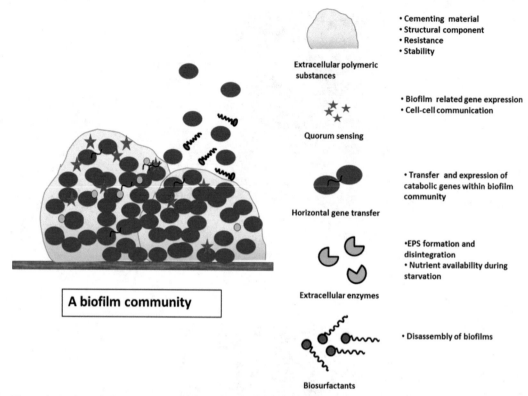

Figure 4. A schematic diagram to explain biofilm associated components, which play critical roles in the biological removal of organic pollutants.

Chapter 13

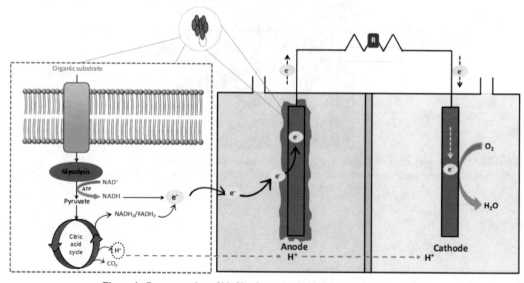

Figure 1. Representation of biofilm formation in bioelectrochemical systems.

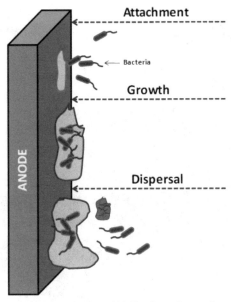

Figure 2. Representation of stages of biofilm formation on electrode surface.

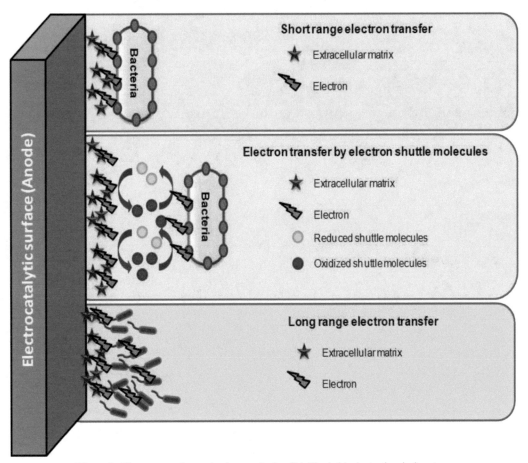

Figure 3. Electron transfer mechanisms and role of biofilm in bioelectrochemical systems.

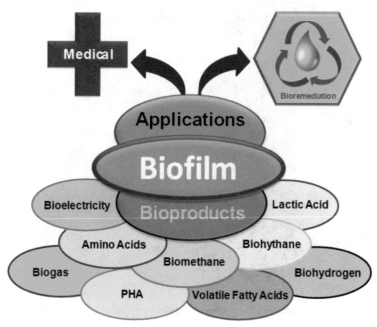

Figure 4. Applications and resource recovery of electroactive biofilms.